TO CONVERT	INTO	MULTIPLY BY
I		
inches	centimeters	2.540
inches	meters	2.540×10^{-2}
inches	millimeters	25.40
inches	mils	1,000.0
inches of mercury	atmospheres	0.03342
inches of mercury	kg/sq cm	0.03453
inches of mercury	pounds/sq ft	70.73
inches of mercury	pounds/sq in.	0.4912
inches of water (at 4°C)	atmospheres	2.458×10^{-3}
inches of water (at 4°C)	inches of mercury	0.07355
inches of water (at 4°C)	kg/sq cm	2.540×10^{-3}
inches of water (at 4°C)	pounds/sq in.	0.03613
K		
kilograms	pounds	2.205
kilograms/sq cm	atmospheres	0.9678
kilograms/sq cm	inches of mercury	28.96
kilograms/sq cm	pounds/sq in.	14.22
kilowatts	foot-lb/min	4.426×10^{4}
kilowatts	foot-lb/sec	737.6
kilowatts	horsepower	1.341
kilowatt-hrs	horsepower-hr	1.341
L		
liters	cu cm	1,000.0
liters	cu feet	0.03531
liters	cu inches	61.02
liters	gallons (U.S. liq.)	0.2642
liters/min	cu ft/sec	5.886×10^{-4}
M		
meters/min	cm/sec	1.667
meters/min	feet/min	3.281
meters/min	feet/sec	0.05468
meters/sec	feet/min	196.8
meters/sec	feet/sec	3.281
mils	centimeters	2.540×10^{-3}
mils	feet	8.333×10^{-5}
mils	inches	0.001
minutes (angles)	degrees	0.01667
minutes (angles)	radians	2.909×10^{-4}
minutes (angles)	seconds	60.0
P		
poise	gram/cm sec	1.00
poundals	grams	14.10
poundals	kilograms	0.01410
poundals	pounds	0.03108
pounds	grams	453.5924
pounds	kilograms	0.4536
pounds	poundals	32.17
pounds of water	cu feet	0.01602
pounds of water	cu inches	27.68
pounds of water	gallons	0.1198
pounds/cu ft	grams/cu cm	0.01602
pounds/sq ft	atmospheres	4.725×10^{-4}
pounds/sq ft	inches of mercury	0.01414
pounds/sq ft	kg/sq meter	4.882

TO CONVERT	INTO	MULTIPLY BY
pounds/sq ft		$\times 10^{-3}$
pounds/sq in.		04
pounds/sq in.		
pounds/sq in.		
pounds/sq in.		
R		
radians	degrees	57.30
radians	minutes	3,438.
radians	seconds	2.063×10^{5}
radians/sec	degrees/sec	57.30
radians/sec	revolutions/min	9.549
radians/sec	revolutions/sec	0.1592
revolutions	degrees	360.0
revolutions	radians	6.283
revolutions/min	degrees/sec	6.0
revolutions/min	radians/sec	0.1047
revolutions/min	rev/sec	0.01667
revolutions/sec	degrees/sec	360.0
revolutions/sec	radians/sec	6.283
revolutions/sec	rev/min	60.0
S		
slug	kilogram	14.59
slug	pounds	32.17
square centimeters	sq inches	0.1550
square feet	sq cm	929.0
square feet	sq inches	144.0
square inches	sq cm	6.452
square inches	sq feet	6.944×10^{-3}
square inches	sq millimeters	645.2
square millimeters	sq cm	0.01
square millimeters	sq inches	1.550×10^{-3}
T		
temperature (°C) +273	absolute temperature (°C)	1.0
temperature (°C) +17.78	temperature (°F)	1.8
temperature (°F) +460	absolute temperature (°F)	1.0
temperature (°F) −32	temperature (°C)	5/9
tons (long)	kilograms	1,016.
tons (long)	pounds	2,240.
tons (long)	tons (short)	1.120
tons (metric)	kilograms	1,000.
tons (metric)	pounds	2,205.
tons (short)	kilograms	907.1848
tons (short)	pounds	2,000.
tons (short)	tons (long)	0.89287
tons (short)	tons (metric)	0.9078
W		
watts	Btu/hr	3.413
watts	Btu/min	0.05688
watts	horsepower	1.341×10^{-3}
watts	horsepower (metric)	1.360×10^{-3}
watts	kilowatts	0.001
watt-hours	Btu	3.413
watt-hours	foot-pounds	2,656.
watt-hours	horsepower-hr	1.341×10^{-3}
watt-hours	kilogram-meters	367.2
watt-hours	kilowatt-hr	0.001

BASIC
FLUID POWER

PRENTICE-HALL INTERNATIONAL, INC., *London*
PRENTICE-HALL OF AUSTRALIA, PTY. LTD., *Sydney*
PRENTICE-HALL OF CANADA, LTD., *Toronto*
PRENTICE-HALL OF INDIA PRIVATE LTD., *New Delhi*
PRENTICE-HALL OF JAPAN, INC., *Tokyo*

BASIC FLUID POWER

DUDLEY A. PEASE

Hydraulic Product Manager
Westinghouse Air Brake Company

Prentice-Hall, Inc.

Englewood Cliffs, N.J.

Library of Congress Catalog Card Number: 67-16389
Printed in the United States of America

ACKNOWLEDGMENTS

Many people have contributed to the writing of this textbook and I sincerely hope that all of these fine people will share with me the educational values that it may impart to those interested in fluid power.

The assistance from the manufacturers of fluid power equipment, if listed, would have to include that given by Ted Pearce, Executive Vice President of National Fluid Power Association, and by many of its well over a hundred members.

Another long list of contributors would be those who over the years have shared their knowledge with me: George Altland, Frank Mackin, Wes Master, Art Ritterford, John Pippenger, Walter Kudlaty, Zeke Lansky, Jack Ellis, Jim Fisher, Rex Burnham, Ernst Weidman, and many others.

Still another long list would be the students who have shared classroom experiences with me at Henry Ford College, Milwaukee School of Engineering, Wayne State University, Moorhead State College, Tuskegee Institute, Bemidji State College, and Kenosha Technical Institute.

In addition to those contributing information, I am especially grateful to Mrs. Howard Cohn for her excellent work in the organizing and typing of the manuscript, to James Querin for helping with the illustrations, and to my wife, Violet, and daughter, Bonnie, for their patience and understanding.

PREFACE

by
THEODORE PEARCE
Executive Vice President
National Fluid Power Association
Fluid Power Society

The rapid development of fluid power as a chief means—along with electrical and mechanical systems—of transmitting power is a modern phenomenon. Fluid power technology is based on scientific principles known for hundreds of years, but only since World War II has it emerged as a technological entity. It has surged to the forefront without a supply of engineers specifically educated in fluid power, without a reserve of skilled workers and technicians trained therein, and with only a handful of qualified teachers.

Practically all of the education available in fluid power until the mid-1960's was industry-sponsored, aimed at teaching companies' customers in the use of the companies' products. A few companies, as well as the U.S. Navy, developed excellent training manuals, which have been widely used. More highly technical and advanced manuals are available for engineers' and designers' use, but until the publication of this text, none was aimed specifically at the needs of vocational and industrial teachers and their students.

Several significant events have led to the preparations of this text. In early 1963, the National Fluid Power Association, the Fluid Power Society, the American Vocational Association, and the National Association of Industrial Teacher Educators organized and co-sponsored the Council for Fluid Power Education. The council's purposes were to investigate needs of fluid power education and to recommend positive programs to relieve shortage of personnel trained in this technological discipline. Its first recommendation was that a source of teachers of teachers be developed, since no teacher education institutions at that time included fluid power in their industrial and vocational education curricula. Subsequently, NFPA and FPS sponsored and financed a summer institute in fluid power education in 1964

at Wayne State University. A group of twenty carefully chosen teacher educators attended.

One of the studies undertaken by this group of teacher educators was the development of an instructor's handbook and bibliography of text material and visual aids in fluid power. Not intended as a text or curriculum guide, the manual simply referred the reader to recommended source material for any given aspect of fluid power technology. It highlighted, however, the need for a single text, combining the best from many unrelated sources.

In December, 1964, the Council on Fluid Power Education urged that the next steps to improve fluid power education should be the writing of a text adaptable to vocational and industrial education classes, and further special institutes for high school and technical institute teachers. During the summer of 1965, through a $234,000 grant made to the Fluid Power Society under the provisions of the Vocational Education Act of 1963, seven summer institutes attended by 165 teachers, were held throughout the United States. Simultaneously, through a grant made to the Fluid Power Society by George Carlson, Chairman of the Board, Minnesota Rubber Company, the society offered to underwrite the expense of a qualified author to develop a basic text.

A pioneer in fluid power education, Dudley Pease accepted this assignment. An organizer and charter member of the Fluid Power Society, he combines academic competence with a practical background in industrial experience. The fluid power curricula he has developed at the Kenosha Technical Institute, and the related laboratory facilities, are among the best in any part of the world.

The Fluid Power Society and its members and the many persons who contributed to and will benefit from this work can take justifiable pride in its completion and publication.

CONTENTS

BASIC
FLUID POWER

CHAPTER *I*

INTRODUCTION

Historical Background

We traditionally think of hydraulics in connection with water hydrology—the use of water to produce power by means of water wheels, or turbines. Likewise, air has been used for centuries to turn windmills or to propel ships.

But traditionally, the use of water or air to produce power depended on the movement of vast quantities of fluid at relatively low pressures. Also, we depended on nature to supply the pressure. Only in the past hundred years have we devised pumps and compressors to develop artificial pressures vastly greater than those nature provides, and correspondingly, we need to deal in vastly lower quantities of fluid as the operating pressure is increased. And here is the heart of the matter so far as advanced fluid power systems are concerned: the higher the pressure, the less the flow required to produce power, and the system becomes increasingly efficient.

Fluid power technology began about three hundred years ago with the discovery, in 1650, of Pascal's law. Simply stated, this law says that pressure in a fluid at rest is transmitted equally in all directions. A hundred years later Bernoulli developed his law concerning the conservation of energy in a flowing fluid. Another hundred years were to pass before these laws were applied to industry.

By 1850 the Industrial Revolution in Great Britain was well advanced. Electrical energy had not yet been developed to power the tools of industry, but fluid power became prominent as the means of powering presses, cranes, winches, and extruding machines. In fact, by the late 1860's cities like London and Manchester had central industrial hydraulic distribution systems similar in nature to central steam generating and distribution systems common today in some large cities. In these systems, large steam engines drove hydraulic water pumps delivering water at relatively high pressures through pipes to factories which used the water for powering machines. But the emergence of electricity in the late nineteenth century caused fluid power to be neglected. Our brain power was concentrated on refining the technology of electricity rather than on refining the application of fluid power.

However, it became apparent by 1900 that electrical systems had certain disadvantages, as did mechanical systems for transmitting power. Engineers again looked to fluid power for the answers to problems which other systems could not solve.

An important milestone in fluid power applications came in 1906, when a hydraulic system was developed to replace electrical systems for elevating and controlling guns on battleships, first installed on the USS *Virginia*. It is important also to note that by now oil was replacing water in fluid power systems.

In 1926 the United States developed what we call the *direct hydraulic system*—a type of packaged system with the pump, controls, and actuator in a self-contained unit on the machine or vehicle using it.

Since then, the applications of fluid power have

become virtually limitless, and include systems on machine tools, automobiles, earth-moving machines, farm equipment, ships and locomotives, airplanes, and even in space vehicles.

Fluid power is power transmitted and controlled through use of a pressurized liquid or gas. *Hydraulics* is the engineering science pertaining to liquid pressure and flow. *Pneumatics* is the engineering science pertaining to gaseous pressure and flow.

Modern fluid power performs work through direct or central power systems and is unrelated to the academic definitions common to civil engineering or to cross-country pipeline systems. Civil engineering involves a technology in hydroelectric systems using dams, wiers, etc., whereas fluid power deals with fluids as the power media operating directly with an actuating cylinder or fluid motor to perform work.

Fluid power jacks up an automobile, drills our teeth, operates computers, launches space ships, controls submarines, mines coal and ores, moves earth, harvests crops, and, in general, makes our everyday living easier and more enjoyable.

Over 90 per cent of all machine tools are controlled or operated with fluid power. Farm machinery is constantly improving with fluid power. Transportation, printing, materials handling, construction, chemical, mining manufacturing, processing, aerospace, and many other industries are finding more and more uses for fluid power.

Numerical control, military applications, and high performance machines are also important to the new and fast growing fluid power industry.

The fantastic growth of this new and exciting technology is opening many new opportunities for trained manpower in all phases of fluid power. Engineers, technicians, mechanics, sales personnel, servicemen, operators and supervisors are badly needed.

A critical shortage of trained teachers in the field of fluid power has opened interesting opportunities in education. Qualified teachers are needed for universities, colleges, technical institutes, vocational schools, and at the high school level.

Why Fluid Power?

Academically, fluid power provides flexible and easy control of variable force, distance, and speed. Fluid power can be varied from a delicate touch of a few ounces to a gigantic force of 36,000 tons or more. Tape-controlled machines can provide accuracies in measurements as low as plus or minus one ten-thousandth of an inch with amazing repeti-

tion. Fluid power can provide constant torque at speeds of nearly 100 miles per hour within a few inches or give a creeping speed of a fraction of an inch per minute.

Advantages of Using Fluid Power

1. Fluid power provides flexibility in the control of machines.
2. Fluid power provides an efficient method of multiplying forces.
3. Fluid power provides constant torque at infinitely variable speeds in either direction with smooth reversals.
4. Fluid power is compatible with other means of control, such as electrical, electronic, or mechanical.
5. Fluid power is accurate.
6. Fluid power gives fast response to controls.
7. Small forces can be amplified to control large forces.
8. Oil fluid power provides automatic lubrication for less wear.
9. Air fluid power is clean and safe from fire hazards.
10. Fluid power provides freedom in machine design.
11. Fluid power is simple and provides ease of installation and maintenance.
12. Fluid power is economical.
13. Fluid power is efficient and dependable.
14. Fluid power gives predictable performance.
15. Fluid power is readily available.

How Does Fluid Power Work?

A classic illustration showing fluid power in operation is the hoist used by automobile service stations to raise a car for servicing. This application of fluid power generally uses both a liquid and a gas.

The hoist operates on the principle that air under pressure is added on top of a confined quantity of oil within a long cylinder set in the ground. The cylinder is fitted with a long movable plunger that is sealed from the atmosphere. As the pressure is added to the system, the plunger moves upward, raising the automobile.

The hoist shown in Fig. I–1 works on a simple physical law that was discovered about 1650 by a French scientist named Blaise Pascal. The law states, "A pressure added to a confined fluid is transmitted undiminished throughout the fluid. It acts on all

AN AIR PIPE runs through the bottom of the plunger up to the top cylinder in this type of lift. Oil flows freely into the plunger and between its outer wall and the cylinder proper. An air valve is the only control. When it is held open to admit air from the compressor line, pressure builds up between the closed upper end of the plunger and the oil beneath, raising the plunger. When the air is shut off, the plunger is held at that height. To lower it, the valve is opened on the side to the atmosphere, letting the air in the plunger bleed off slowly.

Fig. I-1. Hoist. (*Courtesy of Globe Hoist Co.*). I-1a. An applied force on a small area produces a large force on a large area.

surfaces in a direction at right angles to those surfaces."

Translating the meaning of this fundamental law into a mathematical expression gives the following equation:

Pressure equals Force divided by the Area.

$$P = \frac{F}{A}$$

In other words, the pressure added increases the force per unit of area, or force equals pressure times area, $F = P \times A$.

Thus, it can be seen that as air enters the space above the oil in the hoist it creates a force, or push, against the area of the large plunger, causing the plunger to move upward.

If the hoist plunger has a cross-sectional area of 60 sq in., then a pressure of 100 psi of air would create a lifting force of 6000 lb.

$$F = P \times A$$
Force = 100 psi × 60 sq in.
Force = 6000 lb (3 tons)

When the serviceman wants to lower the hoist,

he merely bleeds air from the space above the oil, which lowers the pressure and the hoist (see Fig. I–2).

Components of the System

1. Air compressor to increase the pressure of air taken from the atmosphere (usually to about 120 psi).
2. Electric motor to drive the air compressor (usually about 2 hp).
3. A receiver tank to store the air energy until it is needed to raise the hoist. (This permits a smaller compressor and drive motor.)
4. Pipe lines to transmit the air power.
5. Control valves to let air into the hoist or out of the hoist.
6. Hoist unit consisting of the cylinder, plunger, seals, and other fittings.
7. Oil to lubricate the moving parts of the hoist and to help transmit the power to raise the hoist.

All fluid power systems use the principle of Pascal's law to create a force which can provide linear and rotary motion to perform work.

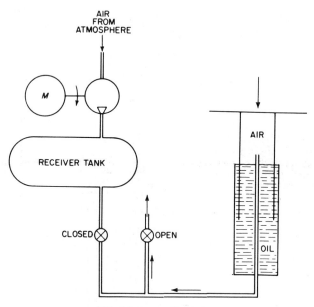

Fig. I-2. Lowering hoist.

A 50-ton Force Raises Utility Poles

The compact power unit shown in Fig. I–3 is portable and can be used in areas inaccessible to vehicle-mounted devices, and is generally used to replace poles.

The unit consists of a 3 hp gas engine, hydraulic pump, control valves, and two small cylinders. The pump operates at 10,000 psi, and the system is capable of creating 50 tons of force.

The controls allow cylinders to operate together or alternately to rock the pole (see Fig. I–4).

It is apparent from this sample application that fluid power can be small and light in weight, and yet can provide tremendous forces. Portable fluid power is doing many jobs on the farm, in lumbering camps, in construction projects, etc.

A New Concept in Farm Machinery

Figure I–5 shows a fluid power combine with design flexibility for placing the power components. Hydraulic drives provide full power for the wheels at low speed. It is also possible to obtain infinitely variable speeds independently of wheel drives.

Fluid Power Provides Remote Control for Giant Earth Movers

The giant tandem earth movers shown in Fig. I–6 use more than a dozen hydraulic circuits from a central hydraulic power source. One operator can control all functions of the machine from the driver's

Fig. I-3. A 50-ton force raises utility poles. (*Courtesy Owatonna Tool Co.*)

Fig. I-4. Controls allow cylinders to operate together or alternately to rock pole.

TOGETHER   ALTERNATELY

Fig. I-5. Fluid power combine. (*Courtesy The Dynex Co.*)

Fig. I-6. Tandem. (*Courtesy of Euclid*)

Fig. I-7. Bin gates. (*Courtesy of Parker-Hannifin.*)

seat. Front and rear scrapers, bowls, aprons, and ejectors are all independently operated with remote controls. Because hydraulic fluid is incompressible, controls operate smoothly with quick response.

The machine is powered with a 444 hp engine and carries 94,000 lb of earth. Both bowls can be loaded to capacity in 1.5 minutes.

Fluid Power Uses Air to Control Concrete Mixing

Fluid power has a tremendous role in the handling of materials. Figure I–7 shows three bin gates which are used in a large concrete mixing plant. Air power is used because of the precise control needed to inch the gates open and shut and to meter the flow of aggregate on to a conveyor belt.

Air power is also an excellent source of power for emergency operations in case of electrical failure.

High Speed Air Power Rivets Automobile Headlights

Riveting headlight assemblies is made simple with fast speed of operation by using eight air circuits. Fast cycling air valves operate six work cylinders and two ejector cylinders. Fluid power using air can provide rapid cycling for high production machines. Most air circuits operate on plant air pressure which is limited to approximately 100 psi; this limits air power to the lower force applications.

Fluid power helps to provide higher production with less cost of manufacture.

Fluid Power Steering and Brakes

Fluid power provides a safe and dependable power source for vehicle steering and braking systems.

The pump operates off the engine and keeps the system charged. A special valving arrangement provides fluid power for both steering and braking and keeps the accumulator charged. (The accumulator stores pressure energy that is available for use when the engine is off.)

All the hydraulic fluid from the pump flows through the brake valve. When the accumulator pressure drops to its low limit, the brake valve diverts a small amount of fluid from the system to charge the accumulator. The brake valve automatically drives the fluid back to the system when the accumulator is fully charged (see Fig. I–8).

Fluid Power for Numerical Control

The three-axis tape-controlled milling machine in Fig. I–9 is designed for numerical control and provides both profile and contour milling.

Dimensions from drawings and machining data

Fig. I-8. Full power hydraulic brake valve. (*Courtesy of The Bendix Corp.*)

Fig. I-9. Numerical control. (*Courtesy of Socony Mobil Oil Co.*)

are coded and put on a control tape. In contour milling where complex curves are involved, the numerous calculations require the use of an electronic computer. As the machining commands punched in the tape are fed into the machine, they are translated into motion and force for hydraulic control operations. Hydraulic actuation provides a very suitable link between the control tape signals and the actual machining motions because of ease of control, rapid response, variable speed, and amplification of force.

Fluid Power Builds Tractor Engines

The automatic transfer machine shown in Fig. I–10 handles over seventy tractor engine blocks per

Fig. I-10. Transfer machine. (*Courtesy of Socony Mobil Oil Co.*)

Fig. I-11. Stabilizing beer filling level. (*Courtesy of Oilgear Co.*)

hour. Castings entering the machine transfer automatically from station to station for clamping, milling, drilling, reaming, tapping and inspection operations. Fluid power provides efficient and dependable control. Emphasis is placed on good design, lubrication, and preventative maintenance to keep loss of production to a minimum. The failure of a single machine component might shut down the production line or even an entire plant's operation.

Hydraulic power plays a very important part in keeping machines operating smoothly and accurately for maximum production.

Fluid Power Bottles or Cans Beer

Figure I–11 shows an important use of fluid power in the making of beverages.

Highly carbonated beverages must be handled gently. Throttling flow of beer into filling machine reservoirs causes foam and irregularity of liquid height, making accurate filling of cans and bottles difficult. Fluid power provides the solution. Shown "on the job" in Fig. I–12 are six hydraulic two-way, "Any-Speed" transmission driven by 3 hp electric

motors, mounted on standard reservoir bases. Transmissions drive beer pumps through dual "V" belts. Beer is pumped from storage tanks to bottle or can-filling machine reservoirs on the floor above. Liquid level in filling reservoirs is controlled automatically by a low-pressure CO_2 signal: as the level rises, the pressure rises; as the level falls, the pressure falls. The CO_2 signal is relayed to controls on the transmissions, which automatically-instantly increase, decrease, reverse, or stop the beer pumps. Where former throttling systems caused foam and irregular levels, this system holds the reservoir level to $\pm \frac{1}{16}$ in. under continuous operation, assuring accurate filling of each can or bottle. Beer pumps are flushed under water pressure without disconnecting drives. Several years of continuous service have proven drives to be extremely dependable under all conditions, with little or no attention.

World's Largest Self-propelled Machine Uses Fluid Power

Figures I–13, I–14, and I–15 show the rear crawlers of Peabody Coal's 9000 ton, 115 cu yd,

Fig. I-12. 6 power units. (*Courtesy of Oilgear Co.*)

Fig. I-13. Stripping shovel. (*Courtesy of Oilgear Co.*)

Fig. I-14. Power unit. (*Courtesy of Oilgear Co.*)

Fig. I-15. Power unit. (*Courtesy of Oilgear Co.*)

"B–E" 3850–B Stripping Shovel. It is 90 ft taller than the Statue of Liberty, wider than an 8 lane highway, and has a 420 ft reach. A shows two of four single-acting cylinders that support, automatically level, and equalize loads. The cylinders are 54 in. in diameter, have an 80 in. stroke, and operate at 3000 psi. Each cylinder weighs 65 tons. B shows single-acting cylinders 19 in. in diameter with a stroke of 90 in., they control the 400 tons of force required to turn the crawlers through a 30 deg arc. They also operate at 3000 psi. C is the steering tie-rods that link crawlers. D is the lower frame of the machine, which is automatically held level within 2 deg. The shovel can move at an average speed of $\frac{1}{4}$ mile per hour Notice the size of just one crawler compared to the man pictured in the lower right-hand corner of the photo.

Fluid Power Mines Coal

With the aid of several hydraulic cylinders and jacks, the continuous coal mining machine shown in Fig. I–16 digs and loads coal at the rate of two tons per minute. The combination of a ripping head and loading equipment marks a long step in coal mine mechanization.

The ripping head and rear conveyor can be swung from side to side through an arc of about 45 deg, which makes it possible for the machine to get around corners when moving from one position to another. During operation, the ripping head is driven into the base of a coal seam and then swung upward. Hydraulic cylinders provide the necessary force. The hydraulic controls can be actuated from one location to operate the machine.

Fluid power is responsible for better, safer, and more efficient methods of mining coal.

Fluid Power Produces Toothbrush Handles, Pocket Combs, and Washing Machine Tubs

Smaller machines of the type shown in Fig. I–17 turn out such items as toothbrush handles, pocket combs, and other plastic products. This large injection molding machine produces larger items, such as washing machine tubs.

Granular plastic material is fed from the hopper (upper right) into a cylinder, where it is softened by heating. The huge dies are closed by the small cyinders above and below the large main ram shown in the center. As the dies close, the large ram's cylinder is prefilled with fluid from the reservoir and then locked closed with hydraulic force from a small holding pump. Flow from the main pump is then diverted to the injection-plunger cylinder (extreme right), which forces the plastic material into the die cavity with tremendous force. After curing, the dies are opened by means of the small auxiliary cylinders above and below the main ram.

A Hydraulically Controlled Goliath

The high-pressure hydraulic forging press in Fig. I–18 is one of the largest self-contained presses in existence. It develops 2000 tons of force and provides high-speed operation with 20 hydraulic pumps, each having a capacity of about 35 gallons per minute. The pumps are turned by ten double-end shaft electric motors rated at 150 hp each. Figure I–19 shows the power units which are separate from the press itself.

A Comparison of Methods of Power Transmission

The three general methods of transmitting power are electrical, mechanical, and fluid power. Transmitting power refers to the time rate of doing work,

Fig. I-16. Coal mining machine. (*Courtesy of Socony Mobil Oil Co.*)

Fig. I-17. Plastic injection molding machine. (*Courtesy of Socony Mobil Oil Co.*)

and work may be defined as overcoming resistance through any distance.

Electrical power uses nuclear energy, water turbines, or steam turbines to produce the voltage and current. Voltage is the intensity factor, and current is the action factor.

Mechanical power uses gravity, heat, combustion, steam, electricity, or fluid power to produce an intensity factor, which is torque, and an action factor of speed.

Fluid power uses compressors, pumps, or intensifiers to produce an intensity factor of pressure and

Fig. I-18. 2000 ton self-contained forging press. (*Courtesy of Denison Engineering.*)

Fig. I-19. Power units. (*Courtesy of Denison Engineering.*)

an action factor of cubic feet of gas per minute or gallons of liquid per minute (cfm or gpm).

Thus, horsepower for the three methods can be stated as follows:

$$\text{Electrical horsepower} = \frac{\text{voltage} \times \text{current}}{*746 \text{ watts}}$$

(Watt is a unit of electrical power, and 746 watts equal one horsepower.)

$$\text{Mechanical horsepower} = \frac{\text{torque} \times \text{speed}}{*5252}$$

$$\text{Fluid power} \begin{cases} \text{Hydraulic horsepower} \\ = \dfrac{\text{pressure} \times \text{gpm}}{*1714} \\ \text{Pneumatic horsepower} \\ = \dfrac{\text{pressure} \times \text{cfm}}{\dagger\text{constant}} \end{cases}$$

In each of the above equations, increasing either the intensity factor or the action factor can vary the resulting power.

1. Force—measured in pounds, tons, grams, or kilograms.
2. Length—measured in inches, feet, yards, rods, miles, etc.
3. Time—measured in seconds, minutes, hours, etc.

Other quantities can be expressed in terms of these three. For example, velocity is defined as distance (length) divided by time. Thus, velocity may be miles/hour, feet/minute, feet/second, inches/day, etc. Mathematcalily stated the equation reads.

$$V = \frac{d}{t}$$

where V = velocity, d = distance, and t = time. Work and energy have the same units.

$$\text{Work} = \text{energy} = F \times d$$

where W = work, F = force on the load, d = distance the load moves. Notice that if the load does not move when a force is applied, $d = 0$, $F \times 0 = 0$;

COMPONENTS USED

	Electrical	Mechanical	Fluid
Distribution or transmission	Wires, Cables	Wheels, shafts, gears, levers, cams, wedges, screw, pulleys, beams, belts, links, clutches, etc.	Pipes, tubes, hoses
Intensity control	Switches, transformers, regulators, relays, distributors	The form of parts transmitting torque depends on the space, weight, size, shape, etc.	Pressure valves, directional valves, flow control valves
Action device	Solenoids, motors	Wheels, pistons, levers, gears, linkage, etc.	Fluid motors, rams, and cylinders
Energy reserve	Batteries and condensers	Flywheels	Accumulators and receivers
Safety devices	Fuses, circuit breakers	Safety latches, strength of part	Relief valves, pressure switches, blowout disc

Mathematics and Principles of Physics Should Be Reviewed

The fluid power technology is somewhat related to all systems of power transmission, because it is flexible to apply and compatible with other means of control. It is, therefore, essential that students in fluid power education understand the general laws of mechanics and the units of measurements involving forces, lengths or distances, and time.

*5252 and 1714 are derived from a unit of one mechanical horsepower, which equals 33,000 ft-lbs per minute.

†Constant is derived from 33,000 ft-lbs per minute, but varies with temperature conditions.

no work is done. In order to have work done, the load must move (see Fig. I–20).

Power is expressed as $\frac{W}{T} = \frac{\text{work}}{\text{time}} = \frac{F \times d}{T}$. Notice that $\frac{d}{t}$ = velocity. We could write the power equation as $P = FV$.

The units of power are most commonly $\frac{\text{ft}-\text{lb}}{\text{min}}$ or $\frac{\text{ft}-\text{lb}}{\text{sec}}$.

Before the internal combustion engine and electric motor were widely used, a test was performed with a horse as the power source. It was found that a good draft horse could do 33,000 ft–lb of work in one

Fig. I-20. Work = Force × Distance.

minute's time or 550 ft–lb of work in one second's time. Thus originated the term *horsepower*. 1 horsepower

$$(hp) = 33,000 \frac{ft-lb}{min} = 550 \frac{ft-lb}{sec} \text{ (see Fig. I–21)}.$$

Example of use of terms. A man unloads steel stock from a flatbed to a cart, a distance of 4 ft. He unloads 1 ton of steel in 1 hour's time.

Work done = $F \times d$ = 2000 lb × 4 ft = 8000 ft–lb

$$\text{Power used} = \frac{W}{T} = \frac{F \times d}{t} = \frac{2000 \text{ lb} \times 4 \text{ ft}}{1 \text{ hr}}$$

$$= 8000 \frac{ft-lb}{hr}$$

$$\text{hp used} = \frac{P}{33,000 \frac{ft-lb}{min}}$$

$$= \frac{8000 \frac{ft-lb}{hr}}{33,000 \frac{ft-lb}{min}}$$

To make the units cancel out, change hours to minutes.

$$\frac{8000 \frac{ft-lb}{hr} \times \frac{1 \text{ hr}}{60 \text{ min}}}{33,000 \frac{ft-lb}{min}} = \frac{8000 \frac{ft-lb}{60 \text{ min}}}{33,000 \frac{ft-lb}{min}}$$

Fig. I-21. Derivation of equation for H. P. formula

$$= \frac{\frac{8000}{60}}{33,000} hp = 0.004 \, hp$$

We have just seen that moving an object through a distance is work. Oil has weight; therefore, in order to move it work must be done. When work is done, power is consumed.

$$P = \frac{F \times d}{t}$$

The amount of power consumed depends upon the quantity of oil, the force required to move it, and the time.

$$P = \frac{gal}{min} \times \frac{lb}{in.^2}$$

Changing gal to in.³ and in. to ft to get units of power, we have

$$P = \frac{gal}{min} \left(\frac{231 \text{ in}^3}{gal}\right) \times \frac{lb}{in^2} \left(\frac{1 \text{ ft}}{12 \text{ in}}\right)$$

$$= \frac{231 \text{ ft}-lb}{12 \text{ min}}$$

$$Hp = \frac{\frac{231 \text{ ft}-lb}{12 \text{ min}}}{33,000 \frac{ft-lb}{min}} = 0.000583 \, hp$$

The hydraulic horsepower formula becomes

$$hp = gpm \times psi \times 0.000583$$

Example. Oil being moved in a system is 12 gpm at a pressure of 1500 psi. What horsepower is being consumed?

$$hp = gpm \times psi \times 0.000583$$

$$= 12 \times 1500 \times 0.000583 = 10.5 \, hp$$

Another useful measurement is torque, which is used in relation to turning or rotary motion. Torque is measured in lb–ft or lb–in.

$$T = F \times d$$

Torque = force × lever arm or radius

$$P = \frac{2\pi RTN}{33,000 \times 12}$$

Clearing the fraction gives

$$hp \times 33,000 \times 12 = 2\pi RTN$$

Equating for unity, we have

$$1 \times 33,000 \times 12 = 2 \times 3.1416 \times 1 \times T \times 1$$

$$T = \frac{33,000 \times 12 \times 1}{2 \times 3.1416 \times 1 \times 1} = \frac{396,000 \times 1}{62,832 \times 1}$$

$$= \frac{63,025 \times hp}{1}$$

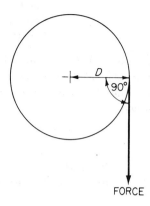

Fig. I-22. Torque = Force × Distance.

$$T = \frac{63{,}025 \times \text{hp}}{\text{rpm}}$$

(Formula for torque)

Notice in Fig. I–22 that the factors F and d are the same as for the work formula ($W = F \times d$). The difference lies in that the d for the work formula is the distance that the load moves while the force is being applied, whereas the d in the torque formula represents the length of the lever arm on which the force is acting. In order to have work done, there must be motion. Torque requires only that a force be applied to *try* to cause a turning effect.

Example. A mechanic tries unsuccessfully to loosen a nut. He exerts a force of 70 lb on the wrench handle, which is 18 in. long. What torque was developed?

$$T = F \times d = 70 \text{ lb} \times 18 \text{ in.} = 1260 \text{ lb–in.}$$

or, change inches to feet

$$= 1260 \text{ lb}$$
$$- \text{in.}\left(\frac{1 \text{ ft}}{12 \text{ in.}}\right)$$
$$= 105 \text{ lb} - \text{ft}$$

It is not possible to get out of a machine all the energy put into it, because of friction of moving parts, heat losses, leakage, etc. If 100 units of energy are put into a machine and 75 units of energy are available from the machine, it is 75 per cent efficient.

Mathematically,

$$\frac{w_0}{w_i} \times 100\% = \text{efficiency}$$

where w_0 = work output.

w_i = work input.

Also, $w_i = w_0 + w_l$, where w_l = work lost due to heat,

friction, etc. Because power is a function of work, we can also say:

$$\frac{P_0}{P_i} \times 100\% = \% \text{ efficiency}$$

where P_0 = power output.

P_i = power input.

$P_i = P_0 + P_l$

where P = power lost.

A review of the basic rules of algebra and trigonometry will be helpful toward a better understanding of applying fluid power.

BASIC ALGEBRAIC FORMULAS

A plus 0 equals A.

$A \times 1$ equals A.

A plus B equals B plus A.

$A \times B$ equals $B \times A$.

A plus (B plus C) equals (A plus B) plus C.

$A(BC)$ equals $(AB)C$.

$A(B$ plus $C)$ equals AB plus AC.

A^n equals $A \times A \times A \times A \times \cdots \times A$ n times.

If $\sqrt[n]{A}$ equals Y, then Y^n equals A.

A^{-n} equals $\dfrac{\text{I}}{A^n}$; $A^{m/n}$ equals $\sqrt[n]{A^m}$.

$A^m \times A^n$ equals $A^{m \text{ plus } n}$; $(A^m)^n$ equals A^{mn}.

$A^n \times B^n$ equals $(AB)^n$.

A/B plus C/D equals DA/DB plus BC/DB equals

$\dfrac{DA \text{ plus } BC}{DB}$.

AB/AC equals B/C.

Work accomplished by fluid power may be linear or nonlinear. This means that loads may move in a straight line in relationship to an actuator, or the work may take a circular or irregular path of travel. When it is essential to compute the effects of loads moving on an inclined plane or when levers and linkage are used, trigonometry is a useful tool.

Fig. I-23. Triangle.

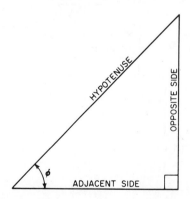

BASIC TRIGONOMETRIC FORMULAS

$\sin \phi = a/c$	opposite side/hypotenuse
$\cos \phi = b/c$	adjacent side/hypotenuse
$\tan \phi = a/b$	opposite side/adjacent side
$\csc \phi = c/a$	hypotenuse/opposite side
$\sec \phi = c/b$	hypotenuse/adjacent side
$\cot \phi = b/a$	adjacent side/opposite side

Reciprocal relations

$\cot = 1/\tan$
$\sin = 1/\csc$
$\cos = 1/\sec$
$\tan = 1/\cot$
$\csc = 1/\sin$
$\sec = 1/ocs$

Angle	sin	cos	tan	cot	sec	csc
45°	$1/\sqrt{2}$	$1/\sqrt{2}$	1	1	$\sqrt{2}$	$\sqrt{2}$
30°	$1/2$	$\sqrt{3}/2$	$1/\sqrt{3}$	$\sqrt{3}$	$2/\sqrt{3}$	2
60°	$\sqrt{3}/2$	$1/2$	$\sqrt{3}$	$1/\sqrt{3}$	2	$2/\sqrt{3}$

Sample problem. Find the amount of force necessary to move 7000 # (pounds) from a standing stop up an incline of 15 deg (0.04 coefficient of friction).

Fig. I-24. Example problem.

Force = (weight × sin θ) + (friction × weight × cos θ)

7000 × 0.259 + 0.04 × 7000 × 0.966

Answer: 2083.48 lb.

Many good textbooks are available for a review of basic mathematics and laws of physics. The foregoing discussion merely highlights some of the important tools helpful in the field of fluid power.

Questions

1. What is the basic law that is important in applying fluid power and when was it discovered? By whom?
2. What theorem pertains to velocity energy, pressure energy, and friction energy?
3. How is fluid power used by the military?
4. List ten fields of application where fluid power can be more effectively used than other power sources.
5. Compare the use of fluid power to a mechanical system, listing the advantages and disadvantages of each.
6. Explain in your own words how the hoist works to lift your car for servicing in a gas station.
7. What advantage of using fluid power is indicated by the application of raising utility poles?
8. What advantages does fluid power give to the design of farm machinery?
9. Generalize on the difference between using pneumatic fluid power and hydraulic fluid power.
10. What effect has fluid power had on automation?
11. Why is hydraulic power especially useful with heavy work?
12. Mathematically express horsepower for the three general methods of transmitting power.

SECTION **I**

HYDRAULICS

CHAPTER **1**

BASIC PRINCIPLES
OF HYDRAULICS

The term *hydraulics* as treated in this chapter pertains to *power transmitted and controlled through the use of a pressurized liquid. Because liquids and gases are both fluids,* it should be understood that *fluid power* is the general term for both hydraulics and pneumatics.

Hydraulics, according to Webster's dictionary, is the branch of physics having to do with the mechanical properties of water and other liquids and the application of these properties in engineering.

Modern hydraulic power systems most generally use petroleum oils or other liquids which provide automatic lubrication for all integral moving parts of the components making up the system. Hydraulic systems may use water, petroleum oils, synthetic fluids, blends and mixtures, or even liquid metals for the media in transmitting fluid power. All liquids used for hydraulic power have certain inherent physical properties which are important for an understanding of hydraulic machines.

Liquids have characteristics similar to those of solids, except that liquids can be pumped and transmitted through pipes and tubes to perform the work.

Liquids possess qualities that are common to both gases and solids, and may be thought of as "fluid solids" from the viewpoint of transmitting fluid power. A comparison of gases, liquids, and solids and their molecular structure is shown in Fig. 1–2.

The molecules in gases possess more kinetic energy (the energy of movement) because they are constantly in motion at high velocity. The gas molecules must be compressed more closely together before forces are transmitted. The transmittal of these forces is in all directions.

The molecules in liquids possess less kinetic energy than those in gases. They are more closely packed, like the molecules in solids. The liquid molecules transmit forces with quick response because they are almost incompressible. The transmittal of these forces is in all directions.

The molecules in solids possess less kinetic energy than those in liquids. They stay in a closely packed condition, and hence act in unison to transmit a force in *only one direction.*

The illustrations above show the main advantage of using fluid power over mechanical systems, which is the ability of fluids (either gases or liquids) to transmit forces in all directions. Thus it is apparent that fluids can be transmitted through pipes, tubes, and controls to end up as energy to perfrom useful work.

Molecules in liquids are flexible, but rigid enough to maintain their fixed distance between each other, except when acted upon by extremely large external forces. Large metal presses would be an example of a hydraulic machine where the compressibility

Fig. 1-1. Incompressibility of liquids.

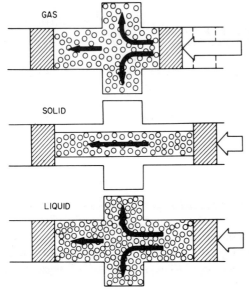

Fig. 1-2. Molecular structure of gases, liquids and solids for comparison.

components of the system regardless of their shape or size.

Liquids have weight. Density of a liquid refers to the weight per unit volume:

$$\text{Density} = \frac{\text{weight}}{\text{volume}} \qquad \rho = \frac{Wt}{V}$$

The density of water = 62.4 lb/ft³. In other words, if a cubic foot of water at 39° For 4°C were accurately weighed with the weight of the container

Fig. 1-3. Hydraulic press. (*Courtesy of Denison Engineering.*)

factor would be important to consider. Big presses such as that shown in Fig. 1–3 have large quantities of liquid under pressures of 5000 pounds per square inch, or higher. This makes it imperative to decompress the liquid before releasing the press.

The accepted compressibility factor for most hydraulic power systems is one-half of one per cent of the liquid volume under pressure for each 1000 pounds per square inch. This is illustrated in Fig. 1–4.

Removing the weight and reducing the pressure will cause the oil volume to return to its original value almost instantly.

Liquids seek their own level regardless of the shape of the container (see Fig. 1–5). When liquids are subjected only to the force of gravity, they seek their own level. This important physical property allows the hydraulic fluid to fill the cavities in all

Fig. 1-4. Compressibility of a liquid.

Fig. 1-5. Various container shapes.

subtracted, it would weigh 62.4 lb. Its density would be expressed as 62.4 lb per ft^3.

Example. What is the density of a quart of hydraulic oil weighing 1.5 lb and occupying a volume of 57.8 in.^{3}?

$$\rho = \frac{\text{Wt}}{\text{V}} = \frac{1.5 \text{ lb}}{57.8 \text{ in.}^3} = 0.0259 \text{ lb/in.}^3$$

Changing in.^{3} to ft^3,

$$\frac{0.0259 \text{ lb}}{\text{in.}^3} \left(\frac{1728 \text{ in.}^3}{\text{ft}^3} \right) = \frac{44.8 \text{ lb}}{\text{ft}^3}$$

The specific gravity (Sp. Gr.) is a comparison of the weight of an oil or other liquid to the weight of an equal volume of water. The specific gravity of water as a standard equals 1.

$$\text{Sp. Gr.} = \frac{\text{weight of liquid}}{\text{weight of equal volume of water}}$$

$$= \frac{\rho \text{ of liquid}}{\rho \text{ of water}}$$

Example. If a cubic foot of hydraulic oil weighs 45 lb, what is its specific gravity?

$$\text{Sp. Gr.} = \frac{\rho \text{ of liquid}}{\rho \text{ of water}} = \frac{45 \text{ lb/ft}^3}{62.4 \text{ lb/ft}^3} = 0.723$$

Notice that all the units cancel out, making specific gravity a dimensionless number. The oil in this case has a specific gravity of 0.723 compared to that of water which is equal to 1.

Knowing the specific gravity of any hydraulic fluid, one can easily compute the weight or density.

$$\text{Density (lb/ft}^3) = \text{Sp. Gr.} \times 62.4 \text{ lb/ft}^3$$

Example. If a hydraulic oil has a specific gravity of 0.90, what is its density in lb/ft^3?

$$\text{Density lb/ft}^3 = 0.90 \times 62.4 \text{ lb/ft}^3$$

$$\rho = 56.16 \text{ lb/ft}^3$$

When a force is applied to the confined liquid of a hydraulic system, the liquid will take the flow path of least resistance. Figure 1–6 shows that input power is lost when leakage in the hydraulic system piping occurs.

The Relationship of Force and Pressure in a Hydraulic System.

Knowing the difference between *force* and *pressure* is one of the most important steps toward an understanding of how hydraulic systems work.

Hydraulic cylinders are used to create a force which can push or pull a load, thus providing linear motion. Hydraulic motors are used to create a

Fig. 1-6. Fluid taking path of least resistance (power peak).

Fig. 1-7. Linary rotary motion.

force in a circular path, thus providing rotary motion (see Fig. 1–7).

In both linear and rotary motion the forces created may be measured in pounds, grams, tons, kilograms, or any unit of weight. The unit of force in each case is associated with the work being done by the cylinder or hydraulic motor.

It should also be understood that the *output forces are being transmitted through the use of a pressurized liquid* which is being supplied to the cylinder and motor by the hydraulic pump.

The principle of all hydraulic systems is based on Pascal's law, which states, "Pressure exerted on a confined liquid is transmitted undiminished in all

21

ONE UNIT OF AREA

Fig. 1-8. Principle of Pascal's Law.

directions and acts with equal force on all equal areas" (see Fig. 1–8).

Hydraulic liquids transmit an applied force undiminished. A force applied on one unit of area of a confined liquid will be transmitted equally and undiminished to each unit of area of the confining container regardless of the shape of the container. Fig. I–8 shows how an applied force can be transmitted through the use of a hydraulic fluid which is confined between two movable pistons.

In Fig. 1–9 the area of the second piston has been increased to twice the area of the first piston. With the fluid contained between the two movable pistons, the same 100 lb of applied force is now doubled. Through the application of this basic principle, a small child can easily lift an automobile with a hydraulic jack.

It is apparent from Fig. 1–10 that the more area on a piston, the more force that piston will exert with the same applied force. In this simple system, each unit of area in the vessel is under an applied force of 100 lb. This could also be stated that the pressure in the system is 100 lb per unit of area.

Fig. 1-9. Pascal's Law illustrated with two pistons.

Fig. 1-10. Four cylinders.

Thus, pressure may be defined as force per unit of area. Mathematically, pressure can be expressed as "Pressure is equal to force divided by area."

$$\text{Pressure} = \frac{\text{force}}{\text{area}}, \qquad P = \frac{F}{A}$$

The units of area may be square inches or square feet. Generally, square inches are used for most calculations concerning fluid power.

The units of force are pounds or tons, with pounds being the most commonly used. Stated again, *pressure* (pounds per square inch, or psi) is equal to *force* (in pounds) *divided* by *area* (in square inches) or $P = F/A$.

This is the most important law in the application of fluid power, and it is necessary to keep force and pressure distinguished at all times. The following

FORCE ON EACH
sq. in. = 0.433 lb

Fig. 1-11. Cu. ft. of water including 1 sq. in. column.

explanation will provide further emphasis on this important difference.

A cubic foot of water (see Fig. 1–11) weighs 62.4 lb. It could be stated that a cubic foot of water exerts a force of 62.4 lb on the bottom of the container. This could also be stated that the weight of one cubic foot of water exerts a pressure of 62.4 lb per square foot. To find pressure in pounds per square inch it is necessary to divide the number of square inches in the base ($12 \times 12 = 144$) into the total weight or force.

$$P = \frac{\text{force}}{\text{area}} = P = \frac{62.4 \text{ lb}}{144 \text{ sq in.}}$$

Thus, pressure (psi) $= 0.433 \dfrac{\text{lb}}{\text{in.}^2}$

Fig. 1-12. Force diagram for Problem Ex. 1.

In order to calculate pressure, the force, either applied or transmitted, must be known, and the area on which the force is acting upon must be known.

Example 1. A force of 200 lb is being exerted by a hydraulic cylinder (see Fig. 1–12). The area of the piston is 2 in.². What is the pressure in the system?

Solution. Known: Force = 200 lb, area = 2 in.².
Find: Pressure (psi)

The formula states $P = F/A$

When the known values are substituted into the formula, it becomes

$$P = \frac{F}{A} = \frac{200 \text{ lb}}{2 \text{ in.}^2}$$

Therefore, the pressure in the system is 200 lb per 2 sq in. or 100 lb per square inch (psi).

Example 2. A hydraulic cylinder (see Fig. 1–13) exerts a force of 500 lb. If the cylinder has a bore of 3 in., what is the pressure in the system?

Solution. In order to find the pressure, use the formula

$$P = \frac{F}{A}$$

The force is given as 500 lb, but the area is not given. However, the diameter of the bore (also the diameter of the piston) is given as 3 in. From geometry, $A = \pi r^2$. This formula will yield the area. However, one must divide the diameter by 2 in order to use the formula. In hydraulics the diameter, or bore size, is most easily found, so it would be advantageous to have a formula for area expressed in terms of the diameter:

$$A = \pi r^2 \qquad \text{[} r \text{ (radius) is equal to } \tfrac{1}{2} \text{ of the diameter]}$$

$$r = \frac{d}{2}$$

Fig. 1-13. Force diagram for Problem Ex. 2.

When writing the formula, instead of putting in r, put in what r is equal to, namely $d/2$. The formula now becomes

$$A = \pi(\frac{d}{2})^2 = \pi d^2/4 \text{ and } \pi \text{ is a constant whose}$$

value is 3.1416, so

$$A = \frac{3.1416 \, d^2}{4} = 0.7854 \, d^2$$

This provides a formula for area expressed in terms of the diameter:

$$A = 0.7854 \, d^2 = 0.7854(3 \text{ in.})^2$$
$$= 0.7854(9 \text{ in.}^2) = 7.0686 \text{ in.}^2$$
$$P = \frac{F}{A} = \frac{500 \text{ lb}}{7.0686 \text{ in.}^2} = 70.74 \frac{\text{lb}}{\text{in.}^2}$$

Because it is a mathematical expression, the equation for pressure can be stated in other forms:

$$P = \frac{F}{A}, \qquad F = PA, \quad \text{and} \quad A = \frac{F}{P}$$

Example 3. If it were required to find the area of a cylinder exerting 200 lb of force under a pres-

Fig. 1-14. Diagram for Problem Ex. 3.

sure of 100 psi (see Fig. 1–14), the formula $A = F/P$ could be used.

$$A = \frac{F}{P} = \frac{200 \text{ lb}}{100 \, \frac{\text{lb}}{\text{in.}^2}} = \frac{2}{\frac{1}{\text{in.}^2}} = 2 \text{ in.}^2$$

Example 4. Given a cylinder with a piston area of 2 in.² operating at a pressure of 200 psi (see Fig. 1–15), what force could the cylinder exert?

Solution. $F = PA$

$$F = 200 \, \frac{\text{lb}}{\text{in.}^2} (2 \text{ in.}^2) = 400 \text{ lb}$$

In using Pascal's law, it is important to remember the units of each expression:

Force = F = lb

Pressure = P = lb/in.²

Area = A = in.²

It may be an aid in remembering the three forms of Pascal's law to use the "force triangle" method illustrated in Fig. 1–16. To use it, simply put your thumb on the factor you wish to find. Do the indi-

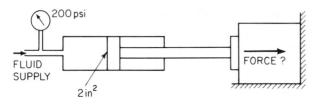

FLUID SUPPLY

FORCE ?

200 psi

2 in²

Fig. 1-15. Diagram for Problem Ex. 4.

(a)

Fig. 1-16. (a) & (b) Force triangle.

cated operation with the remaining factors. For example, to find pressure, cover P. The remaining factors appears as F/A, so divide the force by the area and pressure will be the result. To find force, cover F. The remaining factors appear as PA, so multiply pressure by area to obtain the force.

Figures 1–17a and 1–17b use the following symbols: EM, the electric motor which drives the pump; P, the hydraulic pump which delivers the fluid to the system; T, the tank for storing, supplying, and conditioning the fluid; R, the relief valve to protect the system from overpressure; G, pressure gauges to determine psi at different points of the system; C, actuating cylinders which perform linear work; S, shutoff valve.

Study the diagram with the following questions in mind:

1. With the pump in operation, which cylinder extends first and at what pressure? Second? Third?
2. What would happen if valve S were opened before the cylinders were fully extended?

There is a great deal of hydraulic knowledge that can be obtained from this illustration, so study it carefully.

To answer the first question, the force triangle, Fig. 1–17b, provides the relationship. (Disregard any frictional losses.) It is easy to see that when a pres-

Fig. 1-17. (a) & (b) Three cylinders.

Fig. 1-18. Intensification.

FORCE $\quad F = P \times A \quad$ PRESSURE $= \frac{F}{A} \quad$ RESULTANT FORCE

$F = 1000 \times 10 \qquad P = \frac{10,000}{2} \qquad F = P \times A$

$F = 10,000^* \qquad\qquad P = 5000\,psi \qquad F = 5000 \times 100$

$\qquad\qquad\qquad\qquad\qquad\qquad\qquad\qquad F = 500,000^*$

sure of 600 psi is attained cylinder 3 will extend. Cylinders 2 and 3 in turn will follow at pressures of 1200 psi and 2400 psi, respectively.

If valve S were opened before cylinder 1 was fully extended, pressure energy would be lost, because the fluid would follow the path of least resistance. Consequently, the work being done by the cylinder would stop.

Hydraulic fluids under pressure can also be used to multiply forces, as shown in Fig. 1–18.

Force $\quad F = P \times A \qquad\qquad$ Pressure $= \dfrac{F}{A}$

$\qquad F = 1000 \times 10\text{in.}^2 \qquad P = \dfrac{10,000\ \text{lb}}{2\ \text{in.}^2}$

$\qquad F = 10,000\ \text{lb} \qquad\qquad P = 5000\ \text{psi}$

Resultant Force

$\qquad F = P \times A$

$\qquad F = 5000 \times 100\ \text{in.}^2$

$\qquad F = 500,000\ \text{lb}$

Figure 1–19 shows that 1000 psi working against an area of 10 sq in. produces a force of 10,000 lb. Using this force of 10,000 lb against a smaller area of 2 sq in., we obtain a resultant pressure of 5000 psi. Using the 5000 psi in turn against a large area of 100 sq in. produces a resulting force of 500,000 lb (250 tons).

This example shows the important differences between force and pressure. Further, it points up the fact that through the use of hydraulics a force of any magnitude may be obtained.

Important Concepts and Principles Affecting Flow in Hydraulic Systems

Pressure energy in a hydraulic system is a form of potential energy and is, therefore, essentially static in character. This form of energy is referred to as hydrostatic pressure or pressure head. Hydrostatic pressure is the pressure energy in a liquid which is at rest or relatively motionless. This pressure energy may be developed by the weight of the liquid under pull of gravity, or it may be caused by the action of a hydraulic pump or other mechanical device.

Potential energy = weight × height

Example. What is the pressure at the bottom of a tank 40 ft high filled with oil having a specific gravity of 0.93? (see Fig. 1–19a).

$$\text{Sp. Gr.} = \frac{\text{Weight (oil)}}{\text{Weight (water)}}$$

$$0.93 = \frac{x}{62.4\ \text{lb/ft}^3}$$

Wt (oil) = Sp. Gr. × Wt (water)

Wt (oil) = 0.93 × 62.4 lb/ft³

Wt (oil) = 58.03 lb/ft³

Since a cubic foot of the oil in question weighs 58.03 lb, the pressure developed by one foot of height would be calculated as follows:

$$\underset{(psi)}{\text{Pressure}} = \frac{\text{lb force}}{\text{Area in.}^2}$$

The number of square inches in the base of the cube in Fig. 1–19b is 12 × 12 = 144 sq in. Dividing this area into the force due to the weight of the oil equals 0.4 psi.

$$\text{Pressure} = \frac{58.03\ \text{lb/ft}^3}{144\ \text{in.}^2} = .403$$

Pressure = 0.4 psi

Now that the pressure has been found for one foot of height, the total pressure head can be found by multiplying 0.4 psi × total height.

$P_h = \text{height} \times 0.4\ \text{psi}$

$P_h = 40 \times 0.4\ \text{psi}$

$P_h = 16\ \text{psi}$ (pressure due to head of liquid

$\qquad\qquad$ if height is 40 ft)

The pressure energy due to the height of the oil in the tank could now be used to perform work. Suppose the cylinder illustrated in Fig. 1–20 were fitted with a movable piston having 24 sq in. How many pounds of force can be exerted with 16 psi?

Fig. 1-19. (a) Specific gravity; (b) Cubic feet of oil.

$$F = P \times A$$
$$F = 16 \times 24$$
$$F = 384 \text{ lb}$$

However, it should be pointed out that the height of the oil in the tank would have to be maintained for the full stroke of the piston. In addition, there could be no losses due to friction or other factors if the work done is 384 ft lb. (Work = force × distance.)

It is apparent from this discussion that a power pump would be much more efficient to maintain the pressure energy to do the work. A power pump would also take care of any losses which occur, such as those caused by friction, if the proper pump is selected to do the job.

There is another phase of potential energy important to the design of hydraulic machinery. Weighted members of any machine that may supply energy because of their relative position must also be considered. A vertical press, whose platen and dies may weigh several hundred pounds, is an example.

Conversion formulas:

$$\text{Pressure in psi} = \frac{\text{head (ft)} \times \text{Sp. Gr.}}{231}$$

$$\text{Head in feet} = \frac{\text{pressure (psi)} \times 2.31}{\text{Sp. Gr.}}$$

Inches of mercury vacuum = suction lift (ft) × 0.883 × Sp. Gr.

Fig. 1-20. Cylinder using pressure head.

Kinetic energy is another form of energy which involves hydraulic machinery in motion. Kinetic energy of an incompressible mass is given by the following equation:

$$\text{Kinetic energy} = \frac{\text{weight of mass} \times \text{velocity}^2}{2 \text{ gravity (32 ft/sec/sec)}}$$

Hydraulic fluid in motion is a result of pressure energy's being transformed into kinetic energy. It has also been discovered that many factors enter into the design of hydraulic systems because of the behavior of fluid in motion.

Bernoulli's Equation

In general, a study by Daniel Bernoulli of fluids in motion disclosed that when fluids flow through pipes or tubes *under steady-state conditions,* where the velocity is high, the pressure decreases; where the velocity is low, the pressure increases. This concept

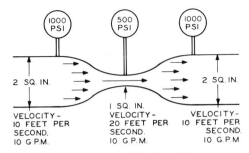

Fig. 1-21. Bernoulli's Equation.

considers a perfect, frictionless fluid and is illustrated in Fig. 1–21.

If no work is done on or by a flowing frictionless fluid, its energy due to pressure and velocity remains constant at all points along the streamline.

Assume that the pump is delivering 10 gpm through the tube, and the pressure is 1000 psi. Also assume the fluid to have a velocity of 10 fps. When the fluid enters the narrow section of the tube, where the area is only one-half that of the larger section, the velocity increases. This happens because the 10 gallons of oil per minute must get through this narrow section. In other words, the same number of molecules of oil must pass through an area only one-half as large. Thus the fluid must travel twice as fast in order to get all the molecules of the oil through this narrow section in the same increment of time. If there are no frictional factors involved, the velocity of flow doubles in the narrow section and the pressure energy decreases by one-half.

Steady-state flow means that the energy contained by the liquid remains constant throughout the tube

Fig. 1-22. Bernoulli's Theorem, Ex. Problem 1.

and that as velocity increases, pressure decreases, or as velocity decreases, pressure increases.

Example.

The reservoir for a hydraulic system shown in Fig. 1–22 is air supercharged at 15 psi gauge. The inlet line to the pump is 6 ft below the oil level. The flow into the inlet port of the pump (point 2) is 30 GPM. Calculate the pressure at point 2 if:

a. It is assumed that there is no head loss from point 1 to point 2.

b. There is a 20 ft head loss from point 1 to point 2.

Solution. Bernoulli's law—If no work is done on or by a flowing frictionless liquid, its energy due to pressure and velocity remains constant at all points along the streamline. In equation form:

$$Z_1 + \frac{144P_1}{\rho_1} + \frac{V_1^2}{2g} = Z_2 + \frac{144P_2}{\rho_2} + \frac{V_2^2}{2g}$$

where Z = head or reference level above an arbitrary base reference level (ft).

P = pressure of fluid (psi).

ρ = density of fluid (lb per cu ft).

V = mean velocity of flow (ft per sec).

g = acceleration due to gravity (32.2 ft per sec²).

Part a.

$$Z_1 + \frac{144P_1}{\rho_1} + \frac{V_1^2}{2g} = Z_2 + \frac{144P_2}{\rho_2} + \frac{V_2^2}{2g}$$

$$6 + \frac{(144)(15)}{53.1} + 0 = 0 + \frac{(144)P_2}{53.1} + \frac{(3.07)^2}{(2)(32.2)}$$

$$6 + 40.7 = 2.715P_2 + 0.1463$$

$$\frac{46.6}{2.715} = P_2$$

$$17.15 \text{ psig} = P_2$$

Part b.

$$\frac{26.554}{2.715} = P_2$$

$$9.78 \text{ psig} = P_2$$

Continuity Equation

Matter such as hydraulic oil cannot be created or destroyed, and since it is incompressible the mass rate of flow into any fixed space is equal to the mass flow out. Thus, the mass flow rate of fluid past all cross sections of a tube is equal.

Example.

In Fig. 1–23, the mean oil velocity at point 1 = 5 fps. Determine the minimum permissible diameter

Fig. 1-23. Continuity Equation, Ex. Problem 1.

at D_2 which will limit the oil velocity to 15 fps. Assume that the fluid is noncompressible.

Solution. Continuity equation—In steady flow, the mass of fluid passing all cross sections of a tube per unit time is the same.

$$\rho_1 A_1 V_1 = \rho_2 A_2 V_2$$

ρ = density of fluid (lb per cu ft)

A = area (ft^2)

V = velocity (ft ps)

$$(\rho_1)(0.785)(5) = (\rho_2)(0.785 D^2)(15)$$

$$D^2 = \frac{(0.785)(5)}{(0.785)(15)} = \frac{1}{3}$$

$$D = 0.577 \text{ in.}$$

Since

Quantity 1 = Quantity 2

Velocity$_1$ Area$_1$ = Velocity$_2$ Area$_2$

Example. Consider Fig. 1–24, where 231 cu in. of oil enters the cylinder every second. When the space is completely filled with oil the piston moves out, and in 3 seconds 693 cu in. of oil will enter the cylinder. The distance the cylinder moves is found by the following:

$$\text{Distance} = \frac{\text{volume}}{\text{piston area}} = \frac{693 \text{ in}^3}{19.25 \text{ in}^2} = 36 \text{ in.}$$

Since the oil is entering at a constant rate, the speed with which the cylinder piston moves is given as follows;

$$\text{Velocity} = \frac{\text{distance}}{\text{time}} = \frac{36}{3} = 12 \text{ in./sec}$$

The general flow equation gives the answer more easily:

$$Q = AV$$

where   Q = rate of flow.

Fig. 1-24. Cylinder showing velocity.

A = area of cylinder.

V = velocity.

Transposing the equation, we have

$$V = \frac{Q}{A}$$

$$V = \frac{231 \text{ cu in./sec}}{19.25 \text{ sq in.}}$$

$$V = 12 \text{ in./sec}$$

Toricelli's Theorem

The liquid velocity at an outlet discharging into the free atmosphere is proportional to the square root of the head (see Fig. 1–25).

$$V = \sqrt{2gh}$$

where V = velocity ft/sec.

        g = acceleration of gravity 32 ft per sec^2.

        h = head in ft.

Fig. 1-25. Tank showing Toricelli's Theorem.

Total pressure head = 0.433 psihft

                 = 0.433 psi × 30 ft

                 (0.433 psi for water)

Total pressure head = 12.99 psi

10 ft	20 ft	30 ft
$v = \sqrt{64 \times 10}$	$v = \sqrt{64 \times 20}$	$v = \sqrt{64 \times 30}$
$v = \sqrt{640}$	$v = \sqrt{1280}$	$v = \sqrt{1920}$
$v = 25.3$ ft/sec	$v = 35.8$ ft/sec	$v = 43.9$ ft/sec

Laminar and Turbulent Flow

Conditions due to the construction or operation of hydraulic machinery may cause irregular flow characteristics of the fluid in the system. In addition the type of fluid used, the size of the tubes or pipes, the smoothness of the internal surfaces of the tubes and pipes, the temperature of the fluid, and the kind of restrictions such as valves, fittings, and other components all have a bearing on the nature of flow within a system.

Laminar flow or streamline flow occurs when the liquid particles flow smoothly in even layers and frictional *losses are at a minimum*. Notice that in Fig. 1–26, which shows laminar flow, there is greater restriction next to the walls of the tube.

So far in our discussion of basic hydraulics, losses resulting from friction have not been considered. They are, however, extremely important to the design

Fig. 1-26. Laminar flow.

of hydraulic systems, and the flow pattern in the diagram shows the effect of friction between the fluid and the walls of the tubing. There would also be some friction loss within the fluid itself because of viscosity. Viscosity is the term for expressing a fluid's resistance to flow and is expressed in absolute units, usually centipoise, or Saybolts Universal Seconds (SUS). (This subject is covered in Chapter 2 of this section.) It is easily understood that some liquids, such as water or gasoline flow much more readily than heavier liquids, such as oil or glycerine.

Experiments have shown that viscosity or resistance to flow in liquids decreases when they are heated. Conversely, as liquids are cooled they offer more resistance to flow, that is, their viscosity increases.

Another small loss would occur because of the kinetic or velocity energy in the fluid. Because the molecules of liquid are flowing, energy due to motion causes some friction, and this is given off in the form of heat. There would be some pressure drop in the system.

It is extremely important to keep all losses in a hydraulic system to a minimum by selecting the proper size of tubing, fittings, and other components that make up the system.

Hagen Poiseuille Law for Laminar Flow

Example. Oil is flowing through a 50 ft straight section of pipe at the rate of 20 gpm. The inside diameter of the pipe is 2 in. Calculate the head loss and pressure drop which will result. Sp. gr. of oil = 0.85. Assume laminar flow oil viscosity = 150 SUS (as shown in Chapter 2).

Solution.

$$h_f = \frac{(\mu)(L)(v)}{(46.5)(D^2)(g)(\rho)}$$

or

$$\Delta P = (0.00000464)\frac{\mu L v}{D^2}$$

ΔP = pressure drop—psi.

h_f = head loss—ft.

L = length of pipe—ft.

ρ = density—lb per cu ft.

v = oil velocity—ft per sec.

g = acceleration of gravity—32.2 ft/sec/sec.

D = diameter of pipe—ft.

μ = absolute viscosity—centipoise.

$$h_f = \frac{(\mu)(L)(v)}{(46.5)(D^2)(g)(\rho)} \quad 150 \text{ sus} = 27 \text{ centipoise.}$$

$$h_f = \frac{(27)(50)(2.04)}{(46.5)(0.1667)^2(32.2)(53)}$$

$$h_f = 1.25 \text{ ft}$$

$$\Delta P = 0.00000464\frac{\mu L v}{D^2}$$

$$\Delta P = (0.00000464)\left[\frac{(27)(50)(2.04)}{(0.1667)^2}\right]$$

$$\Delta P = 0.46 \text{ psi pressure loss}$$

Turbulent flow occurs when the liquid particles flow in a random or erratic pattern, creating inefficiencies, high losses due to friction, and pressure drop. Notice in Fig. 1–27 that frictional losses due to erratic flow are all through the liquid. This means that greater pressure drop would occur because of heat buildup in the system. Turbulent flow occurs when tubes, pipes, fittings, and other components are too small or not properly selected for the job.

Fig 1-27 Turbulent flow.

Reynold's Numbers

Reynold's number is a numerical ratio of the dynamic forces of mass flow to the shear stress due to viscosity. Flow usually changes from laminar to turbulent between Reynold's numbers 2000 and 4000.

Reynold's Equation

Example. Oil is flowing through 1 in. inside diameter pipe at the rate of 100 gpm. The oil viscosity is 150 SUS. Sp. Gr. of oil = 0.85. Is the flow turbulent or laminar? When 10 gpm is flowing through the same pipe, is flow turbulent or laminar?

Solution.

If

$Re < 2000 = $ laminar flow

If

$Re > 4000 = $ turbulent flow

$$Re = \frac{Dv\rho}{\mu e}$$

Re = Reynold's number.

D = pipe diameter—ft.

v = mean velocity—fps.

ρ = density—lb per cu ft.

$\mu e = \dfrac{\mu}{1490} = $ absolute viscosity—lb/sec/ft².

μ = absolute viscosity—centipoise

Note: 150 sus = 27(μ) centipoise

$$Re = \frac{Dv\rho}{\mu e} = \frac{Dv\rho\,1490}{\mu}$$

$$Re = \frac{(0.0833)(40.9)(53)(1490)}{27}$$

$Re = 9960$

Therefore, flow is turbulent for 100 gpm.

$$Re = \frac{(0.0833)(4.09)(53)(1490)}{(27)}$$

$Re = 996$

Therefore, flow is laminar for 10 gpm.

Darcy's Formula

Darcy's formula is used to determine the pressure drop due to flow friction through a tube or pipe.

$$h_f = \frac{fLv^2}{D2g}$$

where $h_f = $ head loss—ft.

f = friction factor.

L = length of tube or pipe—ft.

v = mean velocity of flow—fps.

g = gravity—32 fps².

D = internal diameter of tube or pipe—ft.

In any tube or pipe through which a fluid is flowing, there is a constant loss of pressure.

Fig. 1-28. Darcy's Formula.

Example. Oil is flowing through an 80 ft straight section of pipe at the rate of 60 gpm. The inside diameter of the pipe 0.75 in. Assume a friction factor of 0.05 (f). Calculate the head loss and pressure drop which will result. Specific gravity of oil = 0.85. Assume turbulent flow.

Solution: Darcy's formula—$h_f = \dfrac{fLv^2}{D2g}$

$\Delta P = \dfrac{\rho fLv^2}{144D2g}$ (in terms of psi)

Δp = pressure drop (psi)

ρ = density—lb per cu ft.

f = friction factor.

L = length of pipe—ft.

v = oil velocity—fps.

D = diameter of pipe—ft.

g = acceleration of gravity—32.2 ft/sec/sec.

$$\Delta P = \frac{(62.4)(0.85)(0.05)(80)(43.6)^2}{(144)(0.0625)(2)(32.2)}$$

$\Delta P = 659$ psi pressure drop.

$$h_f = \frac{fLv^2}{D2g}$$

$$h_f = \frac{(0.05)(80)(43.6)^2}{(0.0625)(2)(32.2)}$$

$h_f = 1890$ ft head loss

The basic principles and concepts discussed in this section will be utilized throughout the book in their practical aspect. A great deal of the complexity involving laws and formulas has been removed by the formulation of charts, nomographs, and tables providing engineering data for the design of hydraulic systems. Much of this information will be found in its pertinent section of the book or in the appendix.

Questions

1. Define fluid power.
2. Name the important characteristics of a fluid used to transmit power.
3. Which state of matter contains the least amount of kinetic energy? Which state of matter contains the most kinetic energy?
4. What is the definition of a fluid?
5. What is the density of a hydraulic oil that has a specific gravity of 0.86?
6. Explain why liquids will follow the path of least resistance.
7. What is the difference between force and pressure?
8. What hydraulic device is used to create a force that can push or pull a load?
9. What hydraulic device is used to create a force in a circular path?
10. What force would be developed by a cylinder operating at 600 lb per sq in. if the area is 7 sq in.?
11. What is meant by the hydrostatic principle? What is meant by the hydrodynamic principle?
12. What is meant by pressure head?
13. Explain the meaning of Bernoulli's equation and how it affects fluid flow in a hydraulic system.
14. What is the continuity equation and what implication does it have for fluid flow?
15. Explain what is meant by laminar flow. Explain what is meant by turbulent flow.
16. What causes the friction when fluid is flowing through pipes and orifices?

CHAPTER **2**

FLUIDS FOR HYDRAULIC POWER

The primary function of the liquid in a hydraulic system is to transmit power to perform useful work. The hydraulic fluid must transmit an applied force from one part of the system to another and must respond quickly to reproduce any change in magnitude or direction of the applied force. In order to perform this primary function, the fluid must be relatively incompressible and have flowability.

In addition to this primary function of the hydraulic fluid, there are secondary factors extremely important to the successful operation of the hydraulic system. A number of these important factors include: film strength to seal close clearances between moving parts against leakage; the minimization of wear and friction by providing adequate lubrication in bearings and between sliding surfaces in pumps, valves, cylinders, and other components of the system; resistance to physical and chemical change and prevention of rusting or corrosion; rapid settling and separation of insoluble contaminants that may enter the system; and the removal of heat from the system.

Any number of liquids would fulfill the primary function of a hydraulic fluid. For instance, water is used as the fluid medium in hydraulic systems for coastal oil drilling rigs, forging presses and elevators. This is ideal from the standpoint of availability and cost, but there are also numerous disadvantages. Water promotes rust, freezes and boils at frequently encountered operating temperatures, and has very little lubricating value. Through the addition of soluble oils or other materials, water can be made suitable for special applications where large quantities of fluid are essential or for working hot materials where a fire hazard exists.

Petroleum Base Fluids

Modern advances in the refining and blending of petroleum base oils have been responsible for the production of satisfactory fluids for hydraulic systems. The petroleum refiner is capable of producing oils of almost any viscosity. Additives also provide other desirable characteristics which prolong the life expectancy of hydraulic machines and increase their efficiencies.

The ability of a fluid to be pumped and transmitted through the system is most important. This ability to flow is determined by the oil's viscosity.

Viscosity is the measure of an oil's resistance to flow or its internal resistance to shear. Viscosity is affected by temperature and must be expressed as a certain value at a certain temperature.

The engineer is sometimes faced with design problems where absolute units of measurement and

Fig. 2-1. Concept-viscosity. (*Courtesy of Socony Mobil Oil Co.*)

the metric system are involved. Fluid friction is caused by the movement of molecules with respect to each other; this movement is termed *viscosity*. Figure 2-1 shows the concept of this movement.

As a thin moving plate is drawn over a film of fluid, the fluid adheres to both the moving plate and the stationary surface. Fluid in contact with the moving plate moves at a velocity the same as that of the moving plate, while the fluid in contact with the stationary surface is at a velocity of zero. Between V_1 and V_0 the fluid flows as imaginary layers with each at a velocity that is proportional to the distance above the stationary surface. The force pushing the moving plate is proportional to the viscosity of the fluid.

The measurement of viscosity in the metric system would be as follows: If the moving plane surface is 1 sq cm in area, the film thickness h is 1 cm, the velocity V_1 is 1 cm per sec, and the force is 1 dyne, the viscosity of the fluid would be 1 poise. The poise is the standard unit of absolute viscosity in the centimeter-gram-second system. It is defined as the ratio of the shearing stress to the shear rate of a fluid, and is expressed in dyne seconds per square centimeter; 1 centipoise equals 0.01 poise. The poise can be converted into other common units of measurement for viscosity.

The method used most commonly as a measure of viscosity is SUS, or Saybolt Universal Seconds. The method of determination is shown in Fig. 2-2.

The formula for changing SUS to viscosity in centipoise is

$$\mu = \text{Sp. Gr.} \left(0.22\, \text{SUS} - \frac{135}{\text{SUS}}\right)$$

μ = viscosity in centipoise

Sp. Gr. = specific gravity of fluid

SUS = Saybolt Universal Seconds

VISCOSITY CLASSIFICATION

SAE num-ber	Viscosity Range, SUS (Equivalent Centistokes in Parenthesis)			
	At 0°F		At 210°F	
	Minimum	Maximum	Minimum	Maximum
5W		4,000(880)		
10W	6,000(1,320)*	12,000(2,640)		
20W	12,000(2,640)†	48,000(10,500)		
20			45(5.75)	58(9.7)
30			58(9.7)	70(13.0)
40			70(13.0)	85(16.85)
50			85(16.85)	110(22.75)

*Minimum viscosity of 10W at 0°F can be waived if viscosity at 210°F is not below 40 SUS (4.2 cs).

†Minimum viscosity of 20W at 0°F can be waived if viscosity at 210°F is not below 45 SUS (5.75 cs).

$$\frac{\text{Centipoises}}{\text{Specific gravity}} = \text{Centistokes.}$$

SAE (Society of Automotive Engineers)

Viscosity is generally considered to be one of the most important physical properties of a hydraulic fluid and is the starting point in the selection of a fluid. If the fluid does not have the proper viscosity it cannot perform, regardless of other superior characteristics.

If the viscosity is too low, the following results can be expected (lightweight oils):

1. Less film strength, and thus more wear on moving parts.
2. More leakage.
3. More pressure loss.
4. Lower volumetric efficiencies in pumps and motors.
5. Less precision control and slower responses.
6. Less over-all efficiency.

If the viscosity is too high, the following results can be expected (heavyweight oils):

1. Higher pressure drop due to friction.
2. Excessive heat generation.
3. Sluggish operation.
4. More power consumption.
5. Less mechanical efficiency.
6. Starvation of the pump inlet causing cavitation. (Cavitation is starvation of the pump inlet.)

Viscosity Index

Viscosity index (V.I.) is a measure of the relative change in viscosity for a given change in temperature. An oil with a high viscosity index shows less change in viscosity for a given change in temperature than an oil with a low viscosity index.

Viscosity index has little significance when the operating temperature range is small. This is the reason for selecting a viscosity of an oil at a specific operating temperature range. However, in hydraulic systems where the temperature range is difficult to control, viscosity index is very important: for example, a lift truck working both indoors and outdoors during extreme temperature conditions. Oil having a V.I. rating above 90 is usually selected for hydraulic systems. The viscosity index nomograph shown in Fig. 2–4 may be used to select oils with proper viscosity indexes.

Stability or Resistance to Oxidation

Another important property of a good hydraulic oil is stability, or resistance to chemical change. The three main accelerators of oxidation are high operating temperatures, high rate of mixing with air, and the catalytic effects of metals and contaminants, such as dirt and water.

The normal operating temperature range for most hydraulic systems is between 100°F and 130°F. The rate of oxidation doubles for approximately every 20°F rise in temperature. A high rate of oxidation cannot be tolerated in a hydraulic system because of the varnishes and sludges produced. A high rate of oxidation also increases the acidity content of the fluid, causing further contamination and damage to system components.

The temperature of the oil in the reservoir does not give the true operating temperature of the system. It is better to use the pump discharge side or other points wherever the temperature is the greatest.

It should also be recognized that pressure increases viscosity, which, in turn, will develop more frictional heat, thus raising the operating temperature of a system and increasing the rate of oxidation.

The amount of air that can be held in solution by an oil increases rapidly with increase in pressure. At atmospheric pressure an oil may contain 10 per cent by volume of dissolved air. Air provides the oxygen necessary to promote oxidation.

Contaminants such as cutting oils, greases, dirt, moisture, paint, pipe compound, and insoluble oxida-

tion products themselves all accelerate the rate of oxidation.

Metals such as copper produce a catalytic action for oxidation, especially in the presence of water. Iron and aluminum also provide some catalytic action, increasing oxidation.

Premium grade hydraulic oils contain inhibitors which tend to decrease the factors causing oxidation. They may be of the type that breaks the chain of reactions, thus preventing oxidation, or they may reduce the catalytic effect of the metals.

The ASTM (American Society for Testing Materials) Oxidation Test (see Fig. 2–5) is important as a research tool in determining which oils have proven to be resistant to oxidation.

Rust Prevention

Rust occurs in a hydraulic system because of the presence of water. The source of moisture may be leakage from an oil cooler, or may be condensation from the air which is present in the system. Air may enter through the reservoir breather at ambient temperatures above 72°F. When the system is shut down and the air cools, moisture is given off. Water can also enter the system through the use of coolants during machinery operations.

Suitable inhibitors have been provided to control rust in a hydraulic system. High grade hydraulic fluids contain an additive material, called rust inhibitor, which has an affinity for metal surfaces. It adheres to the metal surfaces, forming a protective coating that resists water, thus preventing a corrosive action between the metal and water (see Fig. 2–6).

Resistance to Foaming

In addition to entrained air in hydraulic oil because of contact with the atmosphere, high velocity mixing of air and oil also takes place in a hydraulic system. This occurs when oil is being discharged over a relief valve and back to the reservoir. Another source of air into a hydraulic system is through the pump inlet line, the pressure of which is generally less than atmospheric pressure.

It is important to bleed air out of all lines and components when first starting up a system. Also, it is important to fill all lines and components when a hydraulic machine is being serviced. Good chemical compounds are available to break out entrained air, quickly separating air from oil.

Premium grade hydraulic oils are available with foam depressants.

Pour Point

Pour point is the characteristic of a fluid that indicates its ability to flow at low temperatures. It is important to consider the extreme temperature conditions under which a system must function, rather than normal conditions. For most installations, the pour point should be at least 20°F below the lowest expected operating temperature.

Anti-wear Properties

In addition to its many other functions, a good hydraulic oil must lubricate all integral moving parts of the system. Petroleum base fluids provide excellent lubricating qualities.

Anti-wear tests are conducted with a vane type of hydraulic pump, as shown in Fig. 2–7. Results indicate that premium hydraulic oils provide excellent lubrication.

Neutralization Number

The neutralization number (see Fig. 2–8) is a measure of the acidity or alkalinity of a hydraulic fluid. This is referred to as the PH factor of the fluid. Fluids with a low neutralization number are recommended to prevent harmful chemical reaction. High acidity causes the oxidation rate in an oil to increase rapidly. The best method to determine the PH factor of an oil is to send a sample of the oil to the supplier. Most suppliers have a modern laboratory facility for making the various tests to determine the condition of a fluid (see Fig. 2–9).

Viscosity

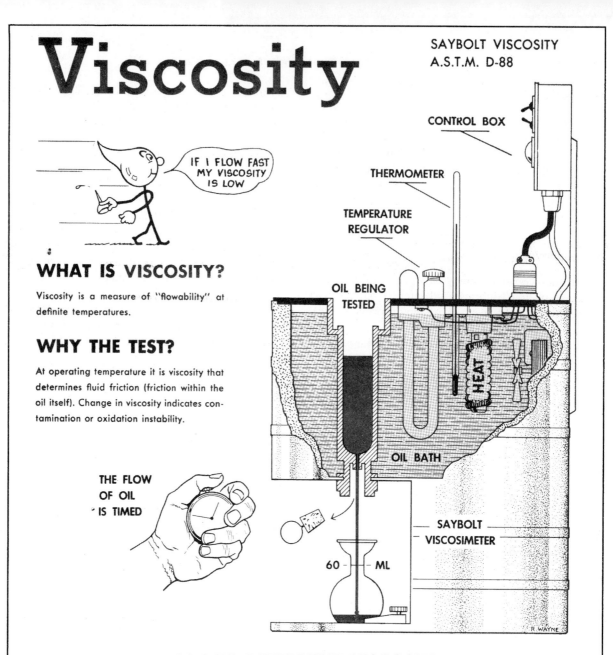

IF I FLOW FAST MY VISCOSITY IS LOW

WHAT IS VISCOSITY?

Viscosity is a measure of "flowability" at definite temperatures.

WHY THE TEST?

At operating temperature it is viscosity that determines fluid friction (friction within the oil itself). Change in viscosity indicates contamination or oxidation instability.

THE FLOW OF OIL IS TIMED

CONTROL BOX

THERMOMETER

TEMPERATURE REGULATOR

OIL BEING TESTED

HEAT

OIL BATH

SAYBOLT VISCOSIMETER

60 — ML

R WAYNE

HOW TO DETERMINE VISCOSITY

The unit of measure is time in seconds required for 60 ML of the oil to flow through a standard orifice under a standard falling head and at a given temperature. 100°F. and 210°F. are common temperatures for reporting Viscosity. Saybolt Furol Viscosity is obtained with the same instrument with a larger orifice producing results approximately one-tenth those of the Universal orifice readings.

	SAE	at 100°F.	at 210°F.
Typical Viscosity	10W	202	48
(Saybolt Universal)	20W	323	57
For SAE	30	538	68
Automotive Ratings	40	850	84
	50	1174	100

Fig. 2-2. Viscometer. (*We gratefully acknowledge the courtesy of United States Steel Corporation in permitting us to reproduce pages from the Lubrication Engineers Manual.*)

Division
United States Steel Company

327 Craft Avenue
Pittsburgh 13, Pennsylvania

ASTM Designation D-88-44

SAYBOLT UNIVERSAL VISCOSITY
(Saybolt Viscosimeter)

Definition:

The viscosity of petroleum products and lubricants is the duration of time that is required to measure 60 ml. through a standard orifice (viscosity is a measure of the flowability at a definite temperature).

Apparatus:

Saybolt Universal Viscosimeter according to A.S.T.M. standard.
60 ml. receiving flask according to A.S.T.M. standard.
Stopwatch capable of measuring .2 second.
Thermometer (A.S.T.M. standard: E-1).

1. bath
2. oil tube

Pipette according to A.S.T.M. standard.
Aluminum pot for heating.
Cork stopper.

Procedure:

A draft free room with temperature maintained between 68 and 86°F. The controls for the viscosimeter are adjusted so that the desired bath temperature is attained. Previous to the introduction of oil into the oil tube, the tube is cleaned with an effective solvent such as xylol, and all excess solvent is removed from the tube and gallery. The oil is shaken thoroughly before it is to be heated in the aluminum pot and is not heated to more than 3°F above the temperature of the test. The oil is then transferred to the oil tube straining it through a 100 mesh wire strainer, with sufficient oil to overflow into the gallery. The bath temperature* is adjusted until the oil temperature remains constant. After removal of the oil tube thermometer, the oil is removed from the gallery until the oil level in the gallery is below that of the oil tube proper. During this procedure the pipette must remain in one position only. The receiving flask is placed below the orifice to direct the stream of oil to strike the neck of the flask. The removal of the cork and the start of the stopwatch are simultaneous. The timer is stopped when the bottom of the menicus of the oil reaches the 60 ml. mark on the neck of receiving flask. The results are recorded to the nearest .1 second below 200 seconds and the nearest whole second for values of 200 seconds or above. The reproducibility is within .5%.

* Allowance for 100°F is plus or minus .05°F and for 140, 175 and 210°F is plus or minus .1°F.

7/15/52

Kinematic Viscosity
A.S.T.M. D-445-53T

WHAT IS KINEMATIC VISCOSITY?

Kinematic viscosity is the property measured when a fixed amount of an oil flows through a capillary tube under the force of gravity. An intermediate measurement in seconds is taken and then converted by the viscometer constant for the final answer. The unit of kinematic viscosity is the stoke or centistoke.

Kinematic Viscosity, $cs = Ct$

Where $C = $ Viscometer Constant

$t = $ Observed flow time in seconds

WHAT IS ABSOLUTE VISCOSITY?

Absolute viscosity is the kinematic viscosity corrected to overcome the variations caused by differences in specific gravity. The centipoise is one hundredth of a poise and is the unit of absolute viscosity more commonly used. The relation between absolute and kinematic viscosity is as follows:

$$\frac{Centipoises}{Specific\ Gravity} = Centistokes, \text{ or } Centipoises = Centistokes \times Specific\ Gravity.$$

WHY THE TEST?

The test is a precise viscosity measurement of fluids.

METHOD OF CHARING SAMPLE

PLACE IN CONSTANT TEMPERATURE BATH

ADJUST HEAD LEVEL 5MM ABOVE STARTING MARK

MEASURE FLOW TIME IN SECONDS FROM START TO STOP MARKS

TYPICAL RESULTS

$225 \times .1200 = \underline{27.0}$ Centistokes

Where 225 seconds = Flow Time

.1200 cs = Viscometer Constant

Fig. 2-3. Conversion chart (absolute viscosity). (*Courtesy of United States Steel Corp.*)

UNITED STATES STEEL LUBRICANTS TESTING LABORATORY
NATIONAL TUBE

Division
United States Steel Company

327 Craft Avenue
Pittsburgh 13, Pennsylvania

KINEMATIC VISCOSITY

PROCEDURE:

The time is measured for a fixed volume of sample, contained in a glass viscometer, to flow through a calibrated capillary under an accurately reproducible head of liquid and at a closely controlled temperature. The viscometer selected should give an efflux time greater than 200 seconds. The kinematic viscosity is calculated from the measured efflux time. The viscometers are calibrated by using standard oils having viscosities established by reference to water in master viscometers or by direct comparison with carefully calibrated viscometers. Temperatures of the bath used must be maintained within $\pm 0.02F$ for test temperatures of 60F or above.

CONVERSION OF KINEMATIC VISCOSITY TO SAYBOLT UNIVERSAL VISCOSITY

Kinematic Viscosity, cs	Equivalent Saybolt Universal Viscosity, sec		Kinematic Viscosity, cs	Equivalent Saybolt Universal Viscosity, sec	
	At 100°F Basic Values	At 210°F		At 100°F Basic Values	At 210°F
2	32.6	32.9	31.0	145.7	146.8
2.5	34.4	34.7	32.0	150.2	151.2
3	36.0	36.3	33.0	154.7	155.8
3.5	37.6	37.9	34.0	159.2	160.3
4	39.1	39.4	35.0	163.7	164.9
4.5	40.8	41.0			
5	42.4	42.7			
6	45.6	45.9	36.0	168.2	169.4
7	48.8	49.1	37.0	172.7	173.9
8	52.1	52.5	38.0	177.3	178.5
9	55.5	55.9	39.0	181.8	183.0
10	58.9	59.3	40.0	186.3	187.6
11	62.4	62.9	41.0	190.8	192.1
12	66.0	66.5	42.0	195.3	196.7
13	69.8	70.3	43.0	199.8	201.2
14	73.6	74.1	44.0	204.4	205.9
15	77.4	77.9	45.0	209.1	210.5
16	81.3	81.9	46.0	213.7	215.2
17	85.3	85.9	47.0	218.3	219.8
18	89.4	90.1	48.0	222.9	224.5
19	93.6	94.2	49.0	227.5	229.1
20	97.8	98.5	50.0	232.1	233.8
21	102.0	102.8	55.0	255.2	257.0
22	106.4	107.1	60.0	278.3	280.2
23	110.7	111.4	65.0	301.4	303.5
24	115.0	115.8	70.0	324.4	326.7
25	119.3	120.1			
26	123.7	124.5	Over 70.0	Saybolt seconds = centistokes × 4.635	Saybolt seconds = centistokes × 4.667
27	128.1	129.0			
28	132.5	133.4			
29	136.9	137.9			
30	141.3	142.3			

NOTE: To obtain the Saybolt Universal viscosity equivalent to a kinematic viscosity determined at $t°F$ multiply the equivalent Saybolt Universal viscosity at 100°F by $1 + (t - 100)$ 0.000064; for example, 10 cs at 210°F are equivalent to 58.9 × 1.0070 or 59.3 sec Saybolt Universal at 210°F.

Viscosity Index

A.S.T.M. D-567
A.S.A. Z11.45

WHAT IS VISCOSITY INDEX?

The Viscosity Index is an empirical number indicating the rate of change in viscosity of an oil within a given temperature range. A low viscosity index signifies a relatively large change in viscosity with temperature, while a high viscosity index shows a relatively small change in viscosity with temperature. Viscosity Index cannot be used to measure any other quality of an oil.

Viscosity Index is calculated as follows:

$$VI = \frac{L - U}{L - H} \times 100$$

U = viscosity at 100F of the oil whose viscosity index is to be calculated.

L = viscosity at 100F of an oil of 0 viscosity index having the same viscosity at 210F as the oil whose viscosity index is to be calculated.

H = viscosity at 100F of an oil of 100 viscosity index having the same viscosity at 210F as the oil whose viscosity index is to be calculated.

THE SPREAD MAKES THE DIFFERENCE IN VISCOSITY INDEX

LOW V. I.   MODERATE V. I.   HIGH V. I.

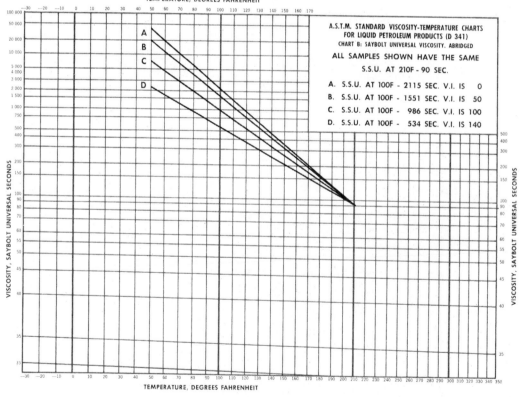

A.S.T.M. STANDARD VISCOSITY-TEMPERATURE CHARTS
FOR LIQUID PETROLEUM PRODUCTS (D 341)
CHART B: SAYBOLT UNIVERSAL VISCOSITY. ABRIDGED

ALL SAMPLES SHOWN HAVE THE SAME
S.S.U. AT 210F - 90 SEC.

A. S.S.U. AT 100F - 2115 SEC. V.I. IS   0
B. S.S.U. AT 100F - 1551 SEC. V.I. IS   50
C. S.S.U. AT 100F - 986 SEC. V.I. IS 100
D. S.S.U. AT 100F - 534 SEC. V.I. IS 140

Fig. 2-4. Viscosity index. (*Courtesy of United States Steel Corp.*)

40

VISCOSITY INDEX

TABLE 100F

V.I.	40	45	50	55	60	65	70	75	80	85	90	95	100	105	110	115	120	125	130	135	140	145	150	155	V.I.
0	138	265	422	596	781	976	1182	1399	1627	1865	2115	2375	2646	2928	3220	3524	3838	4163	4498	4845	5202	5570	5959	6339	0
5	136	261	414	584	763	953	1153	1364	1585	1816	2059	2311	2573	2846	3129	3423	3727	4042	4366	4701	5046	5402	5768	6145	5
10	135	256	405	570	745	930	1124	1329	1543	1767	2002	2246	2500	2765	3038	3323	3616	3920	4233	4557	4890	5234	5587	5950	10
15	133	252	397	557	727	907	1095	1294	1502	1718	1946	2182	2427	2683	2947	3222	3505	3799	4101	4413	4734	5065	5406	5756	15
20	132	247	389	545	710	884	1066	1259	1460	1670	1889	2117	2355	2601	2856	3121	3394	3677	3968	4269	4578	4897	5225	5562	20
25	130	243	380	532	692	861	1038	1224	1418	1621	1833	2053	2282	2520	2765	3021	3284	3556	3836	4125	4423	4729	5044	5368	25
30	129	238	372	519	674	837	1009	1188	1376	1572	1776	1989	2209	2438	2674	2920	3173	3434	3703	3981	4267	4561	4863	5173	30
35	127	234	364	506	656	814	980	1153	1334	1523	1720	1924	2136	2356	2583	2819	3062	3313	3571	3837	4111	4392	4682	4979	35
40	126	230	355	493	638	791	951	1118	1293	1474	1663	1860	2063	2274	2492	2718	2951	3191	3438	3693	3955	4224	4501	4785	40
45	124	225	347	480	621	768	922	1083	1251	1425	1607	1795	1990	2193	2401	2618	2840	3070	3306	3549	3799	4056	4320	4590	45
50	123	221	339	468	603	745	893	1048	1209	1377	1551	1731	1918	2111	2311	2517	2729	2948	3173	3405	3643	3888	4139	4396	50
55	121	216	330	455	585	722	864	1013	1167	1328	1494	1667	1845	2029	2220	2416	2618	2827	3041	3261	3487	3719	3957	4202	55
60	119	212	322	442	568	699	835	978	1125	1279	1438	1602	1772	1948	2129	2316	2507	2705	2908	3117	3331	3551	3776	4007	60
65	118	207	314	429	550	676	806	943	1084	1230	1381	1538	1699	1866	2038	2215	2396	2584	2776	2973	3175	3383	3595	3813	65
70	116	203	305	416	532	653	777	908	1042	1181	1325	1473	1626	1788	1947	2114	2285	2462	2643	2829	3019	3215	3414	3619	70
75	115	199	297	403	514	630	749	873	1000	1132	1268	1409	1553	1703	1856	2014	2175	2341	2511	2685	2864	3046	3233	3425	75
80	113	194	288	391	497	606	720	837	958	1083	1212	1345	1480	1621	1765	1913	2064	2219	2378	2541	2708	2878	3052	3230	80
85	112	190	280	378	479	583	691	802	916	1035	1155	1280	1408	1539	1674	1812	1953	2098	2246	2397	2551	2710	2871	3036	85
90	110	185	272	365	461	560	662	767	875	986	1099	1216	1335	1457	1583	1711	1842	1976	2113	2253	2396	2542	2690	2842	90
95	109	181	263	352	443	537	633	732	833	937	1042	1151	1262	1376	1492	1611	1731	1855	1981	2109	2240	2373	2509	2647	95
100	107	176	255	339	426	514	604	697	791	888	986	1087	1189	1294	1401	1510	1620	1733	1848	1965	2084	2205	2328	2453	100
105	106	172	247	326	408	491	575	662	749	839	930	1023	1116	1212	1310	1409	1509	1612	1716	1821	1928	2037	2147	2259	105
110	104	167	238	314	390	468	546	627	707	790	873	958	1043	1131	1219	1309	1398	1490	1583	1677	1772	1869	1966	2064	110
115	103	163	230	301	372	445	517	592	666	741	817	894	970	1049	1128	1208	1287	1369	1451	1533	1616	1700	1785	1870	115
120	101	159	222	288	355	422	488	557	624	693	760	829	898	967	1037	1107	1176	1247	1318	1389	1460	1532	1604	1676	120
125	99	154	213	275	337	399	460	522	582	644	704	765	825	886	946	1007	1066	1126	1186	1245	1305	1364	1423	1482	125
130	98	150	205	262	319	375	431	486	540	595	647	701	752	804	855	906	955	1004	1053	1101	1149	1196	1242	1287	130
135	96	145	197	249	301	352	402	451	498	546	591	636	679	722	764	805	844	883	921	957	993	1027	1061	1093	135
140	95	141	188	236	284	329	373	416	457	497	534	572	606	640	673	704	733	761	788	813	806	859	880	899	140

LEGEND

Vertical Bold Type - Viscosity Index
Horizontal Bold Type Headings - 210F

JANUARY, 1956

Inhibited Oil Oxidation

A.S.T.M. D-943

BY BUBBLING OXYGEN THROUGH OIL AND WATER WITH CATALYST

WHAT IS INHIBITED OIL OXIDATION?

Inhibited oil oxidation for characteristics of inhibited steam turbine oil is a chemical action between oil and oxygen, producing reaction products of acid and sludge.

WHY THE TEST?

To measure resistance to oxidation by measuring the change in acidity in the oil by the oxygen absorbed.

OIL OXIDATION CAN SCRAP MACHINERY

PETE TROLEUM'S SCRAP

SCHEMATIC DRAWING OF THE INHIBITED OIL OXIDATION TEST

WATER
FLOWMETER
CONDENSER
OIL BATH
OXYGEN
DRYER
OXIDATION CELL

CONDENSER

CATALYST COILS

OIL SAMPLE 300 ML.

WATER 60 ML.

OXIDATION CELL

TEST PROCEDURE

A 300 ml. of sample is placed in the tube and immersed in the bath at 95°C. Three liters per hour of oxygen are allowed to pass continuously through the sample for a period of 1000 or more hours. The acid number determined for neutralization value is the determining factor in the test. Limits may be established between .25 and 2.0 mg. of KOH per gram of oil.

TYPICAL RESULTS

1000 hours to 4000 hours.

Fig. 2-5. A.S.T.M. oxidation test data. (*Courtesy of United States Steel Corp.*)

Rust Prevention

A. S. T. M. D-665-50-T

PASS

FAIL

A GOOD LUBRICATING OIL MUST NOT ONLY LUBRICATE BUT PREVENT RUSTING IN THE PRESENCE OF WATER

WHAT IS THE RUST PREVENTION TEST?

The rust prevention test indicates the ability of lubricating oil to prevent corrosion during the lubrication of ferrous parts in the presence of water.

WHY THE TEST?

Corrosion of mechanical parts, lines, and tanks in connection with lubrication contributes to high maintenance costs.

TEST PROCEDURE

The steel test rod is placed in the beaker of oil and water at 140° F for 24 hours and the results reported as pass or fail. The degree of failure may be light, moderate, or severe.

Fig. 2-6. Rust Prevention. (*Courtesy of United States Steel Corp.*)

Hydraulic Pump Test
UNITED STATES STEEL MODIFIED
SOCONY MOBIL HYDRAULIC SERVICE TEST

WHAT IS THE TEST?

The Hydraulic Pump Test evaluates the oxidation stability and wear characteristics of a lubricant under fixed conditions of pressure, temperature, and time.

WHY THE TEST?

The test measures the conditions encountered by vane type pumps in actual operation.

PROCEDURE

The machine is thoroughly flushed with a solvent and followed by a 24-hour preliminary run of the test oil which is then discarded.

The machine is assembled using new pump parts that have been carefully adjusted and weights recorded of all parts including the filter screen. The tank is filled with 3½ gallons of sample with the Sim-ply-trol adjusted to the desired temperature. After starting, pressure is slowly built up to 1000 PSI over a two to three-hour period by adjusting the relief valve. Machine readings are recorded periodically as indicated on the report tabulation.

During the test run periodic checks are made for viscosity, color, neutralization, and precipitation numbers. At the conclusion of the test the oil is removed for final analysis, weighing, and inspection of pump parts and filter.

November 1955

Fig. 2-7. Vane pump test. (*Courtesy of United States Steel Corp.*)

Neutralization Number

A.S.T.M. D-974-54T

WHAT IS THE NEUTRALIZATION NUMBER?

A number expressed in mg. of potassium hydroxide needed to neutralize the acid in one gram of oil.

WHY THE TEST?

To show relative changes in an oil under oxidizing conditions. It measures development of injurious products in oils.

TEST PROCEDURE

A weighed amount of sample in titration solvent is titrated with a standard alcoholic potassium hydroxide solution to a definite end point. The indicator is alpha-Naphtholbenzein solution. The color change is from orange to green or blue-green.

$$\frac{\text{TOTAL ml of TITRATING SOLUTION} \times 5.61}{\text{WEIGHT OF SAMPLE USED}} = \frac{\text{NEUTRALIZATION}}{\text{NUMBER}}$$

Fig. 2-8. Neutralization number. (*Courtesy of United States Steel Corp.*)

Gravity

WHAT IS SPECIFIC GRAVITY?

The ratio of the weight in air of a given volume of material at a stated temperature to the weight in air of an equal volume of distilled water at a stated temperature.

WHAT IS A.P.I. GRAVITY?

An arbitrarily chosen scale in which the specific gravity of pure water is taken as 10. Liquids lighter than water having values greater than 10 and liquids heavier than water having values less than 10.

$$\text{Degrees A.P.I} = \frac{141.5}{\text{Sp. Gr. } 60/60F} - 131.5$$

WHY THE TEST?

Liquid petroleum products are sold by volume. To arrive at the correct volume, gravity must be known. Gravity is not necessarily an index of quality of a product.

TEST PROCEDURE

The determination of gravity of crude petroleum or its products may be made at any suitable temperature between 0 and 195F.

Pour the sample into a hydrometer jar, removing any air bubbles formed after they have collected on the surface by touching them with a piece of clean filter paper. Lower the hydrometer gently into the sample and when it has settled, depress it about two scale divisions into the liquid and then release. Allow sufficient time for the hydrometer to become completely stationary, simultaneously read the hydrometer to the nearest half scale division and the temperature recorded on the thermoscale of the hydrometer. If a thermo hydrometer is not used, then a separate thermometer is necessary for the temperature reading. The observed gravity at the observed temperature shall be corrected to the gravity at 60F by use of Tables 1 and 3 of the "National Standard Petroleum Oil Tables" as published in the Bureau of Standards Circular C 410.

Fig. 2-9. Gravity chart. (*Courtesy of United States Steel Corp.*)

Fire Resistant Fluids

In many applications of fluid power, hazardous conditions and safety requirements dictate the use of a fire resistant fluid. Hazardous operating conditions found in coal mines or in hot metal working processes are good examples. Safety requirements regarding fluid power systems for aircraft and marine uses call for maximum safety for the prevention of fire.

Manufacturers of fluids have done an excellent job in research and development to keep abreast with the increasing demands of all industry for better and dependable fire resistant hydraulic fluids.

A fire resistant fluid is a fluid, difficult to ignite, which shows little tendency to propagate flame. Such fluids should be used where any hazard exists which may endanger human lives or destroy valuable property.

In general, fluid powered equipment is designed for use with petroleum fluids. The manufacturers of the equipment and of the fire resistant fluid should be consulted for detailed consideration of any installation or conversion problem. When fire resistant fluids are used, the suppliers are the best source of important information concerning design features, test requirements, operational techniques, maintenance procedures, equipment life, or other pertinent data, such as compatibility with system materials.

General Industrial Types of Fire Resistant Fluids

The general types of fire resistant fluids include water glycol, straight synthetics, and water-in-oil emulsions.

The manufacturers of these various types of fire resistant fluids have designed their products to be used in the same standard atmospheric conditions as petroleum oils. Whenever there are extreme atmospheric conditions, such as in foundries where smoke and dirt may be drawn into the hydraulic system, the fluid composition could change. This might also happen in chemical and plating plants, where highly corrosive conditions exist. The change in composition could, over a period of time, be harmful to the hydraulic system. Always consult with the suppliers when adverse environmental conditions must be met.

Water Glycol Fluids

Short-term vapor exposure or handling of water glycol fluids in their complete form has no pronounced toxic effect. Leakage or spillage of these fluids can be easily cleaned by flushing or mopping up with water.

Viscosity Control

Viscosity of the water glycol base fluids depends on the water content, which ranges from 35 to 50 per cent. It is good practice to check the fluid in a system at regular intervals to determine the average rate of water evaporation so that proper viscosity can be maintained. Loss of water also could seriously reduce the effectiveness of fire resistance.

Pure water, such as distilled or clean condensate or deionized water, should be used for make-up. This avoids the possibility of introducing harmful iron, lime, salts, or other foreign materials into the system.

Evaporation may also cause the loss of important inhibitors, which should be replaced by consulting the fluid supplier.

Operating Temperature

The operating temperature of the fluid should not exceed 120°F because of excessive evaporation rate and loss of water and inhibitors.

High temperatures could also cause separation of water from the glycol base. The separated fluids may or may not go back into solution as the temperature is reduced to normal. It is essential to maintain good temperature control to avoid serious damage to pumps and other system components.

Excessive temperature may also cause physical change of the inhibitors, resulting in a gummy substance which does not go back into solution. This may cause loss of protection against fire and may also contaminate the hydraulic system.

Foaming and Aeration

Air retention in water glycol finds is more prominent than with petroleum oils. Reservoirs should be selected large enough with adequate vents and baffles to help eliminate air from the fluid. The fluid returning to the reservoir from the system should be at a maximum distance from the inlet line.

Wear Resistance Characteristics

Wear resistance characteristics of water glycol fluids have been found to be adequate in systems where components are carefully selected to insure compatibility with the fluid. Antifriction bearings and other parts are definitely limited at high-speed, high-load capacities when lubricated with these fluids.

Hydraulic equipment manufacturers, particularly pump manufacturers, should be consulted concerning the use of their equipment with water glycol fluids, since some modification may be desirable for wearing parts.

Specific Gravity and Pump Inlet

The higher specific gravity of water glycols can aggravate pump inlet conditions. Inlet strainers or filters should be amply sized. Fine filtration should be restricted to the discharge side of the pump. Pump inlet lines must also be large enough to prevent excessive velocities of the fluid to the pump. Lines too small may starve the pump, causing damage.

Compatibility Factors

Water glycol fluids attack zinc, cadmium, and magnesium, forming sticky or gummy substances. This residue may cause plugging of strainers or orifices, sticky valves and pump parts, and deterioration or destruction of parts plated or alloyed with these elements.

Check with the fluid supplier for compatibility with the following:

1. All major components.
2. Galvanized pipe.
3. Strainers which have internal parts plated with zinc or cadmium.
4. Die cast reservoir fittings such as vents, fillers, and oil level gauges.
5. Instrumentation components such as thermometer bulbs, gauge bulbs, diaphragms, fittings, level indicators, heater coils, and tubing.

These fluids also have a solvent action on most paints, enamels and varnishes. (Some coatings are now available which resist this action and may be used for internal and external use.) Softening coatings on the inside of sumps or reservoirs must be removed before water glycols are put into the system.

Experience to date indicates that the Buna "N" elastomers and neoprene seals and packings, which are normally used with petroleum oils, are satisfactory for use with water glycol fluids. Silicones, butyl rubbers, nylon, Teflon, Viton, Fluorel, and Kel-f are also suitable materials. (These are trade names for compounds used in sealing devices.)

Asbestos, leather, and cork impregnated materials should be avoided in rotary seals, since these materials have a tendency to swell because of their water-absorption properties.

Changing Fluids in a System

Thoroughness and carefulness should be prime considerations when one is changing fluids in a hydraulic system. Sufficient time, thought, and care can often mean the difference between good, successful operation or complete shutdown and major overhaul. The importance of working closely with fluid suppliers' recommendations cannot be overstressed. Representatives of fluid manufacturers are readily available and are willing to provide recommendations and assistance.

Changing from Petroleum Oil to Water Glycol Fluid

1. The system should be completely drained and cleaned.
2. Drain and blow out the pipe lines. Manually clean the valves and reservoirs.
3. Pumps, accumulators, and cylinders should be actuated to insure removal of all fluid.
4. Strainers should be dismantled, washed, and cleaned. Disposable cartridges should be replaced.
5. Drain and clean the filter housings. Replace or clean the filter elements.
6. Check the component functions before the system is closed and water glycol fluid is added.
7. Close the system and flush with a small amount of water glycol fluid or a suitable flushing compound as recommended by the fluid supplier. Do not use carbon tetrachloride or other chlorinated solvents, as they can react with water by hydrolysis and form corrosive hydrochloric acid. Carbon tetrachloride produces toxic vapors when it comes in contact with flame.
8. Check the system carefully after it is in operation. Leakage at seals is appreciably more expensive with water glycols than it is with petroleum oils.

Changing from Water Glycol to Petroleum Oil

1. The system should be completely drained and cleaned. If all the water glycol is not removed, a sufficient amount of water may remain to cause harmful effects to the replacement oil. As little as 1 per cent water in petroleum oil in a high-pressure hydraulic system can be very detrimental.
2. Pipe lines should be opened and thoroughly

drained. Valves should be opened or removed and drained if there is a possibility of fluid in recesses. Reservoirs should be opened for thorough washing and cleaning.

3. Pumps, accumulators, and cylinders should be actuated to insure removal of all fluid.
4. Strainers should be dismantled, washed, and cleaned. Disposable cartridges should be replaced.
5. Drain and clean the filter housings. Replace or clean the filter elements.
6. Coolers should be thoroughly drained and cleaned. A thin insulating film may adhere to cooler tubes. This film should be removed to maintain original cooler capacity.

The use of steam, high-pressure water jet stream or the wiping of surfaces with a brush, cloth, or other suitable material, and water are effective methods of cleaning surfaces which have been in contact with water glycol fluid. Care must be taken to prevent rust or corrosive action if steam or water is used.

Do not attempt to use carbon tetrachloride or commercial chlorinated solvents, since they may form corrosive reaction products that do not dissolve with water base fluids. Usually a "sticky" or "tacky" gummed surface film results which makes cleaning with water more difficult. Many common water-detergents speed cleaning action, especially in reservoirs. Clear water rinse should follow detergent cleaning.

Circulate a minimum amount of petroleum oil through the system and flush out as much as possible without reopening the system. This oil should not be reused.

Trade Names and Manufacturers of Water Glycol Fluids

Ucon Hydrolube, Union Carbide Chemicals Company.
Houghto-Safe (600 Series), E. F. Houghton & Company.
Mobil Nyvac Fluid, Mobil Oil Company.
Celluguard, Celanese Chemical Company.
Fyre Safe 225, Nalco Oil Company.
Safety Fluid 200, 300, Texaco, Inc.
Quintolubric, Quaker Chemical Corp.

Straight Synthetic Phosphate Ester and Phosphate Ester Base Fluid

These fluids ordinarily do not present any irritating or sensitizing action to the skin. Continued ex-posure tends to dry the skin. Skin tends to be more sensitive to the chlorine-containing fluids than to the phosphate esters. Ordinary care, avoidance of continual skin contact, and normal personal hygiene is advised.

Vapors from the fluid *at recommended temperatures* are not considered harmful. Irritating fumes may be developed when fluid is sprayed on high temperature metals.

Leakage or spillage, like that of petroleum oils, can be cleaned up with the use of cleaning solvents.

Viscosity Control

Generally, no additives are necessary to maintain original viscosity characteristics. Synthetic fluids are compounds and by nature are inherently stable. Changing, draining, or periodic adjustments are not normally required.

Synthetic fluids have a lower compressibility factor than petroleum oils, which contributes to faster circuit response.

Operating Temperature

Normal operating temperatures between 120°F and 130°F are recommended. Specific temperature limitations should be obtained from the fluid suppliers.

If the viscosity index is low, fluid will thin out more rapidly with increased temperatures. This causes an increase in leakage and can also change flow characteristics through control valves.

If the viscosity index is high, the fluid may contain "viscosity index improvers" which may "shear down" with use in pumps and valves, causing the viscosity to lower. Consult the fluid supplier to determine if the fluid will "shear down."

Foaming and Aeration

Air retention in straight synthetics may be more prominent than with petroleum oils. The reservoir, vents, and effective baffles should be adequate, and the pump inlet line should be as far as possible from the system return line. This will provide maximum deaeration.

Wear Resistance Characteristics

Wear resistance characteristics of straight synthetic fluids appear to be similar to those of good grades of petroleum oils.

Specific Gravity

Straight synthetic fluids have a higher specific gravity and are not water soluble. Water tends to go to the top when the fluid is not agitated. Because of the higher specific gravity, pump inlet conditions may be affected. Lower micron filtration should be restricted to the discharge side of the pump. Inlet strainers and pump inlet lines must be of adequate size to maintain proper inlet conditions at the pump.

It is possible that Fuller's earth active-type depth filters can be used to advantage if the synthetic fluids are of the *non*additive type.

Compatibility Factors

No corrosive action is known to occur when straight synthetic fluids are used with the metals normally used in the components of a hydraulic system. Corrosion can occur if water contaminates the system.

The straight synthetic fluids have a strong solvent action on most paints, enamels, and varnishes. Nylon base paints, cured phenolic and epoxy paints are compatible with these fluids. Obtain specific information from the fluid supplier.

Sealing materials normally used with petroleum oils are *not* suitable for use with straight synthetic fluids. Seals should be changed when the system is being converted to this fluid. Suitable sealing materials are butyl rubbers, some silicones, Teflon, Kel-f, nylon, Viton, and Fluorel.

Although neoprene swells in this type of fluid, hoses have been used satisfactorily. However, butyl-lined hoses are recommended.

Changing from Petroleum Oil to Straight Synthetic Fluid

1. The system should be completely drained and cleaned. Very small amounts of residual oils usually do not interfere with the performance of phosphate ester base and chlorinated hydrocarbon base fluids. Resistance to fire is reduced by the amount of oil left in the system.
2. Drain and blow out pipe lines. Manually clean valves and reservoirs.
3. Valves should be opened or removed and drained. Replace gaskets, packings, and seals with those compatible with straight synthetic fluids. It is particularly important to change dynamic seals. Static seals do not necessarily have to be changed if there is no objection to

replacing them each time hydraulic components are serviced.
4. Pumps, accumulators, and cylinders should be removed if necessary for draining purposes. Replace packings, seals, gaskets, and bladders with those compatible with synthetic fluids. Consult seal and packing manufacturers for recommended materials.
5. Check the component functions before the system is closed and straight synthetic fluid is added.
6. Close the system and flush with a small amount of straight synthetic fluid. Do not continue to use the flushing fluid unless the manufacturer approves.
7. Check the system carefully after it is in operation. Leakage at seals is appreciably more expensive with straight synthetic fluids than with petroleum oils.

Changing from Straight Synthetic Fluid to Petroleum Oil

1. The system should be completely drained and cleaned.
2. Pipe lines should be opened and thoroughly drained.
3. Valves should be opened or removed and drained. Replace gaskets, packings, and seals to those compatible with oil.
4. Pumps, accumulators, and cylinders should be removed, if necessary, for draining purposes. Replace packings, seals, gaskets, and bladders with compatible materials. Substitute rubbers recommended for straight synthetic fluids are, in general, not resistant to the solvent or softening action of conventional petroleum base oils.
5. Reservoirs should be opened so that fluid can be thoroughly removed and surfaces wiped clean.
6. Strainers should be dismantled and completely cleaned.
7. Drain and clean filter housings. Replace or clean filter elements.
8. Coolers should be thoroughly drained.
9. Ordinary commercial cleaning solvents will "cut" the straight synthetic fluids and can be used to clean reservoirs or to wipe parts.
10. Circulate a small amount of oil through the system and flush out as much as possible without reopening the system. The flushing oil should not be reused.

Straight Synthetic Fire Resistant Fluids—Trade Names & Manufacturers Phosphate Ester and Phosphate Ester Base

Cellulube, Celanese Corporation.

*Pydraul F*9, 150, 625, and 60, Monsanto Chemical Company.

Skydrol (aircraft and ground support equipment), Monsanto Chemical Company.

Mobil Pyrogard 42, 43, 53, and 55, Mobil Oil Company.

Houghto-Safe, 1000 Series, E. F. Houghton Company.

Shell, SFR B, C, and *D,* Shell Oil Company.

Fyre Safe, 1090, Nalco Oil Company.

Chlorinated hydrocarbon base

Pydraul, A–200, Monsanto Chemical Company.

Water-in-Oil Emulsions

Short-term vapor exposure or handling of emulsion fluids in their complete form has no pronounced toxic effect. Normal precautions for handling and cleaning any spillage are the same as with other hydraulic fluids.

Water-in-oil emulsions are premixed or finished emulsions ready for use in hydraulic systems. The amount of water in various brands usually ranges around 40 per cent and should be maintained according to the fluid manufacturer's recommendation.

A number of manufacturers have concentrates available consisting of the base oil plus emulsifiers and additives. Finished emulsions can be prepared from these concentrates by adding water.

Viscosity Control

Viscosity of the water-in-oil emulsion decreases as the water content is reduced. It is good practice to check the fluid at regular intervals after fluid is first put into operation in order to determine the average water evaporation rate. Check with the supplier to determine proper testing procedures, because loss of water increases fire hazard.

Pure or distilled water free of harmful iron, lime, salts, or other foreign matter should be used for make-up.

Operating Temperature

Fluid operating temperatures should not exceed 120°F. Higher temperatures may cause difficulty because of excessive evaporation of water or oxidation. Products of oxidation can contribute to emulsion instability and the separation of free water from the fluid.

The performance of emulsion fluids depends on their ability to resist phase separation. Emulsions which are trapped in stagnant areas, such as single acting cylinders or accumulators, will, in time, suffer some oil and water separation. In such cases, re-emulsification can be established by circulating through the pump or across relief valves.

Repeated freezing and thawing of emulsions in unprotected storage can cause phase separation. Storage above 100°F can also result in phase separation.

Foaming and Aeration

The foaming and air entrainment characteristics of emulsion fluids are comparable with those of conventional petroleum oils.

Wear Resistance

The qualities of wear resistance of many emulsion fluids have been found to be adequate in systems where components are carefully selected to insure compatibility with the fluid. However, any water separation or reversion of the fluid may greatly reduce wear resistance. Emulsions may exhibit a reduction in viscosity when subjected to the high rates of shear existing in most hydraulic pumps. Consequently, a viscosity level somewhat higher than that of petroleum oils is manufactured into the fluid purposely, and the reduction in viscosity occurs when the emulsion is sheared.

Pump inlet conditions are especially critical with these fluids of higher viscosity, and precautions should be taken to assure proper installation.

Antifriction bearings and other parts are definitely limited at high-speed, high-load capacities when lubricated with these fluids.

Consult hydraulic equipment manufacturers concerning any modification necessary to use this type of fluid.

Specific Gravity

The higher specific gravity of these fluids can affect the pump inlet conditions. Fine filtration should be restricted to the discharge side of the pump. The degree of filtration allowed depends on the fluid used, and the supplier should be consulted.

Filter materials, such as metal screens or discs

and sintered metal filters, are suitable for use with emulsion-type fluids.

Emulsion-type fluids tend to hold dirt and fine metallic particles in suspension more readily than petroleum oils. Filtration should include magnets to remove ferrous materials.

Compatibility Factors

Emulsion-type fluids are not corrosive to the metals normally encountered in industrial hydraulic systems.

Emulsion-type fluids may have a solvent action on some paints, enamels, or varnishes. However, many of the coatings used with petroleum oils are satisfactory for use with water-in-oil emulsion fluids. Fluid suppliers should be consulted for the proper protective coating to use.

Experience to date indicates that Buna "N"-type elastomers and neoprene seals and packings, which are normally used for petroleum oils, are satisfactory for use with emulsion fluids. Silicones, Teflon, Viton, Fluorel, and Kel-f are also suitable materials.

It is generally not necessary to replace static seals until they are disturbed during overhaul. Butyl dynamic seals should be replaced immediately when the system is charged over for use with emulsion fluids.

Changing from Petroleum Oil to Emulsion Fluid

1. The system should be completely drained and cleaned. Very small amounts of residual oils usually do not interfere with the performance of emulsion fluids.
2. Drain and blow out the pipe lines. Manually clean the valves and reservoirs.
3. Pumps, accumulators, and cylinders should be actuated to insure removal of all fluid.
4. Strainers should be dismantled, washed, and cleaned.
5. Drain and clean the filter housings. Replace or clean the filter elements.
6. Check the component functions before the system is closed and emulsion fluid is added.

7. Close the system, fill with emulsion fluid, and circulate for at least eight hours.
8. Drain the flushing charge, blow down the system, and install operating charge of emulsion fluid.

Changing from Emulsion Fluid to Petroleum Oil

1. The system should be completely drained and cleaned.
2. Drain and blow out the pipe lines, and manually clean the lines and reservoirs.
3. Pumps, accumulators, and cylinders should be actuated to insure removal of fluid.
4. Strainers should be dismantled and completely washed and cleaned.
5. Drain and clean the filter housings. Replace or clean the filter elements.
6. Close the system and install a minimum quantity of petroleum oil. Circulate for 16–24 hours.
7. Drain and blow down the system and install the operating charge of petroleum oil.

Water-in-Oil Emulsion Fire Resistant Fluids—Trade Names and Manufacturers

Gulf FR Fluid, Gulf Oil Company.
Houghto-Safe 5000 Series, E. F. Houghton & Company.
Pyrogard C and D, Mobil Oil Company.
Sinclair Duro FR Fluid, Sinclair Refining Company.
Shell Irus Fluid 902, Shell Oil Company.
Sunsafe, Sun Oil Company.
Hul-E-Mul, Hulbert Oil & Grease Company.
Imol, Humble Oil Company.
Dasco, D. A. Stuart Oil Company.
Texaco TL 4-625, Texaco, Inc.
Puro RF, Pure Oil Company.
Permanul, American Oil Company.
Fyre Safe w/o, Nalco Oil Company.
LSE-1256, Atlantic Oil Company.
Pacemaker, Cities Service.
Quintolubric, Quaker Chemical Corporation.

Questions

1. What is the primary function of a liquid in a hydraulic system?
2. What type of fluid is most generally used to transmit power in a hydraulic system?
3. What is viscosity and how is it measured?
4. What effect does temperature have on viscosity?
5. What happens when the viscosity of an oil is too high?

6. Why should normal operating temperatures be controlled below 130°F in most hydraulic systems?

7. What causes catalytic action in a hydraulic oil?

8. How does moisture get into a hydraulic system?

9. What does the neutralization number of a fluid refer to?

10 When are fire resistant fluids used in a hydraulic system?

11. What is meant by compatibility?

12. What effect does a higher specific gravity have on the inlet of a pump?

13. Why is it necessary to use precautions when one is changing from a petroleum base fluid to a fire resistant fluid?

CHAPTER **3**

SEALING DEVICES FOR HYDRAULIC POWER

The success of applying fluid power to any application depends largely on the ability of the sealing devices to prevent both internal and external leakages in the system. Around the year 1650 Blaise Pascal discovered the fundamental laws of physics by which all modern fluid power systems are governed. Unfortunately, his discovered knowledge was not put to practical use until about one hundred and fifty years later, when Joseph Bramah developed the first hydraulically operated press (see Fig. 3–1). The success of this first press, which used water for the transmission of power, is accredited to the invention of the cup packing. Cup packing was the sealing device used to prevent leakage within the cylinder or ram.

Modern fluid power systems depend more and more on the development of sealing materials that will operate with special fluids and at conditions of high pressures and temperatures. The manufacturers of sealing devices have kept pace with the needs of all industry through excellent research and development programs.

Because of this progress in the chemistry of dealing with sealing materials, many new trade names associated with products are constantly being in-

troduced. The manufacturers of sealing devices are the experts in this highly specialized field, and their recommendations are always important in any application.

Fluid power design engineers work very closely with the suppliers of sealing devices to insure proper sealing for rotating shafts, reciprocating rods and stems, and for stationary mating parts of fluid power equipment.

Losses of fluid from hydraulic machinery caused by leakage can be very costly not only in terms of the cost of the fluid but also in production. External leaks in a system can be quickly discovered and fixed, but internal leakage is not so easily detected. Internal leaks slow up machines, causing more rejects of parts, as well as loss in production. Often, lost production from one machine may slow up the entire operation of a plant.

Good maintenance procedures with regard to external leaks are essential to minimize the safety hazards created by oil leaking on the floor and around the machines. Also, oil gets on machined parts, causing an extra cleaning operation. Experience indicates that where maintenance procedures have been

Fig. 3-1. (First Press)—cup packing. A 25,000 ton press made successful through cup-packing. (*Courtesy of Socony Mobil Oil Co.*)

established which minimize excessive leakage, increased production has been the result.

Conditions Affecting the Selection of Sealing Devices

Each application for a seal must be treated as a specific problem because of slight variations in minor conditions of operation. The entire set of conditions encountered in the operation of a component, such as a pump, fluid motor, cylinder, or control valve, must be considered collectively. For example, a given type of seal design might have the ability to stand a given pressure, or it might be able to stand a given speed, but it might not be able to withstand the same given pressure and speed occurring together.

1. Speed is an important factor in determining frictional temperature buildup, which results in wear of both the seal and the shaft. Therefore, surface speed or peripheral speed and shaft size are important considerations. Seals may also be affected by frequency, depending upon the revolutions per minute. This is important when a seal must operate satisfactorily under eccentricity, wobble, and whip conditions of rotation. The same is true concerning end play of shafts. For example, a 1 in. diameter shaft having a speed of 4000 rpm presents a condition more serious than a 4 in. diameter shaft rotating at 1000 rpm, with the same eccentricity, although the surface speed is the same in both cases.

2. Pressure in the system in contact with a sealing device may increase the contact force. This would increase friction, causing heat buildup and more wear. In some cases, seals

may be pressure balanced, which would eliminate increased friction.

Pulsating pressures are troublesome because they set up vibrations in the sealing surface which interfere with the sealing contact.

3. Temperature at the point of seal depends upon the materials in contact, the shaft speed, pressure, amount of lubrication, heat from bearings, heat from fluid, or other sources. The sealing device selected must be unaffected by frictional temperature or the fluid being sealed.

4. Compatibility between sealing devices, the fluid, and other materials that make up the complete system is extremely important. Selecting the wrong sealing materials may result in costly shutdowns and loss of production. Foreign materials, corrosion, excessive heat, and many other conditions can affect sealing operations.

In addition to the factors listed above, there are many machine operating conditions which may also affect good sealing practices. Fast reversals, intermittent operation, long intervals between operations, or any erratic motions are a few examples.

The designer who specifies the type of sealing device for any given application should study all of the problems carefully, using the supplier for consultation and help in making the final decision.

Sealing Materials

Sealing materials may be classed into three general categories: leather, fabricated rubber, and homogeneous.

Leather is the oldest material used for sealing devices and is still very popular for many applications. Leather inpregnated with synthetic rubber may be operated safely at temperatures up to 180°F. Leather seals have low frictional properties and relatively high tensile strength which resists extrusion into clearance spaces. Thus, leather is used for applications where higher system operating pressures are used. It is also used as anti-extrusion rings to bridge clearances for using "0" rings and thus preventing their extrusion into the clearances (see Fig. 3–2).

Leather does not score metal surfaces. However, rough metal surfaces have a tendency to shorten the life of any material used for sealing. A metal surface between 8 and 40 RMS (Root Mean Square) (micro finish) is recommended.

Leather sealing devices are manufactured in "U,"

Fig. 3-2. Anti-extrusion ring for "O" ring. (*Courtesy of Socony Mobil Oil Co.*)

Fig. 3-3. "U" "V" cup and flange shapes of seals.

WASHER

CUP PACKING

FLANGE PACKING

U-PACKING

V-PACKING

O-RING

"V," cup, and flange shapes (see Fig. 3–3), which are in accordance with Fluid Power Standards.

Fabricated rubber sealing devices are composed of synthetic rubber compounds and fabrics. The fabrics reinforce the synthetic rubber to give sealing devices resistance to extrusion. There are numerous combinations of synthetic rubbers and fabrics.

Cotton duck, asbestos, and nylon are three common kinds of fabrics in use. Duck is used for normal operating temperatures, asbestos for high temperature operation, and nylon for greater strength and flexibility.

The synthetic rubbers or polymers used in fabricated seals depend on the type of fluid, temperature, and other operating conditions. Some of the common synthetic rubber bases include chloroprene, Buna "N," Buna "S," Butyl and Viton. Chloroprene and Buna "N" are suited for petroleum oils, Buna "S" for water systems, Butyl for straight synthetic fire resistant fluids, and Viton for high temperature operations.

Fabricated seals have a wider operating temperature range than leather, because the base polymer can be varied. For example, a combination of asbestos and Viton would have an operating temperature range of − 65°F to 400°F.

The metal finishes recommended for fabricated seals is between 16 RMS and 32 RMS.

Fabricated sealing devices are made in "U," "V," cup, and flange in Fluid Power Standard sizes for interchangeability with other material seals.

Homogeneous sealing devices are compounded from many different base polymers of synthetic rubber. They are made in many hardnesses, depending on the shape, application, and intended operating pressure.

The compounding of materials with elastic qualities in a synthetic rubber for a specific fluid or operating condition is complex. Resilient compounds have intricate, semiplastic, colloidal structures which are capable of almost infinite variation because of the hundreds of ingredients from which they are produced.

A hardness range from 8 Shore durometer (see Fig. 3–4) to 40 durometer measured on the "A" scale are commonly available. Other hardnesses are available for special installations.

Homogeneous seals have a wide range of temperature operation similar to that of fabricated seals (−65° to 400°F). When operating pressures in the system exceed 1500 psi, anti-extrusion rings or minimum metal clearances must be used.

The metal surface finish for homogeneous seals should be between 8 RMS and 40 RMS to minimize friction.

The homogeneous seals are made in "V" and "U" cups and "O" ring shapes, according to Fluid Power Standards for interchangeability.

Selecting Shape of Seal

Selecting the seal which is properly shaped for the job depends on how the seal must be applied and how the part to be sealed is constructed. Figure 3–5 shows common uses of the various seals.

Multiple "V's" may be used for an extreme range in operating pressures. They also are adaptable for rotational or reciprocating motion with pistons or

Fig. 3-4. Durometer. (*Courtesy of Chicago Rawhide Mfg. Co.*)

Fig. 3-5. "V" seals. (*Courtesy of Socony Mobil Oil Co.*)

shafts. When "V" seals are installed, the gland nuts should be torqued just enough to seat the seal and then backed off to a point where the sealing is still maintained. Damage may occur from excessive friction if the seals are kept too tight. Often a spring is used to maintain the correct force on the seals at all times. A value of 10 lb per inch of mean diameter is generally used for selecting spring force. "V" seals are considered a balanced seal.

The "0" ring (see Figs. 6a and b) is placed in a groove in such a way that there is enough contact pressure, because of direct compressive stress in the material, to accomplish sealing at low pressures. When greater hydraulic pressure is applied, the material has a tendency to reshape itself in the space away from the pressure, which increases contact pressure.

"0" rings are used as stationary seals or for reciprocating motion on piston or shaft applications. At higher pressures, anti-extrusion rings or back-up washers should be used to prevent the seal from extruding into the clearance space. "0" ring seals are not generally used for rotary applications, except for very low speeds and pressures. They are very popular as static seals, but should not be used where vibration or motion is excessive. "0" rings should be re-

placed when components are disassembled and reassembled during service or repair.

"0" rings are considered a balanced seal.

The "U" seal (see Fig. 3–7) is a form of pressure-actuated seal that depends upon the hydraulic pressure in the system to act against the lip to create contact pressure for sealing. In order for the lip to move freely, the groove space should be slightly greater than the seal. This type of seal is more effective when the groove is properly proportioned to allow lip flexing with little solid compression. When used at operating pressures over 400 psi, it is advantageous to use a leather or fabricated backup washer to prevent extrusion of material into the clearance space.

Seals of this type are commonly used for pres-

COMPATIBILITY CHART

Sealing material	Fluid compatibility	Remarks
Natural rubber (NR)	Water base fluids	Operating temperature range low to 200° F. Compounded to suit fluid requirements.
Leather (treated) (Impregnated with thiokol)	Petroleum base fluids Some synthetics and phosphate ester	Operating temperature range −65 to 265° F. Multiple seals good for high pressure.
Butyl rubber (11R)	Water base fluids and phosphate ester	Operates from −65 to 250°F.
Nitrile rubber (Buna-"N") (Hycar)	Petroleum base fluids Water base and some synthetics	Operates from −65 to 250°F. Does not have corrosive tendencies, but swells in aromatic oils and some synthetics.
Silicone rubbers	Water base fluids Fair with petroleum base Good with some synthetics	Operating temperature range from −120 to 500°F. Not suitable for silicone fluids.
Fluoro-plastics (Teflon) (Kel-F)	Good with almost all fluids used	Operating temperature range from −320 to 500°F. Low friction with good chemical resistance.
Fluoro-elastomers (Viton A)	Good with water, petroleum base fluids and most synthetics	Operating temperature range from −20 to 500°F.
Polyacrylates (Hycar 4021) (Vyram)	Good with chlorinated hydrocarbons	Not recommended for low temperatures. High temperature about 350°F.
Polysulphides (Thiokol)	Good with water, petroleum base, and most synthetic fluids	Operating temperature range from 10 to 160°F Vaporizes above 160°F.
Styrene (SBR)	Good for water base fluids	Operating temperature range slightly higher than natural rubber.

RELATIVE POSITIONS OF "O" RING PACKINGS IN DIFFERENT GROOVES
AT INCREASING PRESSURES

OTHER VERSIONS OF RING SEALS

Fig. 3-6. (a) & (b) "O" rings. (*Courtesy of Socony Mobil Oil Co. and E. F. Houghton Co.*)

RUBBER CUPS   LEATHER CUPS

Fig. 3-9. Cup seals. (*Courtesy of Socony Mobil Oil Co.*)

"V" AND "U" SEALS

A   D

B   E

C   F

RUBBER   LEATHER

Fig. 3-7. "U" seals. (*Courtesy of Socony Mobil Oil Co.*)

Fig. 3-8. Flange seals. (*Courtesy of Socony Mobil Oil Co.*)

RUBBER FLANGES   LEATHER FLANGES

sures as high as 3000 psi for cylinder pistons and rods.

The "U" seal is considered a balanced type of seal.

The flange seal (see Fig. 3–8) is a pressure-actuated seal that effects sealing through the hydraulic pressure in the system. Some of these seals are made with cylindrical sides, or lips, the same dimension as the cylinder rod, but it is considered better practice to make the lip tapered toward the member to be sealed so that the flange seal is somewhat under the rod size. Flange seals are generally applied in the lower pressure ranges.

The flange seal is considered an unbalanced seal.

Cup seals (see Fig. 3–9) are primarily used for sealing cylinder pistons. This seal is also pressure-actuated, with the sealing accomplished by the contact pressure outward against the cylinder bore. It is essential that these seals be clamped in place, and the method of clamping is important. The lip of the seal cannot be too long or compressed too tightly against the cylinder walls. If the lip is too long, repeated stroking of the cylinder causes weakening and the seal breaks down and bits of it extrude into the clearance spaces. Eventually, the pressure breaks the cup all the way through. It is, therefore, important that the cup lip be no higher than is necessary to follow the side play.

This type of seal is considered an unbalanced seal, and its operating pressure depends on the material and conditions under which it operates.

60

Questions

1. What results when a hydraulic system has internal leaks?
2. What effect does the speed of an actuating cylinder or fluid motor have on the selection of sealing devices?
3. What effect does pulsating pressure have on seals?
4. What effect does temperature have on sealing materials?
5. What happens when the sealing material and the fluid are not compatible?
6. Which kind of sealing material is generally used for systems operating at higher pressures?
7. What is an anti-extrusion ring?
8. Which types of sealing materials are used for higher temperature operations?
9. Make a sketch of the various shapes of sealing devices which are normally used with hydraulic systems.
10. What is meant by a static seal?

CHAPTER 4

THE DISTRIBUTION OF HYDRAULIC POWER

The distribution of hydraulic power involves pipe, tubing, hose assemblies, manifolds, and the various fittings which are used to connect the components of a system.

The choice of pipe, tube, or hose depends largely on the system's operating pressure and flow. Other important factors to consider include environmental conditions, type of fluid and operating temperature, shock loads, relative motion between connected parts, practicality, and compliance with certain standards.

All materials used to convey fluid power are commonly classified as *conductors* and the various fittings for connecting components are classified as *connectors*.

Fabricating a hydraulic system requires skill, patience, and pride of workmanship. Personnel engaged in this type of work are required to have a good understanding of basic fluid power and to know how to read circuit diagrams. Skills in the art of measuring, bending, cutting, flaring, or threading of conductors are acquired mainly through experience or practice.

Good tools are also essential for the proper fabrication of a hydraulic system. Using proper tools

for flaring tubing, or threading pipe prevents leakage and aids in the prevention of detrimental contamination to the system. Small particles of metal from ragged threads or flares can cause serious damage to such components as pumps, motors, cylinders, and valves. Making the proper bends and connections can also help to minimize friction and reduce power losses.

This entire area of distributing fluid power is very significant in terms of the successful operation of any hydraulic equipment.

Operating Pressure and Factor of Safety

Conductors and connectors used for hydraulic circuits are selected mainly to operate at the highest pressure level that is reached during a cycle of the equipment operation. The materials used must have a continuous operating pressure rating which will withstand working pressures and provide a margin of safety for short-lived pressure peaks resulting from hydraulic shock.

Hydraulic shock occurs when control valves operate suddenly, stopping or reversing flow which is backed up by large mass forces or high velocities. This condition produces a shock wave due to the kinetic energy in the fluid. Hydraulic shock waves also result from the sudden deceleration, stopping, or reversing of heavy work loads. In either case, the produced shock waves travel through the conductors on the downstream side into various branches at the speed of sound for the fluid, and then bounce back toward the point of origin. This process repeats, with each bounce or reaction traveling a lesser distance than the preceding, until the kinetic energy is dissipated into friction and heat.

Pressure ratings of the various conductors depend mainly on the tensile strength of the materials used and the wall thicknesses of the conductor. The wall thicknesses and safety factors recommended by the Fluid Power Industry Standards are based on calculations using Barlow's formula:

Minimum wall thickness (in.)

$$= \frac{\text{maximum pressure (psi)} \times \text{OD of conductor (in.)}}{2 \times \text{tensile strength (psi)}}$$

(OD = outside diameter. TS = tensile strength)

Example.

a. What would be the maximum pressure for SAE_{1010} Dead Soft Cold Drawn Steel Tubing with an outside diameter of $\frac{3}{4}$ in.; tensile strength of 55,000 psi; and wall thickness of 0.065 in.

Max. psi

$$= \frac{\text{Wall thickness (0.065)} \times 2 \times 55,000 \text{ psi T.S.}}{\text{OD 0.750 in.}}$$

Max. psi = 9533 psi

b. What would be a safe working pressure for this tubing if the factor of safety must be at least a 4:1 ratio?

$$\text{Working pressure (psi)} = \frac{\text{Max. pressure (psi)}}{\text{Factor of safety}}$$

$$\text{WP} = \frac{9533 \text{ psi}}{4}$$

WP = 2383 psi

The formula is adequate for practical purposes in selecting conductor wall thicknesses to withstand the maximum rate of surge peak pressures at the frequencies developed by the cycling of the hydraulic equipment operation. Surge pressures that may be encountered within a system are rarely known and seldom appreciated in their full potential strength. Actual experience has shown that surge peaks of two to four times that of design working pressure are not un-

common and often not suspected by the system designer. When detailed information on surge peaks is not readily available, the importance of a safety factor becomes appreciated, and it becomes a prime consideration in the circuit design.

In addition to normal working pressures and surge pressure peaks, there may be mechanical stresses produced by thermal expansion, abuse, and environmental factors. Evaluation of all phenomena is difficult, so the proper selection of conductors and connectors is usually a compromise drawing heavily on what has been experienced as most successful. Suppliers are most co-operative in helping to solve problems, and their experience may serve well to avoid trouble and provide excellent performance with maximum safety.

However, pressure surges and general operating characteristics within a hydraulic system can be accurately measured by the use of instruments such as a Hydrauliscope (see Fig. 4–1) or with an oscilloscope with power supply and proper accessories. Instruments such as these use a transducer or type of strain gauge that changes the pressure surge into an electronic pulse, which, in turn, is transmitted onto a screen for visual reading. The screen is graduated into divisions that give accurate measurement of the pressure surges within the circuit.

When excessive shock loads are discovered through instrumentation, it is possible to add relief valves, hydraulic fuses, or shock absorbers to prevent serious damage to the hydraulic equipment.

Pressure relief valves appropriately installed are of prime importance for the protection of all system components, including pumps, fluid motors, cylinders, control devices, conductors, connectors, and auxiliary equipment. The relief or safety valve helps to protect a hydraulic system during over pressure conditions that may be impossible to predict. The person fabricating a hydraulic system has a responsibility to recognize the need for additional pressure relief protection that may be caused by a particular installation. The following may serve as a guide in providing safety from over-pressures:

1. Provide ample protection for all parts of the system, using one or more relief valves of proper capacity.
2. Install relief valves to provide priority protection for the pump.
3. Provide an additional relief valve or hydraulic fuse to protect extra long conduit runs.
4. Provide an additional relief valve or hydraulic

Fig. 4-1. Hydrauliscope in action. Kenosha Technical Institute students are shown using a hydrauliscope to measure the pressure surges in a shock-absorber when a 700 lb weight is accelerated to 15 mph and stopped within a couple of inches. The shock absorber absorbs 72,000 in lbs of kinetic energy in stopping the weight. Maximum pressure surge recorded nearly 8000 lbs per sq in.

fuse to protect against trapped fluid which may be subject to shock loads.

5. Use relief valves of the design type best suited to the specific application.
6. Apply shock absorbers where the intensity of surges is beyond the damping characteristics of common relief valves.

Conductors and connectors are available for working pressures to 10,000 psi and higher. The choice of pipe, tubing, or hose is usually a matter of economics. Tubing provides greater flexibility, cleaner appearance, fewer connectors, less leakage, and better reusability. However, pipe is readily available and is generally less in cost. Pipe also has advantages for low-pressure applications, where larger conduits are needed. Figure 4 shows a typical installation using modern techniques to connect system components.

Pressure Drop

Pressure drop is essentially a result of a change in the hydraulic-energy form. Hydraulic pressure energy starting at the pump with an initial potential value has the ability to perfom a specific amount of work. However, as the fluid flows through the conductors or around the bends and turns through the various connectors and components of the system, pressure energy is exchanged for velocity energy and heat. This means that some of the hydraulic energy is lost along the way, because of resistance and friction. Factors which influence loss of energy or pressure include sudden contractions, sudden enlargements, number and kinds of bends, size and smoothness of conductors, as well as the temperature and properties of the fluid itself.

Only a few problems in fluid mechanics can be solved entirely by rational mathematics. Much of the information available involves empirical formulas which are based, at least in part, on experimentally determined coefficients.

Dimensional analysis of flow through a straight conduit can be determined with the Darcy formula, except for the friction factor, which has to be determined by experiment:

$$h_f = \frac{fLv^2}{D2g} \qquad f = \frac{64}{\text{Re}}$$

where D = internal diameter of conduit—ft.
  f = friction factor.
  g = acceleration due to gravity—nominal 32.2 ft/sec².
  h = head—ft.
  h_f = head loss—ft.
  L = length of conduit—ft.
  Re = Reynolds number
  v = mean velocity—ft/sec.

Fig. 4-2. Distribution lines. (*Courtesy of Flodar.*)

Fig. 4-3. Coefficients of orifices.

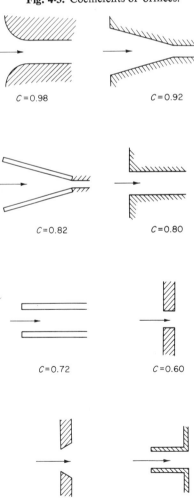

$C = 0.98$

$C = 0.92$

$C = 0.82$

$C = 0.80$

$C = 0.72$

$C = 0.60$

$C = 0.80$

$C = 0.53$

Darcy's formula is used to determine the pressure drop due to flow friction through a conduit.

When the flow is laminar, Poiseuille's law expresses the friction factor of Darcy's formula as a ratio of 64 to the Reynold's number.

Reynolds Number

The Reynolds number is a nondimensional combination of quantities designated by the letter R.

$$Re = \frac{\rho VD}{\mu} \quad \text{or} \quad Re = \frac{VD}{\nu}$$

where Re = Reynolds number.

ρ = mass density of the fluid in lb sec/sq ft.

V = mean velocity in ft/sec.

D = diameter of conductor.

μ = absolute viscosity in lb mass/ft/sec.

$$\mu = 0.000672\,s\left(0.22\,SUS - \frac{135}{SUS}\right)$$

Sample problem.

Calculate the Reynolds number for the flow of oil with the following characteristics: $\rho = 57$, $V = 10$, D = 2 in., $SUS = 300$, s equals 0.9.

$$R = \frac{57 \times 10 \times 0.167}{0.000672 \times 0.9 \times (0.22 \times 300 - 135/300)}$$

Answer. 2379.

$R. < 2000$ = laminar flow

$R. > 2000$ = turbulent flow

Pressure drop through conductors, connectors,

valves, and other components depends on the internal passage design, smoothness of passage, and the effect of edge angles. Because the internal design configuration varies with the manufacturer, information concerning pressure drop must be obtained through bench testing or directly from the manufacturer.

Coefficients of Typical Orifices

$C = 0.98$ $C = 0.92$ $C = 0.82$ $C = 0.80$

$C = 0.72$ $C = 0.60$ $C = 0.80$ $C = 0.53$

In general, the following formulas may be useful in converting the various orifice coefficients into pressure drop or flow:

$$\text{gpm} = C_V\sqrt{\frac{\text{Pressure drop}}{\text{Sp. Gr.}}}$$

$$C_V = \frac{\text{gpm}}{\sqrt{\dfrac{\text{Pressure drop}}{\text{Sp. Gr.}}}}$$

$$\text{Pressure drop} = \left(\frac{\text{gpm}^2}{C_V}\right) \times \text{Sp. Gr.}$$

Conversion equations:

$$\text{lb/hr} = 500C_V\sqrt{\frac{\text{Pressure drop}}{\text{Sp. Gr.}}}$$

$$\text{psi} = \frac{\text{Sp. Gr.} \times \text{head in feet}}{2.31}$$

$$\text{Sp. Gr.} = \frac{\text{Fluid density in lb/ft}^3}{62.34}$$

$$\text{Centistokes} = \frac{\text{Centipoises}}{\text{Sp. Gr.}}$$

$$\text{SUS} = \text{Centistokes} \times 4.635$$

$$\text{Saybolt Furol seconds} = \text{Centistokes} \times 0.470$$

The application of orifice discharge calculations is primarily for determining flow measurement. If the pressure drop through a conductor, connector, or valve is determined, the velocity and volume of flow can be calculated by using flow coefficients.

When exact pressure drops are important to determine, it may be important actually to fabricate and test portions of the hydraulic system. By controlling conditions of fluid velocity, density, size of conductor, and friction, it is possible to maintain laminar flow. When laminar flow conditions are maintained, it is possible to determine accurate pressure drop with Darcy's formula.

Fluid Velocity

Velocity in a hydraulic conductor is obtained at the expense of pressure energy, and the decrease in pressure due to velocity is expressed as follows:

$$\text{Velocity head (ft)}(2.31 \text{ ft} = 1 \text{ psi}) = \frac{v^2}{2g}$$

It is obvious that when no flow occurs the pressure drop is zero. It is also apparent that when a high flow rate occurs there will be a high pressure drop. This flow concept is based on the conservation of mass principle and can be applied to flow through conductors. When the density of the fluid is constant, the following equation may be used:

Inside diameter of conduit (in.)

$$= 0.639\sqrt{\frac{\text{gpm}}{\text{velocity ft/sec}}}$$

Example. What is the velocity of flow in ft/sec if the conductor has an inside diameter 0.500 in. and the flow is 10 gpm?

Substituting, we have

$$0.500 = 0.639\sqrt{\frac{10}{\text{fps}}}$$

$$\left(\frac{0.500}{0.639}\right)^2 = \frac{10}{\text{fps}}$$

$$v \text{ in fps} = 10\left(\frac{0.639}{0.500}\right)^2$$

$$\text{Velocity} = 16.33 \text{ ft/sec}$$

The hydraulic horsepower of a system is determined by the number of gallons per minute multiplied by the operating pressure and divided by a constant.

$$\text{hp} = \frac{\text{gpm} \times \text{psi}}{1714}$$

The horsepower formula indicates that hydraulic systems can operate at high pressure and low volume or at low pressure and high volume to obtain any given horsepower. This is illustrated in Table 4–1.

TABLE 4-1
POWER TRANSMISSION THROUGH CONDUCTORS
20 FT/SEC FLUID VELOCITY

Conductor size I.D.	Horsepower at 1000 psi	Horsepower at 6000 psi
$\frac{1}{4}$	1.8	10.7
$\frac{3}{8}$	4.1	24.3
$\frac{1}{2}$	7.1	42.6
$\frac{5}{8}$	9.6	66.8
$\frac{3}{4}$	16.1	96.3
1	28.6	171.3

Inside diameters of the conductors are important for the determination of flow rate capacity of a system. The velocity of a given flow through conductors will vary inversely as the square of the inside diameter. Conductors between actuators, selector valves, pressure controls, pump, and other components of a

BASED ON FORMULA

$$\text{AREA (SQ.IN.)} = \frac{\text{G.P.M.} \times 0.3208}{\text{VELOCITY (FT/SEC.)}}$$

*RECOMMENDED VELOCITY RANGE FOR INTAKE LINES

*RECOMMENDED VELOCITY RANGE FOR PRESSURE LINES

FLOW – CUBIC INCHES PER MINUTE

FLOW – GALLONS PER MINUTE

NOMINAL STANDARD PIPE SIZE

ACTUAL INSIDE DIAMETER OF CONDUCTOR-INCHES

INTERNAL AREA OF CONDUCTOR SQ. IN.

VELOCITY – FEET PER SECOND

Fig. 4-4. Flow capacities of conductors at recommended flow velocities.

*RECOMMENDATIONS ARE FOR OILS HAVING A VISCOSITY INDEX ABOVE 95 OR 150 SSU AT 100°F AND 50 SSU AT 210°F (ANALINE POINT 190° TO 210°). ASTM #2 GRADE

system should be constructed rigidly to confine a minimum volume of fluid. This is important in order to maintain constant controlled motion and to restrict the effort of varying forces.

Fluid flow velocity is an important factor, and experience has shown that some higher pressure systems can tolerate greater velocities because of the low volume generally required for a given horsepower. The generally accepted design target of 15 ft/sec can in some cases be increased without harmful effects to flow characteristics, but the heat factor of a system can be controlled better at the lower velocities. This is important in machines that operate continuously, such as machine tools.

The nomograph in Fig. 4–4 is useful for calculating the following formula:

$$\text{Area of conductor (sq in.)} = \frac{\text{gpm} \times 0.3208}{\text{velocity fps}}$$

It is a simple operation to use the nomograph when any two of the variables are known. Velocities may be selected arbitrarily on the basis of the recommended values for either pump inlet or discharge.

Example. What is the required inside area of a conductor that will discharge 9 gpm at 15 ft/sec?

Solution. Connect 15 ft/sec on the right-hand scale, with a straight edge, across the center scale to the left-hand scale which reads 9 gpm. When the line crosses the center scale the area reads close to 0.1950 sq in.

To find the inside diameter, use the following equation:

$$D = \sqrt{\frac{\text{Area}}{0.7854}}$$

$$D = \sqrt{\frac{0.1950}{0.7854}}$$

$$D^2 = \frac{0.1950}{0.7854}$$

$$D = \tfrac{1}{2} \text{ in.} = 0.1964 \text{ sq in.}$$
(consult nomograph)

Check the above solution by using the formula:

$$\text{Area (sq in.)} = \frac{\text{gpm} \times 0.3208}{v}$$

$$\text{Area} = \frac{9 \times 0.3208}{15}$$

$$\text{Area} = \frac{2.88}{15}$$

$$\text{Area} = 0.192 \text{ sq in.}$$

When selecting pipe, hose, or tubing as the conductor, it is general practice to use the nearest larger standard size.

The velocity of the hydraulic fluid should not exceed the ranges recommended in the right-hand column for maintaining laminar flow.

Vibration

Another important design and installation consideration is the effect of vibration on a system. Excessive and continuous vibrations cause conductors, connectors, etc. to fracture with fatigue. Vibration can be absorbed effectively if conductors are properly supported at regular intervals. Table 4–2, taken from Industry Hydraulic Standards, should serve as a guide for the placement of tie-down supports.

A good ductile material, such as wood, various plastics, composition materials, etc., can be used as tie-down blocks. Industry standards state that all piping shell be continuous from one piece of appa-

TABLE 4-2

Tube (O. D. in.)	Distance between supports (feet)
$\frac{1}{4}, \frac{5}{16}, \frac{3}{8}, \frac{1}{2}$	3
$\frac{5}{8}, \frac{3}{4}, \frac{7}{8}$	4
1	5
$1\frac{1}{4}, 1\frac{1}{2}$	7

ratus to another, with couplings used only where necessary. Piping shall not be welded to supports and must be removable without dismantling equipment components. When proper installation is made, noise due to vibrations can be minimized. This is an important factor in any circuit.

Chemical Reaction

Corrosion and pitting due to moisture or chemical reactions are destructive agents which attack all components, conductors, and other system parts. Every precaution should be exercised to prevent contamination, which tends to cause malfunction, wear, and a reduction in strength of all materials that make up the system.

Records and Identification

Records and machinery history are important and should be maintained to have ready information available concerning circuit diagrams, source and identification of each component used in the system, spare parts and service information, type of fluid, etc. When systems call for two or more pressure levels, branch circuits, auxiliary systems, or metered flow, etc., the color code pattern shown in Table 4–3 may be helpful to apply for the training of operators and service personnel (see Fig. 4–5).

The Selection of Conductors

The prime considerations for selecting conductors for a hydraulic power system are the type of materials,

Fig. 4-5. Circuit showing use of color coding.

INLET TO PUMP

FLUID UNDER PRESSURE

EXHAUST FLUID

TABLE 4-3

Function	Definition	Diagram color	Zip-a-Tone no.	
Intensified pressure	Pressure in excess of supply pressure	Black	#7R	
Supply pressure	Power actuating fluid	Red	#177	
Charging press. Pilot pressure Reduced pressure	Inlet supercharge above atmosphere Control actuating Auxiliary pressure lower than supply	Intermittent Red	#52	
Metered flow	Fluid at controlled flow rate, other than pump delivery	Yellow	#63	
Exhaust	Return of power and control fluid to reservoir	Blue	#75	
Intake Drain	Subatmospheric pressure Clearance leakage return to reservoir	Green	#16	
Inactive	Fluid which is idle during the phase being represented	Blank	Blank	

capacity, and pressure rating. A number of materials, such as steel, iron, and plastic, in both pipe and tubing, are readily available to adequately handle a wide range of flow capacities.

Industry Recommendations for Systems Below 3000 PSI

If the system operating pressures are below 3000 psi, the following tables, taken from the Hydraulic Standards, give pertinent data important as a guide in the selection of pipe and tubing.

Schedule Pipe

Piping was originally classified by weight as "standard," "extra heavy," and "double extra heavy." This classification has been superseded by classification according to schedule numbers. Hot or cold drawn seamless pipe is recommended for use with hydraulic systems and must be free internally from rust, scale, and dirt.

Schedule numbers run from 40 (formerly standard), to schedule 80 (formerly extra heavy) and to the schedule 160 (formerly double extra heavy).

TABLE 4-4
OPERATING PRESSURES (0 TO 1000 psi)

Flow rate (15ft sec) gpm	Valve size	Pipe schedule	Tubing O. D.	Tubing-wall thickness
1	$\frac{1}{8}$	80	$\frac{1}{4}$	0.035
1.5	$\frac{1}{8}$	80	$\frac{5}{16}$	0.035
3	$\frac{1}{4}$	80	$\frac{3}{8}$	0.035
6	$\frac{3}{8}$	80	$\frac{1}{2}$	0.042
10	$\frac{1}{2}$	80	$\frac{5}{8}$	0.049
20	$\frac{3}{4}$	80	$\frac{7}{8}$	0.072
34	1	80	$1\frac{1}{4}$	0.109
58	$1\frac{1}{4}$	80	$1\frac{1}{2}$	0.120

Safety factor 8 : 1

TABLE 4-5
OPERATING PRESSURES (1000 TO 2500 psi)

Flow rate (15ft sec) gpm	Valve size	Pipe schedule	Tubing O. D.	Tubing-wall thickness
2.5	$\frac{1}{4}$	80	$\frac{3}{8}$	0.058
6	$\frac{3}{8}$	80	$\frac{5}{8}$	0.095
10	$\frac{1}{2}$	80	$\frac{3}{4}$	0.120
18	$\frac{3}{4}$	80	1	0.148
30	1	80	$1\frac{1}{4}$	0.180
42	$1\frac{1}{4}$	160	$1\frac{1}{2}$	0.220

Safety factor 6 : 1. Above $\frac{1}{2}$ in. tubing, welded flange fittings or fittings having metal to metal seals or seals that seal with pressure are recommended.

Tables 4–4 and 4–5 indicate usage of schedule pipe under Hydraulic Standards. Table 4–8 gives the dimensions of schedule 160 pipe and the rated working pressure for each wall thickness.

Industry Hydraulic Standards recommend a 4:1 factor of safety for systems operating above 2500 psi working pressure. (Welded flange fittings, or fittings having metal to metal seals, or seals that seal with pressure are recommended.) Pipe should be selected accordingly from the above tables. Proper area size for a given flow at a recommended velocity can be determined by using the nomograph or its formula $A = \text{gpm} \times 0.3208 \div \text{velocity}$.

Tubing for Higher Pressure Systems

Experience has indicated that on a broad basis tubing is proving to be increasingly more applicable to hydraulic systems than pipe. However, each has its advantages and disadvantages.

Table 4–9 has been tabulated for the selection of tubing for working pressures up to 10,000 psi with a 4:1 safety factor. Very little difficulty exists in the selection of tubing for high pressure systems in the

TABLE 4-6
DEAD SOFT COLD DRAWN STEEL TUBING SAE 1010 RECOMMENDED BY INDUSTRY STANDARDS AND COMMONLY CARRIED IN STOCK BY METAL WAREHOUSES

O. D. (in.)	Wall thickness (in.)	Burst pressure (psi)
$\frac{1}{8}$	0.032	28,000
$\frac{3}{16}$	0.032	18,500
$\frac{1}{4}$	0.035	15,400
$\frac{5}{16}$	0.035	12,300
	0.049	17,300
$\frac{3}{8}$	0.035	10,200
	0.049	14,400
$\frac{1}{2}$	0.035	7,700
	0.049	10,700
	0.065	14,000
$\frac{5}{8}$	0.049	8,600
	0.065	11,500
	0.083	14,600
$\frac{3}{4}$	0.049	7,200
	0.065	9,400
	0.095	13,900
$\frac{7}{8}$	0.065	8,150
	0.083	10,000
1	0.049	5,400
	0.065	7,150
	0.095	10,400
$1\frac{1}{4}$	0.083	7,300
$1\frac{1}{2}$	0.095	6,950

smaller capacity sizes (below $\frac{1}{2}$ in. O.D. tubing).

In the larger sizes of $\frac{1}{2}$ in. O.D. to $1\frac{1}{2}$ in. O.D. (graph), Table 4–10 gives the recommended wall thickness for a given working pressure. Compilations are based on AISIA 4130 Steel.

Plastic Conductors

Plastic hose and tubing are being applied with success to a number of hydraulic systems requiring a lightweight and flexible conductor in the lower pressure ranges.

A number of plastics are available which include nylon, polyethylene, polypropylene, and polyvinyl chloride. Each of these materials has its own advantages and disadvantages for specific applications. Some plastics have the ability to absorb water, swell, and lose strength, which may make it necessary to consult the manufacturer for this important property. Table 4–12 shows the pressure ratings for the materials mentioned above at temperature conditions.

Although testing for a specific purpose is recommended, a 4:1 safety factor is considered good engineering practice in the average hydraulic systems. Table 4–13 gives the standard sizes available in two popular pressure ranges.

In general, plastic tubing can be worked and installed with the ordinary tubing tools. It cuts easily and can be heated and given permanent bends. It can be used with standard metallic compression and flare fittings designed for metal tubing. Many new developments are being made in the nature of tools, fittings, quick disconnects, and other devices especially for plastic fabrications.

Flexible Hoses

Synthetic hose is manufactured from a large number of elastomeric or rubberlike compounds and plastics. Manufacturers are keeping up with the present trend toward more horsepower with higher pressure. Flexible hose and connectors are now available to fulfill the requirements of circuits operating at pressures up to 10,000 psi and higher. Below are a series of tables listing important data for the selection of flexible hoses covering a broad range of working pressures with a 4:1 safety factor.

TABLE 4-7

Tubing composition		Physical properties
Carbon	0.08–0.18	Tensile strength—max. 55,000 ps
Manganese	0.30–0.60	Elongation in 2 in.—min. 35%*
Phosphorus	0.50 max.	Rockwell hardness B65 max
Sulfur	0.55 max.	

*For tubes with O.D. of $\frac{3}{8}$ in. and/or wall thickness of 0.035 a minimum elongation of 30 per cent is permitted.

TABLE 4-8
THE FOLLOWING TABLE SHOWS DIMENSIONS OF STANDARD PIPE AND TUBING FOR THE HIGHER PRESSURE RANGES. (ASA B36. 10-1950) SCHEDULE 160 PIPE

Size (in.)	O. D. (in.)	I. D. (in.)	Wall thickness (in.)	Internal area (in.)	Rated pressure (psi)
$\frac{1}{2}$	0.84	0.466	0.187	0.171	5010
$\frac{3}{4}$	1.05	0.614	0.218	0.208	4670
1	1.315	0.815	0.250	0.522	4280
$1\frac{1}{4}$	1.660	1.160	0.250	1.060	3390
$1\frac{1}{2}$	1.900	1.338	0.281	1.410	3330
2	2.375	1.689	0.343	2.250	3250
$2\frac{1}{2}$	2.875	2.125	0.375	3.55	2940
3	3.500	2.626	0.437	5.41	2810
4	4.500	3.438	0.531	9.32	2655
5	5.563	4.313	0.625	14.65	2530
6	6.625	5.189	0.718	21.20	2440
8	8.625	6.813	0.906	36.70	2365

TABLE 4-9
TUBING FOR PRESSURES UP TO 10,000 PSI (TYPE AISIA 4130 OR EQUIVALENT RECOMMENDED FOR HIGH PRESSURE SYSTEMS)
TENSILE STRENGTH 70.000 PSI (STANDARD NORMALIZED)

Size or O.D. (in.)	I.D. (in.)	Wall thickness (in.)	Internal area (sq in.)	Size or O.D. (in.)	I.D. (in.)	Wall thickness (in.)	Internal area (sq in.)
1/8	0.069	0.028	0.003739		0.482	0.134	0.1825
3/16	0.131	0.028	0.01348		0.437	0.156	0.1500
1/4	0.180	0.035	0.02545		0.375	0.188	0.1104
	0.166	0.042	0.02164		0.312	0.219	0.07645
	0.152	0.049	0.01815		0.250	0.250	0.04909
	0.134	0.058	0.01410	7/8	0.709	0.083	0.3948
	0.120	0.065	0.01131		0.685	0.095	0.3685
	0.084	0.083	0.005542		0.657	0.109	0.3390
	0.060	0.095	0.002827		0.635	0.120	0.3167
5/16	0.214	0.049	0.03597		0.625	0.125	0.3068
	0.196	0.058	0.03017		0.607	0.134	0.2894
	0.182	0.065	0.02602		0.562	0.156	0.2481
	0.146	0.083	0.01674		0.500	0.188	0.1963
	0.122	0.095	0.004072		0.438	0.219	0.1507
3/8	0.277	0.049	0.06026		0.375	0.250	0.1104
	0.259	0.058	0.05269	1	0.834	0.083	0.5463
	0.245	0.065	0.04714		0.810	0.095	0.5153
	0.209	0.083	0.03431		0.782	0.109	0.4803
	0.185	0.095	0.02688		0.760	0.120	0.4536
	0.135	0.120	0.01431		0.750	0.125	0.4418
1/2	0.402	0.049	0.1269		0.732	0.134	0.4208
	0.384	0.058	0.1158		0.687	0.156	0.3707
	0.370	0.065	0.1075		0.625	0.188	0.3068
	0.334	0.083	0.08762		0.562	0.219	0.2481
	0.310	0.095	0.07548		0.500	0.250	0.1963
	0.282	0.109	0.06246		0.250	0.375	0.04909
	0.260	0.120	0.05309	1 1/4	1.060	0.095	0.882
	0.232	0.134	0.04227		1.010	0.120	0.801
	0.187	0.156	0.02746		0.982	0.134	0.7574
5/8	0.509	0.058	0.2035		0.937	0.156	0.6896
	0.495	0.065	0.1924		0.920	0.165	0.6648
	0.459	0.083	0.1655		0.875	0.188	0.6013
	0.453	0.095	0.1612		0.812	0.219	0.5178
	0.407	0.109	0.1301		0.750	0.250	0.4418
	0.385	0.120	0.1164		0.625	0.313	0.3068
	0.357	0.135	0.1001		0.500	0.375	0.1963
	0.312	0.156	0.07645	1 1/2	1.310	0.095	1.348
	0.250	0.188	0.04909		1.260	0.120	1.247
	0.187	0.219	0.02746		1.232	0.134	1.192
3/4	0.606	0.072	0.2884		1.187	0.156	1.108
	0.584	0.083	0.2679		1.125	0.188	0.9940
	0.560	0.095	0.2463		1.062	0.219	0.8866
	0.532	0.109	0.2223		1.000	0.250	0.7854
	0.510	0.120	0.2043		0.875	0.313	0.6013
					0.750	0.375	0.4418
					0.500	0.500	0.1963

Tubing Composition

Carbon	0.28–0.33
Manganese	0.40–0.60
Phosphorus	0.04 max.
Sulfur	0.04 max.
Silicon	0.20–0.35
Chromium	0.80–1.10
Molybdenum	0.15–0.25

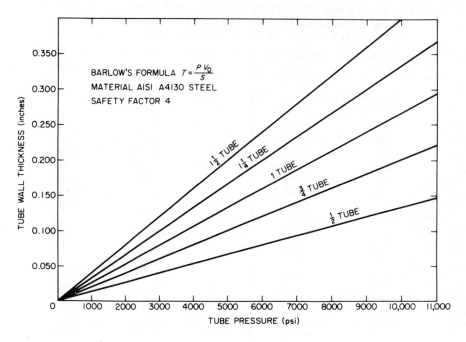

BARLOW'S FORMULA $T = \dfrac{P V_0}{S}$

MATERIAL AISI A4130 STEEL

SAFETY FACTOR 4

TABLE 4-11
STANDARD TUBE RADII–INCHES

Number	Tube size O. D.	Radius bend centerline
2	$\frac{1}{8}$	$\frac{3}{8}$
3	$\frac{3}{16}$	$\frac{7}{16}$
4	$\frac{1}{4}$	$\frac{9}{16}$
5	$\frac{5}{16}$	$\frac{11}{16}$
6	$\frac{3}{8}$	$\frac{15}{16}$
8	$\frac{1}{2}$	$1\frac{1}{2}$
10	$\frac{5}{8}$	2
12	$\frac{3}{4}$	$2\frac{1}{2}$
14	$\frac{7}{8}$	3
16	1	$3\frac{1}{2}$

Approximately 3X the tube O. D. except for heavy-walled
tube, for which the following formula may be useful.

$$\text{Minimum radius bend} = \frac{0.5 \times \text{O.D. in.}}{\%\ \text{elongation in a 2 in.}}$$
length expressed in
decimal

TABLE 4-12
PRESSURE RATINGS FOR PLASTIC TUBING

Operating temperature at room temperature	Nylon		Polyethylene		Polypropylene		Polyvinyl chloride	
	Burst pressure 2200	Work pressure 550	Burst *pressure 250–600	Work *pressure 75–160	Burst *pressure —	Work pressure 20–55	Burst pressure 1800	Work pressure 450
140°	1500	375	125–300	37–80	—	—	1200	300
230°	890	225	—	—	—	—	540	135
300°	310	75	—	—	—	—	90	25

*Depending on tube size and wall thickness

TABLE 4-13

	Up to 250 psi *working pressure*					Up to 500 psi *working pressure*				
O. D.	$\frac{1}{8}$	$\frac{3}{16}$	$\frac{1}{4}$	$\frac{5}{16}$	$\frac{3}{8}$	$\frac{1}{8}$	$\frac{3}{16}$	$\frac{1}{4}$	$\frac{5}{16}$	$\frac{3}{8}$
I. D.	0.096	0.138	0.190	0.242	0.295	0.078	0.110	0.150	0.188	0.225
Minimum bend radius inches	$\frac{5}{8}$	1	$1\frac{1}{4}$	2	3	$\frac{3}{8}$	$\frac{5}{8}$	$1\frac{1}{4}$	3	$2\frac{1}{2}$

TABLE 4-14
ONE WIRE BRAID HOSE
SAE 100 RI

I.D. (in.)	O.D. (in.)	Burst pressure (psi)	Work pressure (psi)	Min bend radii (in.)
$\frac{1}{4}$	$\frac{11}{16}$	11,000	2750	4
$\frac{3}{8}$	$\frac{25}{32}$	9,000	2250	5
$\frac{1}{2}$	$\frac{29}{32}$	8,000	2000	7
$\frac{3}{4}$	$1\frac{3}{16}$	5,000	1250	$9\frac{1}{2}$
1	$1\frac{1}{2}$	4,000	1000	$9\frac{1}{2}$
$1\frac{1}{4}$	$1\frac{13}{16}$	2,500	625	16
$1\frac{1}{2}$	$2\frac{1}{16}$	2,000	500	20

TABLE 4-15
TWO WIRE BRAID HOSE
SAE 100 R2

I.D. (in.)	O.D. (in.)	Burst pressure (psi)	Work pressure (psi)	Min bend radii (in.)
$\frac{1}{4}$	$\frac{11}{16}$	20,000	5000	4
$\frac{3}{8}$	$\frac{27}{32}$	16,000	4000	5
$\frac{1}{2}$	$\frac{31}{32}$	14,000	3500	7
$\frac{3}{4}$	$1\frac{1}{4}$	9,000	2250	$9\frac{1}{2}$
1	$1\frac{9}{16}$	7,500	1875	$9\frac{1}{2}$
$1\frac{1}{4}$	2	6,500	1625	16
$1\frac{1}{2}$	$2\frac{1}{4}$	5,000	1250	20
2	$2\frac{3}{4}$	4,500	1125	22

TABLE 4-16
*THREE WIRE BRAID HOSE

I.D. (in.)	O.D. (in.)	Burst pressure (psi)	Work pressure (psi)	Min bend radii (in.)
$\frac{1}{4}$	$\frac{3}{4}$	28,000	7000	5
$\frac{3}{8}$	$\frac{29}{32}$	22,000	5500	6
$\frac{1}{2}$	$1\frac{5}{64}$	20,000	5000	8
$\frac{3}{4}$	$1\frac{5}{16}$	16,000	4000	10
1	$1\frac{9}{16}$	11,000	2750	12
$1\frac{1}{4}$	$2\frac{1}{16}$	10,000	2500	17
$1\frac{1}{2}$	$2\frac{5}{16}$	8,000	2000	21
2	$2\frac{7}{8}$	8,000	2000	26

TABLE 4-17
*FOUR WIRE BRAID HOSE

I.D. (in.)	O.D. (in.)	Burst pressure (psi)	Work pressure (psi)	Min bend radii (in.)
$\frac{1}{4}$	$\frac{13}{16}$	35,000	8750	5
$\frac{3}{8}$	$\frac{15}{16}$	30,000	7500	6
$\frac{1}{2}$	$1\frac{3}{32}$	25,000	6250	8
$\frac{3}{4}$	$1\frac{7}{16}$	20,000	5000	10
1	$1\frac{3}{4}$	16,000	4000	12
$1\frac{1}{4}$	2	12,000	3000	18
$1\frac{1}{2}$	$2\frac{1}{4}$	10,000	2500	22
2	$2\frac{25}{32}$	10,000	2500	—

TABLE 4-18
*SIX WIRE BRAID HOSE

I.D. (in.)	O.D. (in.)	Burst pressure (psi)	Work pressure (psi)	Min bend radii (in.)
$\frac{1}{4}$	$\frac{15}{16}$	45,000	11,250	5
$\frac{3}{8}$	$1\frac{1}{16}$	40,000	10,000	6
$\frac{1}{2}$	$1\frac{7}{32}$	30,000	7,500	8
$\frac{3}{4}$	$1\frac{9}{16}$	25,000	6,250	10
1	$1\frac{29}{32}$	20,000	5,000	12
$1\frac{1}{4}$	$2\frac{5}{32}$	20,000	5,000	12
$1\frac{1}{2}$	$2\frac{13}{32}$	14,000	3,500	18
2	$2\frac{31}{32}$	12,000	3,000	22

*Wire diameter is of equal importance, and the various manufacturers may have different ways of expressing wire braid or spiral wrapping.

Important Factors for the Application of Flexible Hose

1. Hose assemblies must be of proper over-all length.
2. Hose under variations in working pressure must have enough length to expand and contract.
3. Do not clamp high and low pressure hoses together.
4. Avoid clamping at bends so that the bend radii can absorb change.
5. When there is relative motion between the two ends of a hose assembly, always allow for adequate length of travel.
6. To prevent twisting, hose should be bent in the same plane as the motion of the boss to which it is connected.
7. To prevent twisting in hose lines bent in two planes, clamp hose at the change of plane.
8. Use the proper hydraulic adapters to reduce the number of joints and improve performance as well as appearance.
9. Where the radius falls below the required minimum bend, an angle adapter should be used.
10. Avoid contact with sharp edges or moving parts.
11. Arrange proper positioning of hose and adapters before tightening to avoid distortion.
12. Apply clamps properly and maintain tight to prevent abrasion due to line surge.
13. Be sure to use the proper strength of hose to maintain a good margin of safety.
14. Select the proper size of hose of stay within the recommended velocity range (consult velocity-flow nomograph). (See Figure 4–5).
15. Prevent dirt, chips, or any foreign materials from entering the system during fabrication of the circuit.
16. Be sure all materials used are compatible with the hydraulic fluid designated for the system.

Connectors and Installation Information

General

Hydraulic systems often require pipe, tubing, and hose in the same system. Connections must be transposed from one design configuration to another and satisfy variables such as size, thread type, adaptability, etc.

Connectors are available in standard sizes, design types, and materials that provide excellent results for the installation of hydraulic systems covering practically all fields of applications and operation pressures. However, several primary requirements must be considered in the selecting of hydraulic system connectors.

1. The efficiency of a connector to pass fluid with a minimum pressure drop determines its performance as an important part of the hydraulic circuit.
2. Connectors shall provide leakproof joints under maximum pressures. (This important requirement, however, depends largely upon the cleanliness in the installation.)
3. The degree of workability and ease with which installations can be made or assembled. Connectors may also be considered from a reusability factor for economy.

To help minimize variations in connectors and to establish that certain connections can be universally used, standards were set up giving recommended practices for making connections and joints. The following paragraphs are taken from the Hydraulic Standards and may serve as a guide.

1. Piping between actuating and feed-control devices should be constructed rigidly to confine a minimum volume of fluid. The purpose is to maintain constant, controlled motion and to restrict the effort of varying forces.
2. Solderless pipe connections such as flare, flareless, self-flaring and flange should be used.
3. Performance of any type of fitting used should be equal to or exceed the requirements of MIL–F–5506.
4. All pipe threads should be Dryseal American (National) Standard Taper Pipe Theads (NPTF).
5. Fittings with straight threads and which incorporate pressure seals may be used in place of pipe thread fittings.
6. All piping connections should be designed and installed to permit quick removal and reassembly by means of hand tools.
7. Only steel should be used as fitting materials. Copper alloy fittings should not be used with petroleum-base fluids. Fittings with restricted or stepped-up passages are not recommended.
8. Whenever practicable, each piping run should be integral and continuous from one piece of apparatus to another. Piping runs should be removable without dismantling and without bending or springing the tubing in a manner that would damage it.

9. When high or extra pressure piping is used, all connections should be welded to steel flanges or connections should be used which are equal in performance and ease of assembly. Flanged connections should use sealing devices that seal with pressure.

Threads and Ports

The following thread standards have been accepted by users of hydraulic equipment including pipe, tubing, hose and fittings. Table 4–19 co-ordinates the dimensions of the various conductors and connectors according to thread type and size.

1. Dryseal Standard American Pipe Thread (NPTF).
2. Joint Industry Conference (J.I.C.) Standard Thread with 37° Male Flare.
3. Joint Industry Conference (J.I.C.) Standard Thread with 37° Female Flare.
4. Society of Automotive Engineers (SAE) Standard Male Thread with an "O" ring.

The Dryseal American (National) Standard Taper Threads differ from National Standard pipe threads by engaging roots and crests before flanks, so that spiral clearance occurring in standard threads is avoided. Special taps are used for Dryseal Threads.

This section includes design illustrations of the various fittings, and pointers on how to bend tubing.

Tube bending and fabrication is an art that can only be learned through practical experience. The information given is useful as a guide for students to follow, and to identify tube and pipe fittings commonly used.

Flange and Weld Connections

Flange connections for large components and on some critical applications are sometimes preferred.

TABLE 4-19
STANDARD THREAD TYPES
CO-ORDINATING I.D. OF HOSE WITH TUBING AND PIPE

Hose I.D.	Tube O.D.	Pipe thread	J.I.C 37° flare	SAE straight thread "O" ring
$\frac{1}{4}$	$\frac{5}{16}$	$\frac{1}{4}$	$\frac{1}{2}$-20	$\frac{1}{2}$-20
$\frac{3}{8}$	$\frac{1}{2}$	$\frac{3}{8}$	$\frac{3}{4}$-16	$\frac{3}{4}$-16
$\frac{1}{2}$	$\frac{5}{8}$	$\frac{1}{2}$	$\frac{7}{8}$-14	$\frac{7}{8}$-14
$\frac{3}{4}$	$\frac{3}{4}$	$\frac{3}{4}$	$1\frac{1}{16}$-12	$1\frac{1}{16}$-12
1	1	1	$1\frac{5}{16}$-12	$1\frac{5}{16}$-12
$1\frac{1}{4}$	$1\frac{1}{4}$	$1\frac{1}{4}$	$1\frac{5}{8}$-12	$1\frac{5}{8}$-12
$1\frac{1}{2}$	$1\frac{1}{2}$	$1\frac{1}{2}$	$1\frac{7}{8}$-12	$1\frac{7}{8}$-12
2	2	2	$2\frac{1}{2}$-12	$2\frac{1}{2}$-12

TABLE 4-20
PIPE THREADS

Pipe size (in.)	O.D. (in.)	Length of thread screwed in fitting (in.)	Threads per in.	Taper
$\frac{1}{8}$	0.405	$\frac{1}{4}$	27	$\frac{3}{4}$ in. per ft
$\frac{1}{4}$	0.540	$\frac{3}{8}$	18	″
$\frac{3}{8}$	0.675	$\frac{3}{8}$	18	″
$\frac{1}{2}$	0.840	$\frac{1}{2}$	14	″
$\frac{3}{4}$	1.050	$\frac{1}{2}$	14	″
1	1.315	$\frac{9}{16}$	$11\frac{1}{2}$	″

Sealing in a flanged joint is achieved with a static seal against a ground surface. The flanges are usually welded to the pipe or tube. In the case of flexible hose, the flange is fixed permanently or clamp-connected to hose. Vibration resistance in this type of connection is excellent and reuseability is unlimited. Restrictions of passages are avoided, since ports of both mating parts coincide. The relatively large area which is required to accommodate flange bolts somewhat restricts its use to the larger components. However, excellent results have been obtained with the flanged joint in connection with high pressure hydraulic systems.

TYPES OF CONNECTORS

Connectors may be divided into two classifications, the flare type and the compression type. These two groups are further sub-divided as follows:

TUBING CONNECTORS

FLARE TYPE	COMPRESSION TYPE
37° Flare	Beveled Sleeve
45° Flare	Tapered Sleeve (Bite Type)
Inverted Flare	Threaded Sleeve
	"O" Ring–Split Sleeve
	Nut-shearable Sleeve
	Flex Sleeve Design
	Tube supported Sleeve Design

All of the above designs have specific advantages (disadvantages) and should be selected for hydraulic service on the basis of the manufacturers specifications. The following series of diagrams show the physical characteristics of a few of the popularly used designs.

TUBING CONNECTORS

NUT — SLEEVE — TUBE

37° FLARE

The 37° Flare provides excellent results for connections when tubing is flarable. The free floating sleeve supports tube and dampens vibration. Approved by Underwriters Laboratories., S.A.E., J.I.C., and A.S.M.E. This type of connector is reuseable, does not twist during assembly and requires low assembly torque.

45° FLARE

The 45° Flare may be used with flarable tubing and will withstand pressures up to 5,000 PSI. This design is approved by Underwriters Laboratories, A.S.A., A.S.M.E., J.I.C., and S.A.E. It is reuseable and available at comparatively low cost.

LONG NUT (FOR EXCESSIVE VIBRATION)
SHORT NUT (LITTLE VIBRATION)
TUBE

PIPE THREAD CONNECTOR — NUT — TUBE

45° INVERTED FLARE

The 45° inverted Flare provides protection for seat and threads which are recessed. It is reuseable, low cost, and fits in tight places. This design is also approved by U.L., A.S.A., A.S.M.E., J.I.C. and S.A.E.

STRAIGHT THREAD "O" RING PORT

The straight thread "O" ring design is proving to provide excellent results for high pressure hydraulic applications. This design is approved by S.A.E. and gives a tighter seal as pressure is increased. There is less chance of over-tightening as it gives a positive hit-home feel when assembled. The "O" ring used must be compatible with the fluid. This configuration covers AND 10050 and MS 16142.

STRAIGHT THREAD CONNECTOR

"O" RING

MINIMUM THREAD DEPTH

INTERNAL STRAIGHT THREAD BOSS

MINIMUM DRILL DEPTH

SELF-FLARING

In this design, assembly is made by a ferrule or sleeve being pulled into the fitting body as the nut is tightened. As the ferrule rides into the body, it follows a taper which causes the sleeve to bite into the tubing and provide seal. Good results have been experienced with this design and it is JIC approved.

PIPE THREAD CONNECTOR   NUT   SLEEVE   TUBE   WEDGE INSERT   INTERNAL V-SHAPED RIB

FITTING BODY   FLOATING THREADED NUT   SELF COINED METAL TO METAL SEAL

STRAIGHT THREAD PORT

This is another design of the straight thread port and connection. It is approved by J.I.C. and allows a metal to metal seal.

COMPRESSION

This design of the tapered sleeve (bite type) compression fitting may be used with heavy wall tubing and is approved by U.L., S.A.E., J.I.C., and A.S.M.E. It is reuseable. Good results have been experienced with this design for high pressure applications. However, extreme care should be given to the amount of torque in assembly.

PIPE THREAD CONNECTOR   NUT   FERRULE   TUBE

"O" RING   NUT   SPLIT TAPERED RING   TUBE

COMPRESSION

This design is of the "O" ring and split sleeve type and is approved by J.I.C. and S.A.E. It is reuseable and allows considerable tolerance in the length of tube cut and squareness of cut. The "O" ring must be compatible with the fluid used in the system.

COMPRESSION

This design is proving satisfactory for plastic tubing. It is easy to install, reuseable and comes in a variety of materials including steel and plastic.

PIPE THREAD CONNECTOR   NUT   SLEEVE   PLASTIC TUBE

In view of the difficulties encountered when large or heavy-walled conductors are being bent, it is often advisable to buy prefabricated bends and other weld-type connectors for a better and neater system. The bends and fittings may generally be supplied with the proper radius and have chamfered edges. The inside and outside diameters of the corresponding tube or pipe can be welded by butt welding, resulting in a neat and streamlined appearance.

Hose Fittings

Hose end connectors can be obtained in permanently pressed-on designs, reuseable clamp designs, and reuseable screw-type designs. The reuseable types of connectors require skill and time in making assembly.

It is usually preferable to have both ends of the hose to be connected by union connectors which incorporate a freely turning nut. Nuts are located either in the hose end coupling or in the connector fitting. For short hoses it is possible to have one end rigidly connected; however, during assembly the entire hose must turn. Unions cost more than the rigid fittings, so it is generally desirable to locate the union in the connector rather than in the hose coupling.

A considerable number of hose end connector designs are available, different in the basic principle employed and the type of flare and thread used.

Four Easy Steps for Figuring Offset Bends

Step 1. Determine the total amount of offset required (dimension Y in diagram) and angle of offset. Wherever possible use 45 deg offset bends. This will enable you to figure the total amount of tubing required for a given application, as explained in the following section.

Step 2. Figure the length of tube which is needed to meet your offset requirements (X in dimension diagram) from Table 4–21. For example: Say the amount of offset you require (Y dimension, Step 1) is $2\frac{1}{2}$ in. and the the offset angle is 45 deg. Check the 45 deg column and find $2\frac{1}{2}$ in. The figure next

Fig. 4-6. Swivel fittings are used when there is relative motion between actuater and machine members. This type is available for pressure up to 15,000 lbs psi. (*Courtesy FMC Corp., Chiksan Div.*)

Fig. 4-7. Application of Swivel fittings. (*Courtesy FMC Corp., Chiksan Div.*)

For bending hard or soft tubing — Copper, Aluminum, Brass, Steel and Stainless Steel

1. To place tube in bender, pull out slip joint and move handles of bender so that they are at right angles. Raise the clip and place the tube in the space between the handle slide block and the bending form.

2. Drop clip over tube and push in slip joint. Note that zero mark on bending form will be even with front of handle slide block.

3. Proceed to bend desired angle as indicated by calibrations on bending form. Bends to any angle up to 180° can be made with the Imperial Tube Bender.

4. To remove the bent tube from bender, pull out slip joint and raise the clip. Tube can then readily be removed.

NOTE: Occasionally placing a drop of oil on the pin and bending shoe will assure a smoother working tool.

GUIDE FOR MAKING DIMENSIONAL BENDS

A — MANDREL
B — HANDLE

BENDS FROM HOOK SIDE

The illustration above is for bends measured from the hook-side of bender. For 7/16" and 1/2" benders, line up mark on tube which is to be the center of the bend with the edge of the Forming Wheel. On 1/8" to 3/8" benders, use either the edge of the Forming Wheel or the mark on the lever designed for that purpose.

TO OBTAIN "X" DIMENSION TUBE SHOULD BE PLACED IN BENDER AS ILLUSTRATED IN FIGURES 1 AND 2

The three position start feature furnished on benders 3/8", 7/16" and 1/2" sizes only.

BENDS FROM LEVER SIDE

For 7/16" and 1/2" benders — on bends measured from the lever-side of bender — the tube mark must be lined up with the reference mark on the Forming Wheel as illustrated. For 1/8" to 3/8" benders, use the reference mark stamped on lever.

Fig. 4-8. Operation of lever-type benders. (*Courtesy of Imperial-Eastman*)

OFFSET DIMENSION — DIMENSION CENTER TO CENTER OF BEND "X" — MARK A — MARK B — "Y" — "Z" DIMENSION

TABLE 4-21
OFFSET BEND CALCULATOR

Angle of offset 15°		Angle of offset 30°		Angle of offset 45°		Angle of offset 60°		Angle of offset 75°	
Amount of Offset (Y Dimension)	(X Dimension)	(Y Dimension)	(X Dimension)	(Y Dimension)	(X Dimension)	(Y Dimension)	(X Dimension)	(Y Dimension)	(X Dimension)
1	$3\frac{7}{8}$	1	2	1	$1\frac{13}{32}$	1	$1\frac{5}{32}$	1	$1\frac{1}{32}$
$\frac{1}{8}$	$4\frac{11}{32}$	$\frac{1}{8}$	$2\frac{1}{4}$	$\frac{1}{8}$	$1\frac{19}{32}$	$\frac{1}{8}$	$1\frac{5}{16}$	$\frac{1}{8}$	$1\frac{5}{32}$
$\frac{1}{4}$	$4\frac{27}{32}$	$\frac{1}{4}$	$2\frac{1}{2}$	$\frac{1}{4}$	$1\frac{25}{32}$	$\frac{1}{4}$	$1\frac{7}{16}$	$\frac{1}{4}$	$1\frac{9}{32}$
$\frac{3}{8}$	$5\frac{5}{16}$	$\frac{3}{8}$	$2\frac{3}{4}$	$\frac{3}{8}$	$1\frac{15}{16}$	$\frac{3}{8}$	$1\frac{19}{32}$	$\frac{3}{8}$	$1\frac{7}{16}$
$\frac{1}{2}$	$5\frac{25}{32}$	$\frac{1}{2}$	3	$\frac{1}{2}$	$2\frac{1}{8}$	$\frac{1}{2}$	$1\frac{23}{32}$	$\frac{1}{2}$	$1\frac{9}{16}$
$\frac{5}{8}$	$6\frac{9}{32}$	$\frac{5}{8}$	$3\frac{1}{4}$	$\frac{5}{8}$	$2\frac{5}{16}$	$\frac{5}{8}$	$1\frac{7}{8}$	$\frac{5}{8}$	$1\frac{11}{16}$
$\frac{3}{4}$	$6\frac{3}{4}$	$\frac{3}{4}$	$3\frac{1}{2}$	$\frac{3}{4}$	$2\frac{15}{32}$	$\frac{3}{4}$	$2\frac{1}{32}$	$\frac{3}{4}$	$1\frac{13}{16}$
$\frac{7}{8}$	$7\frac{1}{4}$	$\frac{7}{8}$	$3\frac{3}{4}$	$\frac{7}{8}$	$2\frac{21}{32}$	$\frac{7}{8}$	$2\frac{5}{32}$	$\frac{7}{8}$	$1\frac{15}{16}$
2	$7\frac{23}{32}$	2	4	2	$2\frac{13}{16}$	2	$2\frac{5}{16}$	2	$2\frac{1}{16}$
$\frac{1}{8}$	$8\frac{7}{32}$	$\frac{1}{8}$	$4\frac{1}{4}$	$\frac{1}{8}$	3	$\frac{1}{8}$	$2\frac{15}{32}$	$\frac{1}{8}$	$2\frac{3}{16}$
$\frac{1}{4}$	$8\frac{11}{16}$	$\frac{1}{4}$	$4\frac{1}{2}$	$\frac{1}{4}$	$3\frac{3}{16}$	$\frac{1}{4}$	$2\frac{19}{32}$	$\frac{1}{4}$	$2\frac{5}{16}$
$\frac{3}{8}$	$9\frac{3}{16}$	$\frac{3}{8}$	$4\frac{3}{4}$	$\frac{3}{8}$	$3\frac{11}{32}$	$\frac{3}{8}$	$2\frac{3}{4}$	$\frac{3}{8}$	$2\frac{15}{32}$
$\frac{1}{2}$	$9\frac{21}{32}$	$\frac{1}{2}$	5	$\frac{1}{2}$	$3\frac{17}{32}$	$\frac{1}{2}$	$2\frac{7}{8}$	$\frac{1}{2}$	$2\frac{19}{32}$
$\frac{5}{8}$	$10\frac{5}{32}$	$\frac{5}{8}$	$5\frac{1}{4}$	$\frac{5}{8}$	$3\frac{23}{32}$	$\frac{5}{8}$	$3\frac{1}{32}$	$\frac{5}{8}$	$2\frac{23}{32}$
$\frac{3}{4}$	$10\frac{5}{8}$	$\frac{3}{4}$	$5\frac{1}{2}$	$\frac{3}{4}$	$3\frac{7}{8}$	$\frac{3}{4}$	$3\frac{3}{16}$	$\frac{3}{4}$	$2\frac{27}{32}$
$\frac{7}{8}$	$11\frac{3}{32}$	$\frac{7}{8}$	$5\frac{3}{4}$	$\frac{7}{8}$	$4\frac{1}{16}$	$\frac{7}{8}$	$3\frac{5}{16}$	$\frac{7}{8}$	$2\frac{31}{32}$
3	$11\frac{19}{32}$	3	6	3	$4\frac{1}{4}$	3	$3\frac{15}{32}$	3	$3\frac{3}{32}$
$\frac{1}{8}$	$12\frac{1}{16}$	$\frac{1}{8}$	$6\frac{1}{4}$	$\frac{1}{8}$	$4\frac{13}{32}$	$\frac{1}{8}$	$3\frac{19}{32}$	$\frac{1}{8}$	$3\frac{7}{32}$
$\frac{1}{4}$	$12\frac{9}{16}$	$\frac{1}{4}$	$6\frac{1}{2}$	$\frac{1}{4}$	$4\frac{19}{32}$	$\frac{1}{4}$	$3\frac{3}{4}$	$\frac{1}{4}$	$3\frac{3}{8}$
$\frac{3}{8}$	$13\frac{1}{32}$	$\frac{3}{8}$	$6\frac{3}{4}$	$\frac{3}{8}$	$4\frac{25}{32}$	$\frac{3}{8}$	$3\frac{29}{32}$	$\frac{3}{8}$	$3\frac{1}{2}$
$\frac{1}{2}$	$13\frac{17}{32}$	$\frac{1}{2}$	7	$\frac{1}{2}$	$4\frac{15}{16}$	$\frac{1}{2}$	$4\frac{1}{32}$	$\frac{1}{2}$	$3\frac{5}{8}$
$\frac{5}{8}$	14	$\frac{5}{8}$	$7\frac{1}{4}$	$\frac{5}{8}$	$5\frac{1}{8}$	$\frac{5}{8}$	$4\frac{3}{16}$	$\frac{5}{8}$	$3\frac{3}{4}$
$\frac{3}{4}$	$14\frac{1}{2}$	$\frac{3}{4}$	$7\frac{1}{2}$	$\frac{3}{4}$	$5\frac{5}{16}$	$\frac{3}{4}$	$4\frac{11}{32}$	$\frac{3}{4}$	$3\frac{7}{8}$
$\frac{7}{8}$	$14\frac{31}{32}$	$\frac{7}{8}$	$7\frac{3}{4}$	$\frac{7}{8}$	$5\frac{15}{32}$	$\frac{7}{8}$	$4\frac{15}{32}$	$\frac{7}{8}$	4
4	$15\frac{1}{2}$	4	8	4	$5\frac{21}{32}$	4	$4\frac{5}{8}$	4	$4\frac{1}{8}$
$\frac{1}{8}$	$15\frac{15}{16}$	$\frac{1}{8}$	$8\frac{1}{4}$	$\frac{1}{8}$	$5\frac{27}{32}$	$\frac{1}{8}$	$4\frac{3}{4}$	$\frac{1}{8}$	$4\frac{9}{32}$
$\frac{1}{4}$	$16\frac{7}{16}$	$\frac{1}{4}$	$8\frac{1}{2}$	$\frac{1}{4}$	6	$\frac{1}{4}$	$4\frac{29}{32}$	$\frac{1}{4}$	$4\frac{13}{32}$
$\frac{3}{8}$	$16\frac{29}{32}$	$\frac{3}{8}$	$8\frac{3}{4}$	$\frac{3}{8}$	$6\frac{3}{16}$	$\frac{3}{8}$	$5\frac{1}{16}$	$\frac{3}{8}$	$4\frac{17}{32}$
$\frac{1}{2}$	$17\frac{3}{8}$	$\frac{1}{2}$	9	$\frac{1}{2}$	$6\frac{3}{8}$	$\frac{1}{2}$	$5\frac{3}{16}$	$\frac{1}{2}$	$4\frac{21}{32}$
$\frac{5}{8}$	$17\frac{7}{8}$	$\frac{5}{8}$	$9\frac{1}{4}$	$\frac{5}{8}$	$6\frac{17}{32}$	$\frac{5}{8}$	$5\frac{11}{32}$	$\frac{5}{8}$	$4\frac{25}{32}$
$\frac{3}{4}$	$18\frac{11}{32}$	$\frac{3}{4}$	$9\frac{1}{2}$	$\frac{3}{4}$	$6\frac{23}{32}$	$\frac{3}{4}$	$5\frac{15}{32}$	$\frac{3}{4}$	$4\frac{29}{32}$
$\frac{7}{8}$	$18\frac{27}{32}$	$\frac{7}{8}$	$9\frac{3}{4}$	$\frac{7}{8}$	$6\frac{29}{32}$	$\frac{7}{8}$	$5\frac{5}{8}$	$\frac{7}{8}$	$5\frac{1}{32}$
5	$19\frac{5}{16}$	5	10	5	$7\frac{1}{16}$	5	$5\frac{25}{32}$	5	$5\frac{3}{16}$
$\frac{1}{8}$	$19\frac{13}{16}$	$\frac{1}{8}$	$10\frac{1}{4}$	$\frac{1}{8}$	$7\frac{1}{4}$	$\frac{1}{8}$	$5\frac{25}{32}$	$\frac{1}{8}$	$5\frac{5}{16}$
$\frac{1}{4}$	$20\frac{9}{32}$	$\frac{1}{4}$	$10\frac{1}{2}$	$\frac{1}{4}$	$7\frac{7}{16}$	$\frac{1}{4}$	$6\frac{1}{16}$	$\frac{1}{4}$	$5\frac{7}{16}$
$\frac{3}{8}$	$20\frac{25}{32}$	$\frac{3}{8}$	$10\frac{3}{4}$	$\frac{3}{8}$	$7\frac{19}{32}$	$\frac{3}{8}$	$6\frac{7}{32}$	$\frac{3}{8}$	$5\frac{9}{16}$
$\frac{1}{2}$	$21\frac{1}{4}$	$\frac{1}{2}$	11	$\frac{1}{2}$	$7\frac{25}{32}$	$\frac{1}{2}$	$6\frac{11}{32}$	$\frac{1}{2}$	$5\frac{11}{16}$
$\frac{5}{8}$	$21\frac{3}{4}$	$\frac{5}{8}$	$11\frac{1}{4}$	$\frac{5}{8}$	$7\frac{31}{32}$	$\frac{5}{8}$	$6\frac{1}{2}$	$\frac{5}{8}$	$5\frac{27}{32}$
$\frac{3}{4}$	$22\frac{7}{32}$	$\frac{3}{4}$	$11\frac{1}{2}$	$\frac{3}{4}$	$8\frac{1}{8}$	$\frac{3}{4}$	$6\frac{5}{8}$	$\frac{3}{4}$	$5\frac{31}{32}$
$\frac{7}{8}$	$22\frac{11}{16}$	$\frac{7}{8}$	$11\frac{3}{4}$	$\frac{7}{8}$	$8\frac{5}{16}$	$\frac{7}{8}$	$6\frac{25}{32}$	$\frac{7}{8}$	$6\frac{3}{32}$
6	$23\frac{3}{16}$	6	12	6	$8\frac{15}{32}$	6	$6\frac{15}{16}$	6	$6\frac{7}{32}$

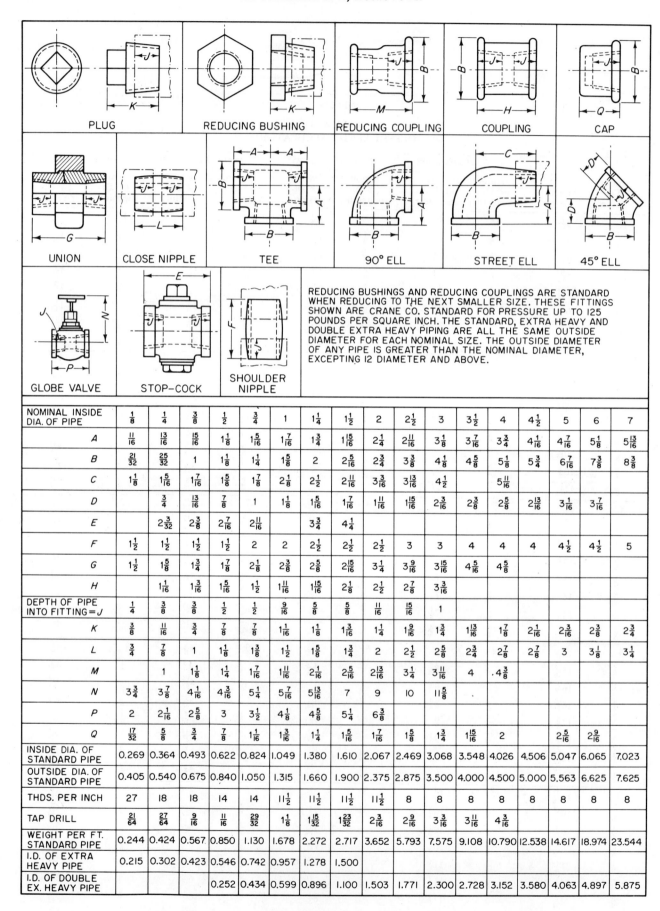

PLUG · REDUCING BUSHING · REDUCING COUPLING · COUPLING · CAP · UNION · CLOSE NIPPLE · TEE · 90° ELL · STREET ELL · 45° ELL · GLOBE VALVE · STOP-COCK · SHOULDER NIPPLE

REDUCING BUSHINGS AND REDUCING COUPLINGS ARE STANDARD WHEN REDUCING TO THE NEXT SMALLER SIZE. THESE FITTINGS SHOWN ARE CRANE CO. STANDARD FOR PRESSURE UP TO 125 POUNDS PER SQUARE INCH. THE STANDARD, EXTRA HEAVY AND DOUBLE EXTRA HEAVY PIPING ARE ALL THE SAME OUTSIDE DIAMETER FOR EACH NOMINAL SIZE. THE OUTSIDE DIAMETER OF ANY PIPE IS GREATER THAN THE NOMINAL DIAMETER, EXCEPTING 12 DIAMETER AND ABOVE.

NOMINAL INSIDE DIA. OF PIPE	1/8	1/4	3/8	1/2	3/4	1	1 1/4	1 1/2	2	2 1/2	3	3 1/2	4	4 1/2	5	6	7
A	11/16	13/16	15/16	1 1/8	1 5/16	1 7/16	1 3/4	1 15/16	2 1/4	2 11/16	3 1/8	3 7/16	3 3/4	4 1/16	4 7/16	5 1/8	5 13/16
B	21/32	25/32	1	1 1/8	1 1/4	1 5/8	2	2 5/16	2 3/4	3 3/8	4 1/8	4 5/8	5 1/8	5 3/4	6 7/16	7 3/8	8 3/8
C	1 1/8	1 5/16	1 7/16	1 5/8	1 7/8	2 1/8	2 1/2	2 11/16	3 3/8	3 13/16	4 1/2		5 11/16				
D		3/4	13/16	7/8	1	1 1/8	1 5/16	1 7/16	1 11/16	1 15/16	2 3/16	2 3/8	2 5/8	2 13/16	3 1/16	3 7/16	
E		2 3/32	2 3/8	2 7/16	2 11/16		3 3/4	4 1/4									
F	1 1/2	1 1/2	1 1/2	1 1/2	2	2	2 1/2	2 1/2	2 1/2	3	3	4	4	4	4 1/2	4 1/2	5
G	1 1/2	1 5/8	1 3/4	1 7/8	2 1/8	2 3/8	2 5/8	2 15/16	3 1/4	3 9/16	3 15/16	4 5/16	4 5/8				
H		1 1/16	1 3/16	1 5/16	1 1/2	1 11/16	1 15/16	2 1/8	2 1/2	2 7/8	3 3/16						
DEPTH OF PIPE INTO FITTING = J	1/4	3/8	3/8	1/2	1/2	9/16	5/8	5/8	11/16	15/16	1						
K	3/8	11/16	3/4	7/8	7/8	1 1/16	1 1/8	1 3/16	1 1/4	1 9/16	1 3/4	1 13/16	1 7/8	2 1/16	2 3/8	2 3/8	2 3/4
L	3/4	7/8	1	1 1/8	1 3/8	1 1/2	1 5/8	1 3/4	2	2 1/2	2 5/8	2 3/4	2 7/8	2 7/8	3	3 1/8	3 1/4
M		1	1 1/8	1 1/4	1 7/16	1 11/16	2 1/16	2 5/16	2 13/16	3 1/4	3 11/16	4	.4 3/8				
N	3 3/4	3 7/8	4 1/16	4 3/16	5 1/4	5 7/16	5 13/16	7	9	10	11 5/8						
P	2	2 1/16	2 5/8	3	3 1/2	4 1/8	4 5/8	5 1/4	6 3/8								
Q	17/32	5/8	3/4	7/8	1 1/8	1 3/16	1 1/4	1 5/16	1 7/16	1 5/8	1 3/4	1 15/16	2		2 5/16	2 9/16	
INSIDE DIA. OF STANDARD PIPE	0.269	0.364	0.493	0.622	0.824	1.049	1.380	1.610	2.067	2.469	3.068	3.548	4.026	4.506	5.047	6.065	7.023
OUTSIDE DIA. OF STANDARD PIPE	0.405	0.540	0.675	0.840	1.050	1.315	1.660	1.900	2.375	2.875	3.500	4.000	4.500	5.000	5.563	6.625	7.625
THDS. PER INCH	27	18	18	14	14	11 1/2	11 1/2	11 1/2	11 1/2	8	8	8	8	8	8	8	8
TAP DRILL	21/64	27/64	9/16	11/16	29/32	1 1/8	1 15/32	1 23/32	2 3/16	2 9/16	3 3/16	3 11/16	4 3/16				
WEIGHT PER FT. STANDARD PIPE	0.244	0.424	0.567	0.850	1.130	1.678	2.272	2.717	3.652	5.793	7.575	9.108	10.790	12.538	14.617	18.974	23.544
I.D. OF EXTRA HEAVY PIPE	0.215	0.302	0.423	0.546	0.742	0.957	1.278	1.500									
I.D. OF DOUBLE EX. HEAVY PIPE				0.252	0.434	0.599	0.896	1.100	1.503	1.771	2.300	2.728	3.152	3.580	4.063	4.897	5.875

37° STEEL FLARE ADAPTERS

Male Connector   Male Elbow   Male 45° Elbow   Male Run Tee

Female Connector   Extra Long Male Elbow   Union Elbow   Female Run Tee

Union   Female Elbow   Male Branch Tee   Female Branch Tee

How to Identify Coupling Ends

Male Pipe Thread   Male 45° SAE Flare   Female Hi-Seal Swivel

Female Pipe Thread   Female 45° SAE Flare Swivel   Flange Head O-Ring

Male 37° JIC Flare   Male Straight Thread O-Ring   Inverted Flare Male Swivel

Female 37° JIC Flare Swivel   Male Hi-Seal   Female Inverted Flare

TABLE 4-22
AREAS OF CIRCLES

$D = \text{diameter}$ $A = 0.7853981634\, D^2$

DIAM.	0	1	2	3	4	5	6	7	8	9	10	11	12	13	14	15	16	17	DIAM.
0		.7854	3.1416	7.0686	12.566	19.635	28.274	38.485	50.266	63.617	78.540	95.033	113.10	132.73	153.94	176.71	201.06	226.98	0
1/64	.000192	.8101	3.1909	7.1424	12.665	19.753	28.422	38.656	50.462	63.838	78.785	95.303	113.39	133.05	154.28	177.08	201.45	227.40	1/64
1/32	.000767	.8352	3.2405	7.2166	12.763	19.881	28.570	38.829	50.659	64.060	79.031	95.574	113.69	133.37	154.63	177.45	201.85	227.82	1/32
3/64	.001726	.8607	3.2906	7.2912	12.863	20.005	28.718	39.002	50.856	64.282	79.278	95.845	113.98	133.69	154.97	177.82	202.24	228.23	3/64
1/16	.003068	.8866	3.3410	7.3662	12.962	20.129	28.866	39.175	51.054	64.504	79.525	96.116	114.23	134.01	155.32	178.19	202.64	228.65	1/16
5/64	.004794	.9128	3.3918	7.4415	13.062	20.253	29.015	39.348	51.252	64.727	79.772	96.388	114.57	134.33	155.66	178.56	203.03	229.07	5/64
3/32	.006903	.9306	3.4430	7.5173	13.162	20.378	29.165	39.522	51.450	64.950	80.019	96.660	114.87	134.65	156.01	178.93	203.43	229.49	3/32
7/64	.009396	.9666	3.4946	7.5934	13.263	20.503	29.315	39.696	51.649	65.173	80.267	96.932	115.17	134.98	156.35	179.30	203.82	229.91	7/64
1/8	.01227	.9940	3.5466	7.6699	13.364	20.629	29.465	39.871	51.849	65.397	80.516	97.205	115.47	135.30	156.70	179.67	204.22	230.33	1/8
9/64	.01553	1.0218	3.5989	7.7468	13.465	20.755	29.615	40.046	52.048	65.621	80.764	97.479	115.76	135.62	157.05	180.04	204.61	230.75	9/64
5/32	.01917	1.0500	3.6516	7.8241	13.567	20.881	29.766	40.222	52.248	65.845	81.013	97.752	116.06	135.94	157.39	180.42	205.01	231.17	5/32
11/64	.02320	1.0786	3.7048	7.9017	13.669	21.008	29.917	40.398	52.448	66.070	81.263	98.026	116.36	136.27	157.74	180.79	205.40	231.59	11/64
3/16	.02761	1.1075	3.7583	7.9798	13.772	21.135	30.069	40.574	52.649	66.296	81.513	98.301	116.66	136.59	158.09	181.16	205.80	232.01	3/16
13/64	.03240	1.1369	3.8121	8.0582	13.875	21.263	30.221	40.750	52.850	66.521	81.763	98.575	116.96	136.91	158.44	181.53	206.20	232.44	13/64
7/32	.03758	1.1666	3.8664	8.1370	13.978	21.391	30.374	40.927	53.052	66.747	82.014	98.850	117.26	137.24	158.79	181.91	206.60	232.86	7/32
15/64	.04314	1.1967	3.9211	8.2162	14.082	21.519	30.526	41.105	53.254	66.974	82.265	99.126	117.56	137.56	159.14	182.28	206.99	233.28	15/64
1/4	.04909	1.2272	3.9761	8.2958	14.186	21.648	30.680	41.282	53.456	67.201	82.516	99.402	117.86	137.89	159.48	182.65	207.39	233.71	1/4
17/64	.05541	1.2577	4.0315	8.3757	14.291	21.777	30.833	41.461	53.659	67.428	82.768	99.678	118.16	138.21	159.83	183.08	207.79	234.13	17/64
9/32	.06213	1.2893	4.0873	8.4561	14.396	21.906	30.987	41.639	53.862	67.655	83.020	99.955	118.46	138.54	160.19	183.40	208.19	234.55	9/32
19/64	.06922	1.3209	4.1435	8.5368	14.501	22.036	31.141	41.818	54.065	67.883	83.272	100.232	118.76	138.86	160.54	183.78	208.59	234.98	19/64
5/16	.07670	1.3530	4.2000	8.6179	14.607	22.166	31.296	41.997	54.269	68.112	83.525	100.509	119.06	139.19	160.89	184.15	208.99	235.40	5/16
21/64	.08456	1.3854	4.2570	8.6994	14.713	22.297	31.451	42.177	54.473	68.341	83.779	100.787	119.37	139.52	161.24	184.53	209.39	235.83	21/64
11/32	.09281	1.4182	4.3143	8.7813	14.819	22.428	31.607	42.357	54.678	68.570	84.032	101.066	119.67	139.84	161.59	184.91	209.79	236.25	11/32
23/64	.1014	1.4513	4.3720	8.8636	14.926	22.559	31.763	42.537	54.883	68.799	84.286	101.344	119.97	140.17	161.94	185.28	210.20	236.68	23/64
3/8	.1104	1.4849	4.4301	8.9462	15.033	22.691	31.919	42.718	55.088	69.029	84.541	101.623	120.28	140.50	162.30	185.66	210.60	237.10	3/8
25/64	.1198	1.5188	4.4886	9.0292	15.141	22.823	32.076	42.899	55.294	69.259	84.796	101.903	120.58	140.83	162.65	186.04	211.00	237.53	25/64
13/32	.1296	1.5532	4.5475	9.1126	15.249	22.955	32.233	43.081	55.500	69.490	85.051	102.182	120.88	141.16	163.00	186.42	211.40	237.96	13/32
27/64	.1398	1.5879	4.6067	9.1964	15.357	23.008	32.390	43.263	55.707	69.721	85.306	102.462	121.19	141.49	163.36	186.79	211.80	238.39	27/64
7/16	.1503	1.6230	4.6664	9.2806	15.466	23.221	32.548	43.445	55.914	69.953	85.562	102.743	121.49	141.82	163.71	187.17	212.21	238.81	7/16
29/64	.1613	1.6584	4.7264	9.3652	15.575	23.355	32.706	43.628	56.121	70.184	85.819	103.024	121.80	142.15	164.06	187.55	212.61	239.24	29/64
15/32	.1726	1.6943	4.7868	9.4501	15.684	23.489	32.865	43.811	56.329	70.417	86.075	103.305	122.11	142.48	164.42	187.93	213.02	239.67	15/32
31/64	.1841	1.7305	4.8476	9.5354	15.794	23.623	33.024	43.995	56.537	70.649	86.333	103.587	122.43	142.81	164.77	188.31	213.42	240.10	31/64
1/2	.1963	1.7671	4.9088	9.6212	15.904	23.758	33.183	44.179	56.745	70.882	86.590	103.869	122.72	143.14	165.13	188.69	213.82	240.53	1/2
33/64	.2088	1.8042	4.9703	9.7072	16.015	23.893	33.343	44.363	56.954	71.116	86.848	104.151	123.03	143.47	165.49	189.07	214.23	240.96	33/64
17/32	.2217	1.8415	5.0322	9.7937	16.126	24.029	33.503	44.548	57.163	71.349	87.106	104.434	123.33	143.80	165.84	189.45	214.64	241.39	17/32
35/64	.2349	1.8793	5.0946	9.8806	16.237	24.165	33.663	44.733	57.373	71.583	87.365	104.717	123.64	144.13	166.20	189.83	215.04	241.82	35/64
9/16	.2485	1.9175	5.1573	9.9678	16.349	24.301	33.824	44.918	57.583	71.818	87.624	105.001	123.95	144.47	166.56	190.22	215.45	242.25	9/16
37/64	.2625	1.9560	5.2203	10.0544	16.461	24.438	33.985	45.104	57.793	72.053	87.883	105.285	124.26	144.80	166.91	190.60	215.85	242.68	37/64
19/32	.2769	1.9949	5.2838	10.1435	16.574	24.575	34.147	45.290	58.004	72.288	88.143	105.569	124.57	145.13	167.27	190.98	216.26	243.11	19/32
39/64	.2916	2.0342	5.3477	10.2318	16.687	24.713	34.309	45.477	58.215	72.524	88.404	105.804	124.88	145.47	167.63	191.36	216.67	243.54	39/64
5/8	.3068	2.0739	5.4119	10.3206	16.800	24.850	34.472	45.664	58.426	72.760	88.664	106.139	125.19	145.80	167.99	191.75	217.08	243.98	5/8
41/64	.3223	2.1140	5.4765	10.4098	16.914	24.989	34.634	45.851	58.638	72.996	88.925	106.425	125.50	146.14	168.35	192.13	217.48	244.41	41/64
21/32	.3382	2.1545	5.5415	10.4994	17.028	25.127	34.798	46.039	58.850	73.233	89.186	106.771	125.81	146.47	168.71	192.52	217.89	244.84	21/32
43/64	.3545	2.1953	5.6069	10.5893	17.142	25.266	34.961	46.227	59.063	73.470	89.448	106.997	126.12	146.81	169.07	192.90	218.30	245.28	43/64
11/16	.3712	2.2365	5.6727	10.6796	17.257	25.406	35.125	46.415	59.276	73.708	89.710	107.284	126.43	147.14	169.43	193.28	218.71	245.71	11/16
45/64	.3883	2.2782	5.7388	10.7703	17.372	25.546	35.289	46.604	59.489	73.946	89.973	107.571	126.74	147.48	169.79	193.67	219.12	246.14	45/64
23/32	.4057	2.3201	5.8054	10.8614	17.488	25.686	35.454	46.793	59.703	74.184	90.236	107.858	127.05	147.82	170.15	194.06	219.53	246.58	23/32
47/64	.4236	2.3623	5.8723	10.9528	17.604	25.826	35.619	46.983	59.917	74.423	90.499	108.146	127.36	148.15	170.51	194.44	219.94	247.01	47/64
3/4	.4418	2.4053	5.9396	11.0447	17.721	25.967	35.785	47.173	60.132	74.662	90.763	108.434	127.68	148.49	170.87	194.83	220.35	247.45	3/4
49/64	.4604	2.4484	6.0073	11.1369	17.837	26.108	35.951	47.363	60.347	74.901	91.027	108.723	127.99	148.83	171.24	195.21	220.76	247.89	49/64
25/32	.4794	2.4929	6.0753	11.2295	17.954	26.250	36.117	47.554	60.562	75.141	91.291	109.012	128.30	149.17	171.60	195.60	221.18	248.32	25/32
51/64	.4987	2.5359	6.1438	11.3236	18.072	26.392	36.283	47.745	60.778	75.382	91.556	109.301	128.62	149.50	171.96	195.99	221.59	248.76	51/64
13/16	.5185	2.5802	6.2126	11.4159	18.190	26.535	36.450	47.937	60.994	75.622	91.821	109.591	128.93	149.84	172.32	196.38	222.00	249.20	13/16
53/64	.5386	2.6248	6.2819	11.5096	18.308	26.678	36.618	48.129	61.211	75.863	92.087	109.881	129.25	150.18	172.69	196.77	222.41	249.65	53/64
27/32	.5591	2.6699	6.3515	11.6038	18.427	26.821	36.787	48.321	61.427	76.105	92.353	110.171	129.56	150.52	173.05	197.15	222.83	250.07	27/32
55/64	.5800	2.7153	6.4215	11.6983	18.546	26.964	36.954	48.514	61.645	76.346	92.619	110.462	129.88	150.86	173.42	197.54	223.24	250.51	55/64
7/8	.6013	2.7612	6.4918	11.7933	18.665	27.109	37.122	48.707	61.862	76.589	92.886	110.753	130.19	151.20	173.78	197.93	223.65	250.95	7/8
57/64	.6230	2.8074	6.5526	11.8885	18.785	27.252	37.291	48.900	62.080	76.831	93.153	111.045	130.51	151.54	174.15	198.32	224.07	251.39	57/64
29/32	.6450	2.8540	6.6337	11.9842	18.906	27.398	37.461	49.094	62.299	77.074	93.420	111.337	130.82	151.88	174.51	198.71	224.48	251.83	29/32
59/64	.6675	2.9010	6.7052	12.0803	19.026	27.543	37.630	49.288	62.518	77.317	93.688	111.630	131.14	152.22	174.88	199.10	224.90	252.26	59/64
15/16	.6923	2.9483	6.7771	12.1768	19.147	27.688	37.800	49.483	62.737	77.561	93.956	111.922	131.46	152.57	175.25	199.49	225.31	252.70	15/16
61/64	.7135	2.9961	6.8494	12.2736	19.268	27.834	37.971	49.678	62.956	77.805	94.225	112.215	131.78	152.91	175.61	199.89	225.73	253.15	61/64
31/32	.7371	3.0442	6.9221	12.3708	19.390	27.981	38.142	49.874	63.176	78.050	94.494	112.509	132.09	153.25	175.98	200.28	226.15	253.59	31/32
63/64	.7610	3.0920	6.9952	12.4684	19.512	28.127	38.313	50.069	63.396	78.295	94.763	112.803	132.41	153.59	176.35	200.67	226.56	254.03	63/64

to this is the amount of tubing required for the offset bend you want (X dimension). In this case it is $3\frac{17}{23}$ in.

Step 3. Determine where you want the center of the offset bend on the tube and make a reference mark (A). Now measure off the X dimension (determined in Step 2) starting from the reference mark and make a second mark (B). You are now ready to make the bends.

Step 4. Align mark A with reference mark R on bender and proceed with first bend. Then align B with R mark and make second bend in proper direction.

Note. When the amount of offset exceeds what is listed on the table, choose an offset from the table which is a multiple of the offset you need. Look this up on the table and multiply the X dimension by the

multiple you used. Example: For an offset of 20 in. for a 45 deg bend, look up 5 in. offset on the table in the 45 deg column and multiply X dimension ($7\frac{1}{16}$ in.) by 4. The resulting X dimension you would use is $28\frac{1}{4}$ in.

How to Figure Total Length of Tubing Required for 45 deg Offset Applications

Determine the X dimension required for a particular application and subtract the amount of offset from this. From the example above (Step 3), $2\frac{1}{2}$ in. offset was required and the X dimension as determined from the table was $3\frac{17}{32}$ in. The difference between these two figures is $1\frac{1}{32}$ in. Simply add this to the vertical distance from the starting point to the finishing point (Z dimension in illustration above).

Questions

1. How can the fabrication of a system minimize friction and reduce power loss?
2. What effect does hydraulic shock have on fluid conductors?
3. What variables determine the wall thickness and safety factor of a conductor for any given operating pressure?
4. What is a hydrauliscope, and what is its use in connection with a hydraulic system?
5. Name several considerations for safeguarding a system from over-pressure conditions.
6. What causes pressure drop through conductors and connectors?
7. What flow velocity is generally recommended for the discharge side of a pump?
8. What is the recommended velocity for the inlet side of the pump?
9. What effect does vibration have on a hydraulic system and how is it controlled?
10. What size inlet line would you use for a 12 gpm pump?
11. What size discharge line would you use for a 12 gpm pump?

CHAPTER *5*

CONDITIONING POWER FLUIDS

Reservoirs, heat exchangers, strainers, filters, and magnets are all important components of a fluid power system to assure smooth and trouble-free operation of machines and equipment. Proper conditioning of the hydraulic fluid is probably one of the most neglected considerations in the design of hydraulic circuits. Consequently, 40 to 60 per cent of all the trouble in hydraulic circuits results from improper care of the hydraulic fluid.

Conditioning the hydraulic fluid means to provide ample storage for the fluid in a system, to maintain its proper operating temperature, and to keep it clean and free of contamination.

Different fields of applying fluid power will require varying degrees of fluid conditioning. For instance, such applications as numerically controlled machines, high performance spacecraft, and nuclear-powered submarines require positive control of the fluids cleanliness and physical conditions at all times. Hydraulic systems of this nature use servo valves and other precision components that have extremely close tolerances between moving parts. Because of these close tolerances, a very small particle of dirt could be

detrimental to the function of a control valve, and, hence, cause a serious mishap. A sharp rise in temperature could also cause various parts to expand and prevent the proper function essential for the control of system actuators. The hydraulic system on a space capsule, as shown in Figure 5–1, is responsible for guiding the craft through space and back to earth safely.

Extreme measures must be taken in these special areas of fluid power application to keep the power fluids clean and stable and at the proper temperature. Many times, costly equipment and human lives depend on the hydraulic system and its ability to function properly.

Hydraulic systems used on mobile equipment are rugged and less sophisticated, with components that can withstand high shock loads under intermittent duty operation. The components such as pumps, motors, and control valves are built to be more compatible with dirt, but there are still limits to the amount of contamination that can be tolerated. Mobile hydraulic circuits often use a pressure cap on the reservoir to seal the fluid from the dusty or dirty

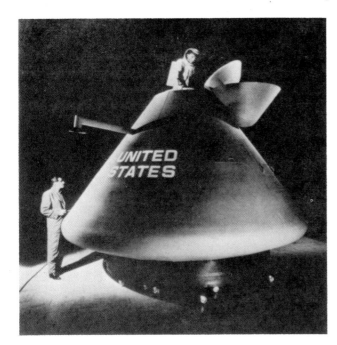

Fig. 5-1. Apollo command module. (*Courtesy of* NASA; *North American Aviation, Inc.;* Reprinted from *APPLIED ASTRONAUTICS An Introduction to Space Flight,* by Ordway, Gardner, Sharpe and Wakeford, Prentice-Hall, Inc.)

atmospheric conditions under which the equipment must work. Scrapers or boots are used on cylinder rods to prevent dirt from getting into the hydraulic fluid as the cylinders operate (see Fig. 5–2).

Space limitations on machinery may dictate the use of smaller reservoirs, which means that temperature control of the working fluid may be critical. Heavy dirt and mud caked on the outside of the reservoir and on the hydraulic hoses and tubing could prevent heat transfer to the surrounding atmosphere; therefore, these areas should be kept clean.

The machine tool industry presents a slightly different problem, because hydraulic machinery must operate continuously. The cleanliness, stability, and temperature of the fluid must be properly maintained for continuity of production and the quality control of machined parts. Long life expectancy of equipment and low maintenance are important factors in preventing down time and loss of production.

It is also easier to control the proper condition of the operating fluid in machine tools because of the accessibility of cooling water and the availability of preventive maintenance. Reservoirs, coolers, filters, and magnets can be maintained and serviced regularly by skilled personnel. This means that machine

designers and original equipment manufacturers (OEM) should provide or recommend the proper conditioning equipment for each installation. Generally, in this type of application, there is little excuse for improper installation of equipment or lack of a good preventive maintenance program to keep the hydraulic systems troublefree and operating smoothly.

Regardless of the field of application, there are certain important factors which must be considered when conditioning equipment is selected for any hydraulic system.

The Hydraulic Reservoir

Reservoir Capacity

The hydraulic reservoir should contain enough fluid that its working level is always maintained during the system's operation. It should also have additional capacity to hold all the fluid in the system during shutdowns. The reservoir capacity is generally between two and three times the capacity of the hydraulic pump. In other words, a pump delivering 10 gallons per minute would need a reservoir with a capacity of about 30 gallons. If a hydraulic system has a great number of cylinders operating simultaneously, more careful consideration should be given to the proper reservoir size by calculating the entire system's volume; for example, an agricultural tractor with fluid power to operate auxiliary equipment, such as a loader, ripper, snow plow, etc. It should also be understood that systems blowing high pressure fluid over the relief valve cause rapid heat build-up, and a proper amount of fluid should be main-

Fig. 5-2. Booted cylinder. (*Courtesy of Socony Mobil Oil Co.*)

Fig. 5-3. (a) Vertical Duplex Hydro-Broach; (b) Circuit.

Vertical Duplex Hydro-Broach. Broaching is roughly comparable to drawing a milled file of suitable shape over the surface to be machined. Each succeeding tooth takes off a small part of the metal to be removed, but, usually, a broach is made long enough and with enough teeth so that the work is finished in one pass. Both internal surfaces (holes of all shapes) and external surfaces (one or several at a time) can be broached. Since many broach teeth are cutting at the same time, greater force is required than when comparable work is done on a milling machine, shaper, or planer. Hydraulic actuation easily provides this force as well as ease of control and adjustment of speed. The cost of broaches and work-holding fixtures limits this method to mass production where a large number of identical parts are machined. The duplex broach shown has two rams, two work tables, etc. At the end of a cutting stroke of one ram, the work table on that side moves away from the broach so that the operator can unload and reload the work pieces safely. During this time, the ram makes its return stroke and the second ram is on its cutting stroke. The ram speeds are adjusted to suit the work and the operator's speed. Cutting is then virtually continuous, with rams moving up and down and work tables moving in and out in synchronism. The main hydraulic circuit for this machine is shown below with control valves in position for the downward work stroke of the right ram. High-pressure oil is supplied by the constant-volume-type main pump through start-stop-reverse, ram-direction, and right-ram valves to the rod end of the right-ram cylinder. Oil under back pressure is forced from the blank end of this cylinder through the ram valves to the blank end of the left-ram cylinder, driving the left ram upward on its return stroke. Oil flows from the rod end of the left-ram cylinder under exhaust pressure to the reservoir. At this time, the work tables are held stationary. At the end of the work stroke, the ram and table control valves are shifted hydraulically by means of a low-pressure pilot circuit (not shown) that has its own pump. The work tables are reciprocated, and the left ram begins its work stroke. The table reciprocating linkages are driven from cranks on a shaft that is rotated by means of a rack on the table-piston rod. Bearings and ways are automatically lubricated with oil under pressure.

tained and circulated within the reservoir to help dissipate its heat. The side wall area of the reservoir must also be adequate to help provide effective transfer of heat through the walls to the surrounding atmosphere.

Actually, a detailed analysis of the heat transfer between the fluid, reservoir, and atmosphere is very complicated and often inaccurate because of changes in convection rate due to local atmospheric conditions. The complications also involve making allowances for radiation, conduction, and convection of the various materials, such as color and thickness of paint, thickness of metal wall, position of surfaces such as surfaces of sides, top, bottom, or ends. For intance, the bottom surface may be a short distance from the floor and have uncertain ventilation, making it extremely difficult to calculate.

Experience has proved, however, that within moderate ambient temperatures, reservoirs generally have the capacity to maintain the proper fluid operating temperatures.

Other features of a well-designed reservoir, as shown in Fig. 5–4, may include the following:

Baffle plate. The baffle plate separates the system return fluid from that going to the pump inlet. This aids in cooling by promoting the flow of fluid along the side walls, and also helps to separate water, entrained air, and dirt from the hydraulic fluid.

Bottom drain. A drain connection on the dished bottom of the reservoir provides a means of draining water and sediment from the system. This is also important when fluid in the system is being changed.

Sight glass. The sight glass provides a visual check on the fluid level. Proper fluid operating levels must be maintained.

Filling cap and hole. A capped hole fitted with a fine mesh screen prevents accidental contamination when the reservoir is being replenished to proper operating level. Make-up fluid should be prefiltered and transferred by means of clean equipment, such as containers, hose, or funnel.

Breather air filter. The air filter must be of adequate size and equipped with a filter element suitable to the local atmospheric conditions under which it must work. Since the fluid level in the reservoir must rise and fall as the cylinders stroke, air comes and goes through the breather. The filter element must be kept clean to allow atmospheric pressure to be in contact with the liquid at all times.

Clean-out cover or door. A large opening to facilitate cleaning, change of strainer, or inspection is a desirable feature. Also, some reservoirs have a protective coating which may have to be removed if one is changing from one type of hydraulic fluid to another.

Return line. The return line should be cut on an angle before it is fitted into the reservoir. This prevents any restriction on the returning fluid by jamming the end of the return pipe against the bottom of the reservoir. The size of the return line should be large enough to prevent excessive back pressures in the system (usually less than 30 psi).

Drain line. An extra drain line must be fitted into the reservoir when back pressures in the return line are too great. The drain line would be used exclusively to allow the free drain of any components fitted with external drain connections.

Pump inlet line. The pump inlet line should be large enough to assure satisfaction of the inlet requirements of the hydraulic pump to prevent cavitation. A slip joint or other means should be employed to facilitate the complete removal of the inlet line and strainer to permit regular inspection or cleaning of the strainer element. Illustration 5–5 shows one

Fig. 5-4. Reservoir. (*Courtesy of Socony Mobil Oil Co.*)

Fig. 5-5. Flange connection for removing pump inlet and strainer. (*Courtesy of Socony Mobil Oil Co.*)

method used to facilitate servicing inlet line strainers.

Drive mounting base. Many installations use the top of the reservoir to mount an additional steel plate to support the drive motor and pump. This is an excellent way to keep the pump mounted closer to the fluid being pumped, and to help the inlet characteristics of the pump.

Casters. Another convenient feature to facilitate the portability of a complete hydraulic power source is to mount the reservoir, drive, pump, and, sometimes, controls as a complete package on casters. A spare power unit of this design is used many times as a stand-by to facilitate quick change of power in cases of emergency or for controlled maintenance.

The design of any reservoir is generally dictated by the type of application for which the system is intended. For example, in a closed loop system (see Fig. 5–6) the fluid transmits power from either side of the pump to change the direction of the fluid motor. Because the same fluid is being circulated constantly, it is necessary only to make up the fluid losses which occur because of clearance spaces between integral moving parts.

Systems of this design generally replenish fluid losses through the use of a second pump, which takes its fluid from a reservoir. The replenishing system should be sized in accordance with the efficiencies of pump and motor drive units. The expected losses

of fluid during operation may be computed through actual testing, or may be obtained directly from the manufacturers of the equipment. The size of the reservoir to be used depends upon the amount of fluid make-up needed to keep the system fully primed at all times.

The proper selection of a reservoir for any hydraulic system depends largely on the field of application and the type of duty under which the system must operate. In some cases, a given length and size of pipe may be very adequate for a reservoir, whereas in other cases much more elaborate equipment is needed.

Heat Exchangers

Coolers or heaters are called heat exchangers and are used to control fluid operating temperatures in a hydraulic system. When high or low ambient temperatures are encountered, such as when machinery is operating close to a hot furnace or when mobile equipment is operating in sub-zero weather, heat exchangers are used to control and maintain proper operating temperatures of the systems' fluid. The

Fig. 5-6. Closed loop system.

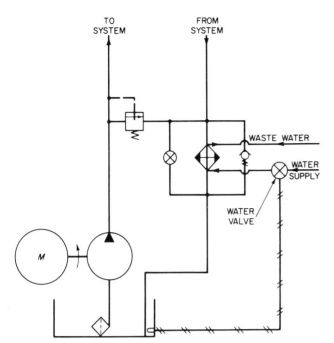

TO SYSTEM   FROM SYSTEM

WASTE WATER

WATER SUPPLY

WATER VALVE

M

Fig. 5-7. Hydraulic application showing heat exchanger.

steady temperature reached by the fluid in a hydraulic system depends on both the *amount of heat generated* and the *heat-dissipating ability of the system.*

Heating units employ hot water, steam, or electricity and are generally restricted to use for startups or for fluid storage and transfer tanks. Electric heaters are generally used for hydraulic equipment or mobile machinery operating in cold temperatures.

Critical and complex machinery may require conditioning of the fluid to proper operating temperatures before startups are made, which would require positive control of both heating and cooling. Generally, the natural heat-generating characteristics of a system after warmups are sufficient or greater than required. Therefore, the greatest concern in designing hydraulic systems is to provide adequate cooling.

Heat is generated naturally in a hydraulic system because it is impossible to maintain 100 per cent efficiency of operation. This is true with any power-transmitting system using fluid, electrical, or mechanical drives.

If heat energy is allowed to accumulate in a hydraulic system, high temperatures result with possible damage to seals, fluid, and moving parts. High temperature also affects the viscosity of the fluid, which may change the operating performance characteristics of a machine or piece of equipment. Desired temperature ranges are usually between 120° and

140°F, and if the natural processes of heat transfer do not maintain the proper temperature range, some method of cooling must be used. Figure 5-7 shows the installation of a heat exchanger on a hydraulic system.

The amount of cooling necessary to maintain the proper temperature in a hydraulic system can be estimated in a number of ways. For example, the heat generated is equal to the total input power less the actual mechanical work done based on a definite time cycle of operation. Thus, the more efficient hydraulic systems require less cooling. It is also an important fact that the kinetic energy in a fluid under pressure is converted into heat when the fluid is passed over a a relief valve or undergoes pressure drop through a component.

Method of Estimating Fluid Temperature Rise in a Hydraulic Sysem

Example 1. When a given quantity of fluid is forced through a restriction, power is required. The power or energy required is converted *into heat* and results in a rise of fluid temperature. For example,

Flow rate = 10 gpm
Pressure drop = 100 psi

The hydraulic horsepower involved is

hp = gpm × psi (drop) × .000583
hp = 10 × 100 × .000583 = 0.583

The conversion factor is

2544 BTU/hr = 1 hp
BTU = British Thermal Unit

One BTU is the amount of heat required to raise the temperature of one pound of water 1°F. Since our hp formula includes flow in gallons per minute, we must reduce the conversion factor to BTU/min.

$$\frac{2544}{60} = 42.4 \text{ BTU/min (1 hp)}$$

The energy converted into heat is

42.4 BTU/min × 0.583 hp = 24.7 BTU/min

The specific heat of oil is approximately

0.42 BTU/#°F

The specific gravity of oil is approximately 0.85. The energy converted into heat (24.7 BTU/min) will raise the temperature of each gallon of oil as follows:

Temp. °F =

$$\frac{\text{BTU/min}}{\text{Specific heat of oil} \times \text{quantity of oil in lb/min}}$$

BTU/min = 24.7
Specific heat of oil = 0.42 BTU per lb per °F
Quantity of oil flowing in lb/min = 71 lb/min
1 gallon of oil = 231 cu in.

1 cu ft = 1728 cu in.

Weight of oil = 62.4 × 0.85 = 53 lb per cu ft

Weight of 1 gallon of oil

$$= \frac{231}{1728} \times 62.4 \times 0.85 = 7.1 \text{ lbs.}$$

$$\text{Temp.} \,^\circ F = \frac{24.7}{0.42 \times 71} = 0.84 \,^\circ F$$

Example 2. A hydraulic pump with a capacity of 20 gpm at 3000 psi discharges 50 per cent of its volume over the relief valve during the feed portion of the machining cycle. If the feed stroke takes 3 minutes, how much heat is generated in British Thermal Units (BTU)?

$$\text{Hydraulic horsepower} = \frac{\text{gpm} \times \text{psi}}{1714}$$

$$\text{hp} = \frac{20 \times 0.50 \times 3000 \text{ psi}}{1714}$$

$$\text{hp} = \frac{30,000}{1714}$$

$$\text{hp} = 17$$

Since there are 42.4 BTU/min in 1 hp, then total heat in BTU

$$= \frac{42.4 \text{ BTU/min} \times 17 \text{ hp} \times 3 \text{ min}}{1 \text{ hp}}$$

Total heat gained in BTU = 2162.4

The result in heat gained due to the high pressure fluid's being discharged over the relief valve shows the importance of selecting the proper cooling equipment based on an analysis of the system's operations.

High fluid velocities, high ambient temperature, mechanical friction, and fluid friction are all factors of heat generation. Experience has shown that heat generation due to these factors is approximately 20 per cent of the input horsepower (prime mover).

Example 3. A hydraulic system has a pump which is driven by a 20 hp electric motor. What is the expected heat generation (heat in BTU)?

Fig. 5-8. Fluid temperature rise in a hydraulic

1 BTU = 778 ft lb

Heat generated = (0.20)(20 hp) = 4 hp

$$4 \text{ hp} \frac{(33,000 \text{ ft lb/min})}{1 \text{ hp}} = 132,000 \text{ ft lb/min}$$

$$132,000 \text{ ft lb/min} \frac{(1 \text{ BTU})}{(778 \text{ ft lb})}$$

$$= 169.4 \text{ BTU/min}$$

Therefore, the system must dissipate 169.4 BTU/min. If the system itself cannot get rid of this much heat, a cooler must be added to control the proper system temperature.

The most popular types of heat exchangers used with hydraulic systems use air or water as the cooling medium. The law of applying heat exchangers states that the heat given off by the hot object must equal the heat gain by the cold object. Generally speaking, if all other factors remain constant, heat transferred in any given heat exchanger will be directly proportional to the temperature difference between the hydraulic fluid and the cooling media.

Air coolers (see Fig. 5–9) are used where water may be too expensive or not readily available. Air-to-oil coolers are generally used on mobile-type equipment. These coolers consist of a core section of oval or round tubes and plate fins, or individual round finned tubes.

The oval core with plate fins has a compact surface which provides good surface efficiency, low air resistance, and minimum power to heat removal ratio.

The round core with plate fins is generally used for fluids having a high rate of heat transfer. It can be used also for systems with high operating pressures. This design operates more efficiently with forced air circulation when used for cooling viscous fluids.

The individual round fin tube core section is comparable with the round core design, except that individual tubes can be removed from the core section.

These coolers are used for applications where the hydraulic fluid enters the cooler at a minimum of 130°F with an ambient air temperature of 100°F or more.

Water-cooled heat exchangers of the shell and tube design are generally used for hydraulic systems when ample water supply is available. A heat exchanger of this type consists of an assembled bundle of tubes inserted into a shell. The tubes are baffled to direct the hydraulic fluid through the shell side of the unit at right angles to the tube bundle. The cooling medium generally flows through the tubes.

There are a number of design variations; however, the straight tube design with fixed tube bundle

Header support channel tapped for rods to hang cooler independent of piping.

Supply and return headers are tubular steel.

Large oil connections for low pressure drop.

Large tubes are brazed into headers. Patented turbulators double heat transfer (see below).

Cooling element designed for low final air temperature.

Fan and shrouds are designed for maximum air delivery with quiet operation.

Fins are mechanically bonded to tubes.

Motor bolted firmly in place on rugged bracket.

Fig. 5-9. Air cooler. (*Courtesy of Vickers.*)

is popular for use with hydraulic systems. The units are available with various baffle arrangements to create single or multiple pass exchangers. The multiple pass designs use less water and can be used more efficiently and at less cost when colder circulating water is available. Tubes are accessible from either end for cleaning.

The main factors for selecting the proper heat exchangers are as follows:

1. Determine the actual heat (BTU/min) generated during the system operations on the basis of a one minute time cycle (total input horsepower minus mechanical work output).
2. Select the hydraulic fluid with careful consideration of: type of fluid, viscosity, density, specific heat, flow rate, inlet temperature.
3. Determine the cooling fluid.
   a. Flow and inlet temperature for water-cooled exchangers.
   b. Ambient temperature for air-cooled units. When water for cooling is readily available the type of cooling system as shown in Figure 5–10 is used to maintain proper fluid temperature.

Heat exchangers installed on the low-pressure side of the hydraulic circuit eliminate the need for high-pressure units. When the pressure differential is exceedingly high between the hydraulic fluid and the coolant fluid, it enhances the possibility of water leakage into the hydraulic system. Heat exchangers should be protected against overpressures as well as against shock loads. Relief valves properly placed help to provide protection for heat exchangers.

A separate low-pressure circuit from the reservoir through the heat exchanger and back to the reservoir

permits the hydraulic fluid to circulate without being affected by flow conditions found in the main circuit.

Excessive pressure drops should be avoided when heat exchangers are selected. They should be inspected periodically and cleaned at regular intervals in order to maintain their efficiency.

Source of Contamination in a Hydraulic System

Modern hydraulic power and control systems must be reliable and dependable, and must provide greater accuracies than ever before. This means that each component used in a hydraulic system must be precisely machined.

The biggest enemy of a precision machine is contamination. Contamination is any substance in a fluid

Fig. 5-10. Water-type water cooler. (*Courtesy of Young Radiator.*)

power system which is detrimental to its proper function. Contamination may be in the form of a liquid, gas, or solid.

The source of contaminants may be any of the following:

1. Assembly or machine fabrication.
2. The system's operational factors.
3. Careless maintenance and service practices.

The contaminants found to result from component assembly or machine fabrication include metal chips, parts of pipe threads, tubing burrs, pipe dope, shreds of plastic tape, lint, bits of sealing material, scale, welding beads, bits of hose, plain dirt, and sludge. New hydraulic machinery may be seriously damaged during startups unless every precaution is observed to free system components from such contaminants.

Recommended Procedures for Users

1. Uncrate equipment when it is received and carefully inspect for any damage that may have occurred during shipment.
2. Carefully store in clean, dry place, making sure ports or openings are sealed tight, until ready for use.
3. Use accepted practices of good workmanship during fabrication or setting up equipment for use, and use clean and compatible materials.
4. Flush system thoroughly (put in blanks for any servo valves during flushing procedures) with a good grade of recommended flushing fluid; then completely drain.
5. Flush system with the hydraulic fluid (filtered) to be used, and then drain completely.
6. Inspect and clean or replace all filters in the system.
7. Fill system with the filtered hydraulic fluid to be used for the machine operation.

This may seem a complex procedure simply to put a hyraulic system into operation, but, compared in cost to expensive down time and loss of production, it is well worth the effort.

During the normal operation of hydraulic equipment, there are many sources of contamination that may cause serious trouble unless proper maintenance practices are used. Contamination due to operation may be caused by water or moisture getting into the system, entrained gases, scale caused by rust, bits of sealing material, metal silvers, particles of the metal due to wear, and sludge or varnishes due to oxidation of the fluid.

Installation of adequate filtration and magnets at vulnerable points throughout the system and a good

preventive maintenance program are the best safeguards against these kinds of contamination. It is also important that the right fluid be used and that its temperature be properly maintained.

Improper maintenance and service practices can also add harmful contamination to the hydraulic system. One of the main sources of contamination is the use of dirty equipment, such as buckets or funnels, to add fluid to a hydraulic system. It has also been found that hydraulic components during assembly or disassembly have been placed on a dirty work bench or wiped off with a dirty rag.

Clean equipment and clean work stations are essential when servicing or maintaining hydraulic equipment. When it is necessary to do field repair or work in contaminated areas, spread an old sheet or wrapping paper on top of the work surface. Wash all parts and dip in clean hydraulic fluid just before assembly.

Good maintenance and service practices can prevent contamination and expensive down time.

Strainers and Filters

The importance of keeping the hydraulic fluid clean and free of harmful contamination cannot be overemphasized.

Strainers and filters are used to remove contaminating particles from the hydraulic system. The terms *strainer* and *filter* are often used interchangeably because they have a common function. The industry has attempted to distinguish between the two types of units according to a basic difference in their construction. Thus, a strainer may be defined as "a device for the removal of solids from a fluid wherein the resistance to motion of such solids is in a straight line." The filter may be defined as "a device for the removal of solids from a fluid wherein the resistance to motion of such solids is in a tortuous path."

Strainers are generally designed of a fine mesh wire screen or a screening element made of specially processed wire of varying thickness wrapped around a metal frame (see Fig. 5–11).

Strainers do not provide the fine filtrating action as do filters; thus they offer less resistance to flow. For this reason, strainers may be used in a pump inlet line to protect the pump from the larger sizes of detrimental particles. Differently constructed hydraulic pumps require different inlet conditions. It is important that the capacity of the inlet strainer, including case and fittings, be sufficient to keep pressure drop at a minimum. The finer the screen mesh, the greater the pressure drop. Pressure drop also in-

FLOW RATES (gpm) FOR HYDRAULIC FLUIDS OF COMMON VISCOSITIES FOR 100 MESH AND PRESSURE DROPS WHEN STRAINER IS CLEAN

Strainer sizes	Approximately 6 sq in.			Approximately 14 sq in.			Approximately 32 sq in.			Approximately 45 sq in.		
psi drop	5	10	15	5	10	15	5	10	15	5	10	15
50 SUS	4.9	7.0	8.5	9.6	13.5	16.6	26	37	45	93	132	161
100 SUS	4.1	5.9	7.2	8.0	11.3	13.9	22	31	38	78	111	136
200 SUS	3.5	4.9	6.0	6.7	9.5	11.6	18	26	32	65	92	113
300 SUS	3.1	4.4	5.4	6.0	8.5	10.4	16	23	28	56	80	98

Fig. 5-11. Strainer. (*Courtesy Schroeder Brothers.*)

ceases during operating as dirt clogs the small screen openings. If the pressure drop reaches a critical limit for a given pump design, cavitation results and the pump life is shortened.

There are a number of arrangements for using strainers in a pump inlet line. If single strainer units are insufficient for the demands of a given pump, two or more can be used, as shown in Fig. 5–12. To help facilitate regular cleaning, strainers need to be installed only hand tight, provided the fittings are constantly submerged.

A 60 mesh screen is generally recommended where actual pressure drop is unknown. However, if the pressure drop at the lowest operating temperature of the fluid used is known, screens as fine as 120 mesh can be used, if of ample capacity.

Filtration is fast becoming an art in itself, and the modern tendency is to measure detrimental particle sizes in microns. A micron equals one-millionth of a meter, or 0.000039 in. The limit of visibility with the unaided eye is about 40 microns; thus, many detrimental particles in a hydraulic system are not normally visible (see Fig. 5–13).

Modern high flow, high pressure, hydraulic systems may require a combination of strainers and filters for protection against detrimental contaminants. The pump is generally the most expensive component in the system and should have the protection of an inlet strainer. It is also a fact that filters applied on the discharge side of the pump or in the return line to the reservoir can be of real significance in preventing faulty operation and loss of production.

The following information is important in order to select the proper filter for any application:

1. The size and physical nature of the detrimental contaminates to be removed.
2. The viscosity, density, operating temperature, and corrosive properties of the hydraulic fluid being used.
3. The materials used for fabricating the system before the filter and after the filter. Seals, fittings, tubing, hose, instruments, and other parts should be analyzed.
4. The operating characteristics of the system, such as shock load, fluid velocity, and direction.

Fig. 5-12. Multiple strainer arrangements. (*Courtesy of Vickers.*)

ACCESS OPENING SHOULD BE PROVIDED
SO STRAINERS MAY BE REMOVED FOR
CLEANING WITHOUT DRAINING OIL FROM TANK

**STRAINER MESH SIZES AND SIZE OF
SOLIDS REMOVED WHEN STRAINER IS CLEAN**

Mesh	Wire	Size in.	Microns
8	0.028	0.097	2500
30	0.012	0.021	540
50	0.009	0.011	282
60	0.0075	0.0092	236
70	0.0065	0.008	205
80	0.0055	0.007	180
90	0.005	0.0061	156
100	0.0045	0.0055	141
150	0.0026	0.0041	105
200	0.0021	0.0029	75
250	0.0016	0.0024	60

(One micron equals 0.000039 in.)
Recommended strainer mesh

700 SUS Hyd. Oil	30 mesh
350 SUS Hyd. Oil	60 mesh
200 SUS Hyd. Oil	100 mesh

5. Pressures and pressure drops that the filter media must encounter during the system's operation.

6. The component in the system that requires the closest tolerances, and, hence, the least contamination.

Filters provide much finer filtration than strainers and may be applied at several different places in the circuit. Where the filter is placed depends a great deal on the application and the design of the hydraulic equipment. Each application must be analyzed to determine the degree of filtration needed, the vulnerable points in the circuit, and the type of filtering media that should be used. Figures 5–14 through 5–17 show examples of where filters may be placed in a hydraulic system.

Filters are rated in terms of "nominal" and "absolute" filtration. Both terms are expressed in microns. The term *nominal filtration* indicates the ability to remove about 98 per cent of the solid particles

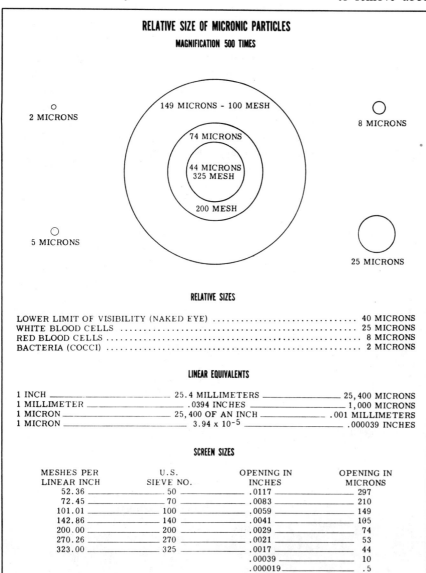

Fig. 5-13. Mesh strainers (magnified 500 times). *(Courtesy of Vickers.)*

Fig. 5-14. By-pass filter; pressure side..

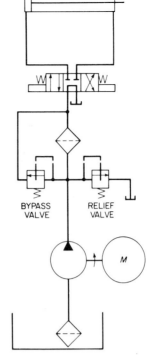

Fig. 5-16. Proportional side; pressure side.

Fig. 5-15. Full-flow filter; pressure side.

Fig. 5-17. Separate filtration system.

T TYPE   POT TYPE   Y TYPE

IN LINE TYPE

PROPORTIONAL TYPE

Fig. 5-18. Basic filter designs.

from the fluid which are equal to or larger than the micron rating of the filter.

The *absolute rating* indicates the size of the filter element pores through which the fluid must flow, and gives the largest possible size solid that can pass through the element. Many laboratory-controlled tests have indicated that irregularly shaped particles and parts of "O" rings, etc., that may be elastic in nature have no true absolute rating. As a filter element removes contaminating particles from the fluid

stream, they build up on the media or element surface. This reduces the size of the pores through which the fluid must flow. Thus, a filter rated 60 microns absolute may have a nominal rating of 35 microns because 98 per cent of the particles 35 microns or larger are removed.

Some of the basic filter designs are illustrated in Fig. 5–18.

Filter accessories include devices which assist in maintenance, such as indicators (see Fig. 5–19) that operate on the pressure differential of the filter to show that a filter element needs to be cleaned or replaced. Many filters are equipped with relief valves (see Fig. 5–20) which are set to open when the differential pressure reaches a predetermined value. Also, pressure gauges on either side of a filter are sometimes used to show the pressure differential across the filter. When the gauges reach a certain pressure difference above that which is normal when the filter is clean, it indicates a dirty filter.

Manual shutoff and by-pass valves are also used to isolate the filter during service (see Fig. 5–21).

Portable units, such as that shown in Fig. 5–22, aid during service and repair or when a system is accidently contaminated to filter large quantities of fluid in order to get the equipment back in service. The portable unit is also excellent in reclaiming hydraulic fluid.

The materials selected for the construction of filter housings depend primarily on the intended service. Aircraft and missile usage require materials of high strength and low weight, such as stainless steels and

PRESSURE DROP CONTROL SPRING

INSPECTION GLASS

INDICATOR OPERATING LEVER

OUTLET PORT

CLEAN-OUT COVER

FILTER IS CLEAN
NEEDS CLEANING
BY-PASSING

DIRT INDICATOR

NOT VISIBLE
By-Pass Port incorporated in Rear end Cartridge Pilot.

REMOVABLE FILTER CARTRIDGE

PISTON RING

INLET PORT

Fig. 5-19. Tell-tale filter.

Fig. 5-20. (a) Full-flow filter with relief valve (*Courtesy of Vickers.*); (b) High-pressure filter (*Courtesy of Schroeder Brothers*).

certain aluminum alloys. Aluminum is generally restricted to medium temperature ranges where noncorrosive conditions are essential. For marine, industrial, and some military applications, carbon steels and copper-nickel alloys are used. Because of the trend in all fields of usage to use full flow filtration, the body must be rated for its intended operating pressure.

There are several basic types of filter media, such as paper, sintered metal powder, woven wire cloth, and certain kinds of ceramic or plastic. In the selection of a filter media, it is important to consider the type of fluid, chemical compatibility, temperature, and the ability to withstand high flow rates.

Paper filter elements are made of special cellulosic materials which are impregnated with a resin to provide greater strength. The paper media may be used as an extended area filter where particles removed depend on the amount of surface exposed and the porosity of the filter cake that is formed. Two popular filters of the type are made by folding sheets of resin-impregnated paper into accordion-like pleats or into "concertina folds" and the forming into the filter housings. Other types are made by stacking

cellulose discs of various designs with alternate layers of cellulose spacers. The fluid is filtered laterally, vertically, and between the discs in an edge-type manner. The design generally determines the efficiency of the filter. Recently it was discovered that by adding glass fiber greater strength can be obtained. Cellulose filter elements can be made with accurate

Fig. 5-21. Manual by-pass.

Fig. 5-22. *Portable Filter.* This filter has its own pump and hose connections. Oil is drawn from a reservoir—from the low point, if possible —is pumped through the filter, and is returned to the clean oil side of the reservoir. The filter has a replaceable filter cartridge, a relief valve, and a pressure gauge. The pressure reading increases as the cartridge becomes loaded with dirt. When a specified pressure is reached, the cartridge should be changed. A portable filter is usually connected to a reservoir for 24 hours at a time. *(Courtesy of William W. Nugent & Co.)*

to 200 microns. This type of filter media provides resistance to corrosion and fatigue. It is free from migration and resists penetration by sharp particles.

Filtration is accomplished by essentially two classes of filters, surface filters and depth filters.

The surface filter (see Fig. 5–23) accomplishes all its filtering action at the surface of the media. The pores are all very nearly of uniform size, and the action approaches the absolute. This class of filter is easy to clean and resists migration, but it has a comparatively low dirt-holding capacity.

The depth filter (see Fig. 5–24) operates throughout the volume of the filter material by presenting many tortuous passages through which the fluid must flow. The pores or passages are not of uniform size, and the entrapment of particles depends on the depth and nature of the various passages. The action of this type of filter is primarily statistical because of the different sizes of the passages and the effect of fluid velocity, which could dislodge trapped particles if rapidly increased. Depth filters operate successfully at lower flow rates and at relatively low pressure drops. The types of filters used to eliminate contaminants from a hydraulic system are shown in Figure 5–25.

Fig. 5-23. *Surface-Type Full-Flow Filter.* This model is suitable for line pressures up to 1500 psi. The replaceable filter element is made of resin-impregnated paper having pores of up to 5-micron size. The paper is pleated and formed into a cylinder as shown. Hydraulic fluid passes through the paper, leaving behind dirt and other solids in a layer on the outer surface. In service, particles down to about one-micron (about 0.00004-in.) size are held back. If pressure drop across the filter becomes too great, the relief valve opens and by-passes unfiltered fluid. *(Courtesy of Socony Mobil Oil Co.)*

control of pore sizes between 2 and 25 microns. They can be used for temperatures as high as 275°F. The large amount of surface area helps to increase the actual operation time between servicing. These filter elements are available for nominal ratings of 0.5 to 100 microns. Absolute ratings range from 5 microns and up.

In general, the sintered metal powder media, including bronze and stainless steel, provide depth filtration with nominal ratings of 2 to 65 microns. The absolute ratings range from 13 to 100 microns. Media migration or self-contamination may be a problem unless proper manufacturing methods are used.

Ceramic or plastic media are also used, but higher pressure drops may be expected as well as less resistance to mechanical and thermal shock.

Woven wire cloth elements are usually made of stainless steel and provide nominal filtration ratings of 2 to 100 microns with absolute ratings from 12

Fig. 5-24. *Depth-Type Size Filter.* Dirty oil surrounds the renewable filter cartridges and flows radially inward through a considerable depth of filter material such as cellulose. Clean oil flows downward and out at the left. These filters are capable of very fine filtration. (*Courtesy of Socony Mobil Oil Co.*)

The use of magnets has been found to be advantageous for removing iron and steel particles from the hydraulic fluid. Magnetic plugs are installed at strategic locations in the reservoir to prevent the particles of steel and iron from getting back up into the system.

Magnetic rings and pencils are also assembled on pump inlet strainers to attract and hold any steel or iron particles trying to get through the wire mesh screen. Magnets help to prevent the sliver-type particles such as broken-off threads from penetrating through the strainer and into the system.

Magnets can be installed easily at any time and may save valuable equipment from being damaged.

Benefits from Proper Conditioning of the Hydraulic Fluid

Using the correct hydraulic fluid, selected especially for the equipment used as well as for the operating conditions involved, is most important. Proper conditioning of the fluid selected must be maintained continually to keep the fluid clean and at proper operating temperature if the hydraulic system is to function properly and give long life. Proper conditioning provides the following benefits:

1. *Reduction of power losses.* The use of properly sized reservoirs and adequate coolers reduces system leakage by controlling the viscosity of the fluid with temperature control. This also helps to minimize friction by maintaining adequate lubrication between moving parts in pumps, fluid motors, and cylinders. When these properties are maintained, power losses are minimized.

2. *More continuous production.* When the hydraulic fluid is properly conditioned, less pressure drop occurs, which maintains good control, fast response, and accurate timing of machine operations. Strainers, filters, and magnets play an important part in keeping the system clean and preventing costly interruptions in production. The cost of installing a filter or magnet in a hydraulic system is incidental, compared to the loss of a machine and the time involved to make costly repairs.

Fig. 5-25. *Edge-Type Size Filter.* The filter element consists of a stack of wheel-shaped metal discs, each separated from the next by a thin metal spacer. The fineness of filtration is determined by the thickness of these spacers. The stack is closed at the bottom and opened to the filter outlet at the top. Dirty inlet oil flows into the space around the element and is forced through the small openings between the discs. All solids larger than the openings are held back. Clean oil flows upward through the stack to the outlet. Stationary cleaner blades extend into the space between the discs (see assembly sketch), and the filter element can be rotated against these for cleaning, using the handle shown. Solids fall to the bottom of the housing and should be removed at regular intervals. (*Courtesy of Socony Mobil Oil Co.*)

3. *Reduction of cost of maintenance.* The life expectancy of hydraulic equipment under normal operating conditions can be many years. Hydraulic equipment has been known to operate 15 to 20 years or more without trouble. The main reason for this kind of service life is the fact that the hydraulic fluid was kept clean and properly conditioned. Detrimental contamination and high temperatures can create excessive maintenance costs, if not controlled. When system fluids are properly conditioned, it requires a minimum amount of maintenance to keep equipment in good operating condition.

4. *Reduction of the cost of hydraulic fluid.* Good tight connections are, of course, needed to prevent external leakage. Temperature control is also important, because hot oil leaks more readily than oil at its proper operating temperature. When hydraulic fluid is overheated, it oxidizes rapidly and loses its important physical properties. This causes a kind of chain reaction which quickly contaminates the entire system, which necessitates a changing of the system's fluid. A good hydraulic fluid does not wear out or lose its desirable properties unless it is overheated or contaminated with dirt. Good control of the hydraulic fluid with proper conditioning equipment increases the life of the fluid and the system's components.

Questions

1. What is meant by conditioning hydraulic fluids?
2. Name the important features found on a well-designed hydraulic reservoir.
3. What is the general rule of thumb for sizing a reservoir's capacity?
4. What is the function of a heat exchanger when used with the hydraulic system?
5. What effect would high temperatures have on the operating characteristics of a hydraulic system?
6. How many BTU per hour would be gained by a system blowing 7.5 gpm over the relief valve at 1500 psi every 18 minutes out of each hour of operation?
7. What are the main factors in selecting the proper heat exchangers for a hydraulic system?
8. How does contamination get into a hydraulic system?
9. What are the contaminants that are generally found in a hydraulic system?
10. How is contamination controlled in a hydraulic system?
11. How are metal chips eliminated from a hydraulic system?
12. In general, what is the difference between a strainer and a filter?
13. How are the sizes of detrimental particles determined?
14. What are some of the important factors to consider in the selection of filtration devices?
15. Explain the difference between a high-pressure full flow filter and a low-pressure full flow filter.
16. Why are indicators used on filters and upon what principle do they operate?
17. Name some of the basic types of filter media.
18. Describe the difference between a surface filter and a depth filter.
19. What are some of the benefits from properly conditioned hydraulic systems?

CHAPTER 6

THE SOURCE
OF HYDRAULIC
POWER

Transmitting hydraulic power requires a source of input energy, one or more pumps to convert this mechanical energy into hydraulic energy, controls, and actuating cylinders or fluid motors which utilize the fluid power to provide linear or rotary motion (see Fig. 6–1).

The input energy source may be an electric motor internal combustion engine, turbine drive, power take-off arrangement or other mechanical device which can impart force and motion to operate the pump. The pump, in turn, must supply hydraulic fluid to the system. A hydraulic pump is defined as a device which converts mechanical force and motion into fluid power. It is the heart of the system and generally one of the more costly components used in the transmission of hydraulic power.

Where electric power is readily available, such as in most industrial plants, electric motors are used to drive the hydraulic pumps. Power units, as shown in Fig. 6–2, are especially popular in the machine tool field of application. Electric motors are readily available in speeds (rpm's) which are compatible with most hydraulic pumps.

Hydraulic power systems found on farm tractors, road machinery, earth-moving equipment, utility vehicles, and various other devices using internal combustion engines use power take-off arrangements to operate the hydraulic pump (see Fig. 6–3).

When power take-off arrangements are not available on machines, auxiliary internal combustion engines are generally used to supply the input energy for the hydraulic system. The operating speed (rpm) of the engine is selected to be compatible with that of the recommended speed (rpm) of the hydraulic pump.

The illustration in Fig. 6–4 shows a method of directly coupling the hydraulic pump to the main drive. Engine speed in rpm must be compatible with that of the pump's rated speed.

Aircraft and other equipment using high speed prime movers, such as turbines or jet engines, require especially designed high speed pumps, or employ the use of speed reduction devices (see Fig. 6–5). Speed reducers coupled to the main drive change the economical operating speed of the prime mover to an economical operating speed of the hydraulic pump.

Belt and chain drives are also used to transfer input force and motion from a prime drive for the operation of a hydraulic pump. Operating speeds (rpm) are adjusting by selecting the proper pulley diameters or sprocket tooth ratios.

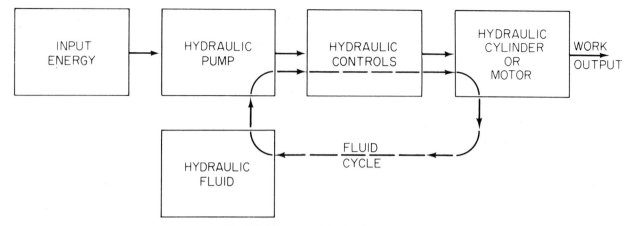

Fig. 6-1. Elements of hydraulic power.

Fig. 6-2. Power units showing electric motor and hydraulic pump. (*Courtesy of Oilgear Co.*)

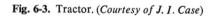

Fig. 6-3. Tractor. (*Courtesy of J. I. Case*)

Fig. 6-4. Direct couples engine and pump.

Fig. 6-5. Power arrangement taken from jet engine.

Speed reduction by means of gears can be expressed as the ratio of the number of gear teeth on one gear to the number of teeth on the other. When N represents the number of teeth on the large gear and n the number on the smaller gear, then the speed reduction is given by

$$\text{Speed reduction} = \text{gear ratio}\, \frac{N}{n}$$

Example.

$$\text{Gear ratio} = \frac{N}{n}$$

$$\text{Gear ratio} = \frac{32}{16}$$

$$\text{Gear ratio} = 2 : 1$$

Example. Consider A and B as a system, and C and D as a system (speed of B equals speed of C). This results as follows:

$$\text{Speed reduction} = \text{gear ratio} = \frac{N_1}{n_1} \times \frac{N_2}{n_2}$$

$N_1 = 16$ teeth

$n_1 = 4$ teeth

$N_2 = 20$ teeth

$n_2 = 5$ teeth

$$\text{Speed reduction} = \text{gear ratio} = \frac{16}{4} \times \frac{20}{5}$$

$$\text{Speed reduction} = \text{gear ratio} = \frac{320}{20}$$

$$\text{Speed reduction} = \text{gear ratio} = 16 : 1$$

$$\text{Speed ratio} = \frac{N_1}{n_2}$$

$$\text{Speed ratio} = \frac{28}{7} = \frac{4}{1}$$

Pulleys and belts are also used for connecting hydraulic pumps to the main drive. Generally, the

Fig. 6-6. (a) Spur gear; (b) Gear train; (c) Sprocket and chain; (d) Pulley and V Belt.

N

n

32 TEETH   16 TEETH

7 TEETH   28 TEETH

N_2 N_1 n_2 n_1

A

C

B

D

$N_1 = 16$ TEETH $N_2 = 20$ TEETH
$n_1 = 4$ TEETH $n_2 = 5$ TEETH

V–BELT

6"   24"

d   D

PULLEY

"V" belt is employed to provide adequate strength and area of contact for the friction required to prevent slippage (see Fig. 6–6d).

In general, pulley and belt drive speeds can be expressed in terms of pulley diameters.

$$\text{Speed ratio} = \frac{D}{d}$$

$$\text{Speed ratio} = \frac{24}{6} = 4:1$$

In most applications of hydraulic power, the pump is connected directly to the input drive shaft by a flexible coupling. A number of different designs are available for a wide choice of power ratings and operating speeds and for a given allowable misalignment.

Important safety rules for the installation of guards covering rotating machinery should be strictly observed, regardless of the type of power drive used.

In general, pump manufacturers limit the amount of side loading when pumps are driven by belts, chains, and other devices. It is more efficient, safer, and less wearing on the equipment when direct coupling arrangements can be used.

Input Horsepower

The input horsepower for hydraulic power systems can be determined by the following equation:

$$\text{hp} = \text{gpm} \times \text{psi} \times 0.000583$$

Example. What horsepower electric motor would be needed to drive a hydraulic pump whose actual delivery is 20 gallons per minute at 1500 psi?

$$\text{hp} = 20 \text{ gpm} \times 1500 \text{ psi} \times 0.000583$$

$$\text{hp} = 17.4$$

It would not be practical to buy an electric motor that has to be specifically built, so the logical choice would be a standard 20 hp electric motor. It is good practice to select a large enough electric motor to handle emergency overloads. Some industrial plants

have a standard practice of selecting the next largest size motor automatically to prevent any future problems.

Output power of a hydraulic pump is expressed as a function of flow volume and pressure. Some engineers use the simple rule that one hydraulic hp equals the delivery of one gallon per minute at 1500 psi. The previous example problem indicates that this is close for a quick estimate of power requirements.

Another version of the hydraulic horsepower formula is shown in the following:

$$\text{Hyd. hp} = \frac{\text{gpm} \times \text{psi}}{1714}$$

Example. A farm tractor is equipped with a 40 hp internal combustion engine. What portion of this hp will be needed to operate a front end hydraulic loader if the pump delivers 12 gpm at 2000 psi?

Solution.

$$\text{hp} = \frac{\text{gpm} \times \text{psi}}{1714}$$

$$\text{hp} = \frac{12 \times 2000}{1714}$$

$$\text{hp} = 14$$

The input horsepower requirements of any hydraulic system depend mainly on the maximum needs at any one time during a cycle of the equipment operation. The operating pressure at any given moment in a hydraulic system depends primarily on the resistance due to the work load. The number of gallons per minute being delivered to a hydraulic cylinder or motor at any given moment determines how fast the work is being done. Therefore, the input power requirements must be adequate to handle the maximum pressure and volume flow at any given moment in a hydraulic system.

In comparison with mechanical systems, one of the main advantages of using hydraulic power is the fact that a well-designed system provides the correct amount of operating horsepower at all times.

Pump Classification

There are two general classifications of pumps which are being used for the transmission of hydraulic power:

1. Nonpositive displacement pumps.
2. Positive displacement pumps.

Nonpositive displacement pumps are primarily velocity-type units which have a great deal of clearance between rotating and stationary parts of the

Fig. 6-7. Coupling.

pump. These pumps are generally used for low-pressure, high-volume flow applications. Because of the large clearance space, these pumps are not self-priming. In other words, the pumping action has too much clearance space to seal against atmospheric pressure. The displacement between inlet and outlet is not positive. Therefore, the volume of fluid delivered by the pump depends on the speed at which the pump is operated and the resistance at the discharge side. As the resistance builds up at the discharge side, fluid slips back into the clearance spaces, or, in other words, follows the path of least resistance. When the resistance gets to a certain value, no fluid at all will be delivered to the system and the volumetric efficiency of the pump will drop to zero for a given speed. Increased speed will increase pressure. Figure 6–8 shows an impeller type nonpositive displacement pump which is generally used for transferring large quantities of liquid at a low pressure.

Figure 6–9 shows the axial flow design of the nonpositive displacement pump design which handles up to thousands of gallons of liquid per hour.

Figure 6–10 shows the centrifugal pump which has the higher potential of the nonpositive displacement type pumps.

Fig. 6-8. Impeller unit. (*Courtesy of Vickers Co.*)

Fig. 6-9. Axial flow unit. (*Courtesy of Vickers Co.*)

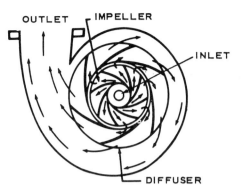

Fig. 6-10. Centrifugal pump-single stage. (*Courtesy of Vickers Co.*)

Fig. 6-11. Centrifugal pump-multistage.

The main uses for nonpositive displacement pumps are for transferring fluids, coolant supply, and to supercharge the inlet of other pumps.

When centrifugal pumps are multistage units (see Fig. 6–11) they can be used for higher operating pressures. They operate on the principle that fluid taken in at the inlet of the first stage is delivered to the inlet of the second stage and so on through each successive stage of the pump. In this manner the potential of the entire pumping unit is substantially increased. In other words, it is a number of centrifugal pumps hooked up in a series.

These units are used for large hydraulic systems using water, such as forge presses where hot molten metal presents a fire hazard.

Positive displacement pumps are most generally used for hydraulic power. A positive displacement pump has a very close clearance between rotating and stationary parts. As the classification implies, pumps of this design provide a specific amount of fluid to the hydraulic system for each revolution that they are turned.

The volumetric efficiency of any pump is the quantity of fluid delivered from the outlet port, compared to the theoretical displacement of the pump.

$$\text{Vol. eff.} = \frac{\text{actual output (gpm)} \times 100}{\text{theoretical output (gpm)}}$$

Theoretical displacement depends primarily on the geometry of the pump by design (cubic inches of fluid displacement per revolution) and the operating speed (rpm).

Theoretical output (gpm)

$$= \frac{\text{displacement (cu in.)/rev.} \times \text{rpm}}{231 \text{ cu in.}}$$

The actual delivery of the pump depends on how well the pump is made and is computed by actual laboratory testing.

Over-all efficiency is expressed as follows:

$$\text{Pump over-all eff.} \% = \frac{\text{output hyd. hp}}{\text{input hp to pump}} \times 100$$

Mechanical efficiency as well as volumetric efficiency have an important bearing on over-all efficiency.

Mechanical efficiency is the ratio of the calculated power required to drive a hydraulic pump compared to the actual input power.

$$\% \text{ mech. eff.} = \frac{\begin{array}{c}\text{theoretical torque required}\\ \text{to drive the pump (lb in.)}\end{array}}{\begin{array}{c}\text{actual torque supplied to}\\ \text{pump (lb in.)}\end{array}} \times 100$$

Theoretical input torque is

$$\frac{\text{Torque}}{\text{(lb in.)}} = \frac{\text{displacement in cu in./rev.} \times \text{psi}}{2\pi}$$

Actual input torque is calculated by laboratory testing:

$$\frac{\text{Actual torque}}{\text{(lb in.)}} = \frac{\text{brake hp} \times 63,025}{\text{rpm}}$$

Brake horsepower can be calculated as follows:

$$\text{Brake hp} = \frac{\text{hyd. output hp} \times 100}{\text{pump vol. efficiency}}$$

Example.

1. What is the volumetric efficiency of a pump that displaces 0.7 cu in. of fluid per revolution if the actual delivery is 4.5 gpm at 1800 rpm?

$$\text{Vol. efficiency} = \frac{\text{actual output (gpm)}}{\text{thereotical output (gpm)}} \times 100$$

Actual output given = 4.5 gpm

$$\text{Theoretical output (gpm)} = \frac{\frac{\text{dis. cu in.}}{\text{rev.}} \times \text{rpm}}{231}$$

$$= \frac{0.7 \times 1800}{231}$$

$$= \frac{1260}{231}$$

$$= 5.4 \text{ gpm}$$

so

$$\text{Vol. eff.} = \frac{4.5}{5.4} \times 100$$

$$\text{Vol. eff.} = 83 \text{ per cent}$$

2. What is the over-all efficiency of a hydraulic pump that is driving a fluid motor at 2000 rpm with a torque of 550 lb in. The pump is driven by a 20 hp electric motor.

$$\text{Pump over-all efficiency} = \frac{\text{output hyd. hp}}{\text{input hp}} \times 100$$

$$\begin{array}{c}\text{(Output hp)}\\ \text{Fluid motor hp}\end{array} = \frac{\overset{\text{(lb in.)}}{\text{torque.}} \times \text{rpm}}{63,025}$$

$$= \frac{550 \times 2000}{63,025}$$

$$= 16$$

The input hp is given as 20 hp. Substituting, we obtain

$$\text{efficiency} = \frac{16 \text{ hp}}{20 \text{ hp}} \times 100$$

Over-all efficiency = 80 per cent

3. What is the mechanical efficiency of a hydraulic pump with a displacement of 0.7 cu in. per revolution and operating at 2000 psi? The pump is operating at 1800 rpm and the brake horsepower is 7.2.

$$\% \text{ mech. efficiency} =$$

$$\frac{\begin{array}{c}\text{theoretical torque required}\\ \text{to drive pump (lb in.)}\end{array}}{\begin{array}{c}\text{actual torque supplied to}\\ \text{pump}\end{array}} \times 1000$$

$$\text{Theoretical torque required} = \frac{\frac{\text{dis./rev.} \times \text{psi}}{\text{cu in.}}}{2\pi}$$

$$\text{Theoretical torque} = \frac{0.7 \text{ cu in.} \times 2000 \text{ psi}}{6.28}$$

$$\text{Theoretical torque} = 223 \text{ lb in.}$$

$$\frac{\text{Actual torque}}{\text{(lb in.)}} = \frac{\text{bhp} \times 63,025}{\text{rpm}}$$

$$\frac{\text{Actual torque}}{\text{(lb in.)}} = \frac{7.2 \times 63,025}{1800 \text{ rpm}}$$

$$\text{Actual torque} = 246 \text{ lb in.}$$

$$\text{Mechanical efficiency} = \frac{223 \text{ lb in.}}{246 \text{ lb in.}} \times 100$$

$$\text{Mechanical efficiency} = 90 \text{ per cent}$$

4. What is the brake horsepower of a hydraulic pump delivering 36 hyd. hp if the pump has an efficiency of 87 per cent?

$$\text{Brake hp} = \frac{\text{hyd. output hp}}{\text{pump eff.}} \times 100$$

$$\text{Brake hp} = \frac{36}{87} \times 100$$

$$\text{Brake hp} = 41.3$$

Pumping Theory

When a positive displacement pump is operated, trapped fluid is displaced from the inlet to the outlet because of the close tolerance between mating parts. In Fig. 6–12, as the gears turn, fluid trapped between the gear teeth and the housing is carried from the inlet side to the discharge side of the pump. When the pump is first started, the fluid displaced is air, and as the air is taken from the inlet pipe a vacuum is created. When the pressure within the inlet line to the pump becomes less than the atmospheric pressure, fluid is forced by the atmospheric pressure through the strainer and up into the pump inlet.

Thus, it becomes obvious that fluid supplied to the pump is dependent upon atmospheric pressure to raise it up into the pump inlet.

Since atmospheric pressure is 14.7 psi at sea level, it is important to control other variables, such

Fig. 6-12. Positive displacement pump.

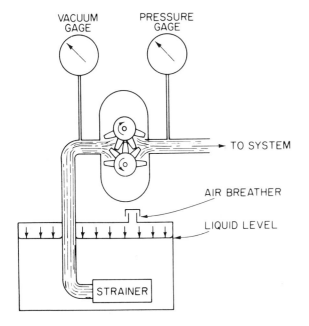

as pipe inlet size and length of inlet pipe within the working range of atmospheric pressure.

The recommended fluid velocity for good inlet conditions of the pump should be maintained between 2 and 5 fps.

$$\text{Fluid velocity fps} = \frac{\text{gpm} \times 0.3208}{\text{inlet pipe area}}$$

Example. What area of pipe is needed for the inlet of a 20 gpm pump? (The value for velocity must be selected between 2 and 5 fps.)

Substituting, we have

$$\text{Velocity 3 fps} = \frac{20 \text{ gpm} \times 0.3208}{\text{area of pipe (sq in.)}}$$

$$\text{Area of pipe (sq in.)} = \frac{20 \text{ gpm} \times 0.3208}{3 \text{ fps}}$$

$$\text{Area} = 2.208 \text{ sq in.}$$

To find the diameter from the area:

$$\text{Diameter (in.)} = \sqrt{\frac{\text{Area sq in.}}{0.7854}}$$

$$\text{Diameter/in} = \sqrt{\frac{2.208}{0.7854}}$$

$$\text{Diameter} = \sqrt{2.8}$$

$$\text{Diameter}^2 = 2.8$$

$$\text{Diameter} = 1.7 \text{ in}$$

The nearest nominal standard pipe size would be $1\frac{1}{2}$ in. diameter pipe with an area of 2.03 sq in. This would increase the velocity rate a little, but it would still remain close to 3 fps.

Pump inlet conditions are also affected by the viscosity, specific gravity, fluid temperature, and the type of fluid being used in the system. It is essential that all these variables be carefully considered to keep inlet conditions of the pump under control for efficient operation. When control of proper conditions is impossible because of the nature of the application or environmental factors, then the pump inlet should be pressurized.

The pump inlet can be pressurized either through the use of a liquid or a gas. Pressurization of the hydraulic pump inlet may be necessary for the following reasons:

1. When the hydraulic application occurs in high altitude areas, where atmospheric pressure is less than normal.
2. If the pump is driven at high rpm and atmospheric pressure cannot act quickly enough to fulfill the inlet requirements.

OIL LEVEL

SPECIFIC GRAVITY = 0.85

2" i.d.

6'

ELECTRIC MOTOR

HYDRAULIC PUMP

Fig. 6-13. Pressurized inlet using liquid.

greater than that of the atmosphere. This may be done by pumping the gas into the space or by relying on the natural thermal expansion of the trapped gas as the hydraulic liquid heats up as a result of the hydraulic system's operations. Pressure caps similar to the automobile radiator cap are generally used for pressurized reservoir conditions (see Fig. 6–14).

Pump Design Types

Pumping units are available for practically any use conceivable to modern man. In fact, it has been stated that more patents have been granted for pump designs than for any other mechanical device invented.

Hydraulic pumps are available for volumetric capacities of a few cubic inches per minute or for hundreds of gallons per minute. The pressure ratings of these pumps range from zero to 10,000 psi, and in some cases higher. Pump operating speeds

Fig. 6-14. Reservoir pressurization with gas.

3. When the specific gravity of the hydraulic fluid is greater than that of water.
4. To provide greater machinery stability for smoother finishes, especially if the work loads are erratic by nature.
5. To allow hydraulic fluids to operate at higher temperatures without vaporization.
6. To assure safety of operation for high performance machinery, or to protect human lives.

If the hydraulic fluid itself is used to pressurize the pump inlet, a second pump is generally used. This second pump may be an independent system or can be attached to the same input shaft of the main pump. The inlet supply pressure to the main pump varies with the application but is usually maintained below 100 psi because of the effect on shaft seals or other low pressure parts. Figure 6–13 illustrates the use of an auxiliary pump to supply the main pump when atmospheric pressure is incapable of forcing an adequate supply of hydraulic fluid to the pump.

When a gas is used to pressurize the pump inlet, the reservoir is completely enclosed. The space above the liquid in the tank is maintained under a pressure

Fig. 6-15. Comparison of 20 hp electric motor to 20 hp hydraulic pump. (*Courtesy of Vickers Co.*)

Fig. 6-15. (*cont'd*)

Fig. 6-16. Fixed pump. (*Courtesy of Dynex.*)

increasing in almost all fields of usage. More and more emphasis is being placed on better efficiencies, faster response, greater accuracy and dependability with less maintenance.

Because of these demands, pump manufacturers are constantly upgrading their designs and developing new pumping techniques to meet these new performance requirements.

Hydraulic pumps are positive displacement pumps which can be further divided into two categories: fixed delivery or variable displacement.

A pump of the fixed delivery design provides a specific volume displacement per revolution which cannot be varied.

Example. A fixed delivery pump with a displacement of 2 cu in. per revolution would deliver 3600 cu in. of fluid at 1800 rpm.

It is apparent from the example that the output volume of a fixed delivery pump depends on the displacement per revolution and the speed (rpm) at which the pump is operated (see Fig. 6-16).

A variable volume pump is a pump in which the displacement per cycle can be varied (see Fig. 6-17). Pumps of this design provide infinitely variable volume from zero to the maximum volumetric capacity.

Example. A variable volume pump with a maximum displacement per revolution of 2 cu in. could deliver any increment of fluid between 0 and 3600 cu in. of fluid at a constant speed of 1800 rpm.

Some variable volume pumps have the ability to provide infinitely variable volume in either direction

Fig. 6-17. Variable pump. (*Courtesy of Racine Hydraulics.*)

are increasing as the pressure ratings go up and the volumetric capacities decrease.

Some aircraft and missile pumps operate at 12,000 rpm with a capacity of about 10 gpm and a pressure rating over 3000 psi. This would be nearly 20 hp for a pump weighing less than 10 lb. Imagine a 20 hp electric motor and its size and weight compared to the aircraft and missile unit (see Fig. 6-15).

Hydraulic pump operating pressures are steadily

through the pump. Pumps of this design are called reversible variable units, or over-the-center pumps (see Fig. 6–18).

Hydraulic power pumps are also categorized by their construction type, such as lobe element, external gear, internal gear, screw, vane, axial piston, radial piston, and reciprocating plunger. There are

also a number of construction types that have variations from these basic designs.

Fixed Delivery Hydraulic Pumps

The simplest and most easily understood pump is a hand pump of the design commonly used in a hydraulic jack. This is the type that is used to raise barber and beauty chairs. Handpumps were among the first designed hydraulic pumps and, surprisingly, operate at rather high pressures.

Figure 6–19 shows the principle of operation. A lever arm is used to provide mechanical advantage with man as the input power source. If 100 lb of force were applied on the 20 in. handle, a 2000 lb force would result on the pumping piston. The resultant force of 2000 lb is applied to a small diameter pumping piston, as shown in Fig. 6–20. Since pressure is force per unit of area, a very high hydraulic pressure of 10,000 psi builds up and is transmitted undiminished through the conductor or hose to the large diameter ram or cylinder to raise the load.

$$\frac{\text{Force of small piston}}{\text{Area of small piston}} = \frac{\text{Force of large piston}}{\text{Area of large piston}}$$

$$\frac{f = 2000}{a = 0.2} = \frac{F = 20,000}{A = 2.0}$$

$$10,000 \text{ psi} = 10,000 \text{ psi}$$

Thus, the pressure at any given moment in a hydraulic system is proportional to the load. *It should also be obvious from the previous relationship that a pump does not pump pressure; it pumps fluid.* The pressure in a hydraulic system is due to the resistance of the load.

Example. What would the pressure (psi) be in

Fig. 6-18. Over the center pump. (*Courtesy of N. Y. Air Brake.*)

Fig. 6-19. Principle of operation-hydraulic hand pump. (*Courtesy of Duff-Norton (Ram-Pac).*)

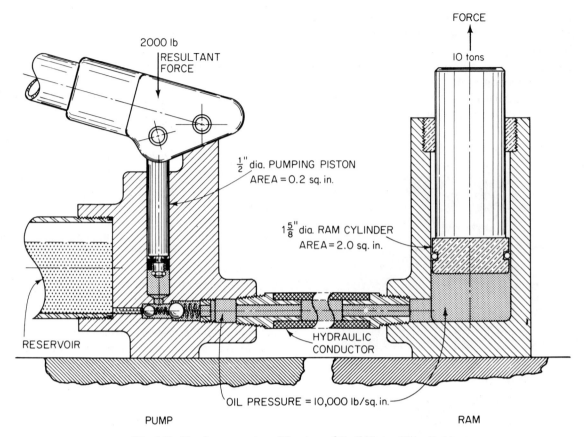

Fig. 6-20. Hand pump system. (*Courtesy of Duff-Norton (Ram-Pac).*)

the previous system if the load were reduced to 8 tons (8 tons = 16,000 lb).

$$\frac{P}{\text{(psi)}} = \frac{F}{A}$$

$$\frac{P}{\text{(psi)}} = \frac{16,000 \text{ lb}}{2 \text{ sq in.}}$$

$$\frac{P}{\text{(psi)}} = 8000 \text{ psi}$$

What handle effort would be required?

Handle force × mech. ad. = total input force on
(lb)   small pumping piston

$$\frac{F \times 20 \, (MA)}{a \text{ (area sm. piston)}} = \frac{\text{force (lb) output}}{\text{area of lg. piston sq in.}}$$

Substituting, we have

$$\frac{F \text{ (lb)} \times 20}{0.2 \text{ in.}^2} = \frac{16,000 \text{ lb}}{2 \text{ in.}^2}$$

Transposing gives

$$2 \text{ in.}^2 \; F/\text{lb} = \frac{0.2 \text{ in.}^2 \times 16,000 \text{ lb}}{20 \, (MA)}$$

$$2 \text{ in.}^2 \; F/\text{lb} = \frac{3200}{20} = 160$$

$$F/\text{lb} = \frac{160}{2}$$

(Handle) Input force = 80 lb

Because the small pumping piston cylinder is $\frac{1}{10}$ the area size of the large cylinder, it has to make ten 1 in. strokes for every one inch of travel for the large output piston. This means that the pump must be stroked in order to raise the load any great distance.

In order to accomplish repetitive stroking with-

Fig. 6-21. Force triangle.

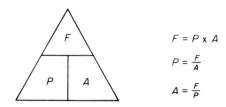

$F = P \times A$

$P = \frac{F}{A}$

$A = \frac{F}{P}$

Fig. 6-22. Inlet stroke. (*Courtesy of Duff-Norton (Ram-Pac)*.)

out losing pressure, hydraulic pumps isolate the discharge stroke from the inlet stroke by means of internal valving. Figure 6–22 shows check valves for this purpose.

On the *inlet stroke,* as the handle is raised a vacuum is created in the cylinder beneath the pumping piston. Atmospheric pressure pushing on the fluid in the reservoir causes the fluid to overcome the slight spring force on the inlet check ball, and fluid enters the piston cavity. As soon as the upward travel of the handle stops, the inlet ball seats by the spring force.

On the *discharge stroke* the handle is forced downward with enough force to overcome the work resistance and transmit fluid under pressure through the discharge check ball and conductor into the large cylinder to raise the load. As the handle stops on the downward stroke, the spring force plus the pressure due to the suspended load causes the discharge check valve to close, and the pump is ready to repeat the cycle (see Fig. 6–23).

When it is necessary to lower the load, a needle-

or bleeder-type valve is used. When the valve is opened, fluid is forced back into the reservoir by gravity or force due to the load on the large cylinder (see Fig. 6–24).

Hand pumps are also available as double acting and two-stage or two-speed units. Double acting hand pumps deliver fluid to the system for each stroke of the handle. The two-speed hand pump in Fig. 6–25 uses two pumping pistons and a relief valve for each piston outlet.

Two-speed hand pumps are popular for applications where actuator resistances and speeds vary.

Hydraulic Power Pumps

Hydraulic power-driven pumps are commonly used for almost all machines and devices using fluid power. There are many different construction types available.

External Gear Pumps

A typical external gear pump consists of a drive gear and a driven gear enclosed within a precision

Fig. 6-23. Discharge stroke. (*Courtesy of Duff-Norton (Ram-Pac)*.)

Fig. 6-24. Bleeder valve. (*Courtesy of Duff-Norton
(Ram-Pac).*)

machined housing. The close fit between the meshing
gears and the housing maintains a seal between inlet
and outlet sides of the pump. When the gears rotate,
teeth unmesh on the inlet side, allowing hydraulic
fluid from the reservoir to fill the resulting space.
The fluid is then transferred around the periphery of
both gears to the discharge side. As the gear teeth
remesh on the discharge side, hydraulic fluid is forced

Fig. 6-25. Two-speed hand pump. (*Courtesy of Duff-
Norton (Ram-Pac).*)

Fig. 6-26. Gear pump and ASA symbol. (*Courtesy
of Webster Electric.*)

through the discharge port and into the system (see
Fig. 6–26).

The direction of rotation for all gear-type hydraulic pumps is determined by facing the shaft end
of the pump. Manufacturers generally stamp an
arrow into the pump housing indicating either left-
hand or right-hand rotation. This is important in
most pumps because integral passages for draining
fluid from bearings, seals, and glands are connected
back to the inlet side to insure flow for lubrication
and seal. If the pump is operated in the wrong direction of rotation, the inlet and output ports reverse
their primary function, and the pump may be
destroyed. (Some gear pumps can be connected for

AS A PUMP

AS A FLUID MOTOR

Fig. 6-27. Gear designs–as a pump and as a fluid motor.

either direction of rotation, and this would be indicated by the manufacturer's literature.)

It is also possible to use some gear designs as either pumps or fluid motors. In other words, the unit could be driven by an electric motor to supply fluid to the system, or it could be supplied with hydraulic fluid under pressure to perform rotary motion. When these units are used as fluid motors, an external drain is generally needed. This is especially true if rotary motion occurs in both directions of rotation. Figure 6–27 illustrates the use of a gear unit for a pump or a motor. When driven by a prime mover, it is a pump; and when supplied with fluid under pressure, it is a motor.

The lobed element pump could be considered an external gear pump with special meshing elements or teeth, both elements being driven by external gearing. The external driving gears maintain proper timing of the lobed elements.

Most gear pumps are spur, helical, or herringbone designs.

Gear pumps can be used in fluid power systems for pressures up to 3000 psi and capacities up to 150 gpm. Spur-type gear pumps are generally less ex-

pensive but are somewhat more noisy than helical or herringbone designs. Herringbone gear pumps are the quietest and provide the least number of pulsations to the system. This design also eliminates thrust against the shafts and housings, which increases bearing life. Herringbone gear pumps must be connected to run in the direction which allows the apex of the vee to make first contact with the mating gear.

Pumps designed with external gear are the most popular in the low operating pressure ranges. They provide good life expectancy, are low in cost, have relatively good efficiency, and are less likely to be damaged by contamination.

External gear pumps may be either single or double units. Double pump or motor units can be used in series or in parallel hookups.

Internal Gear Designs

A popular construction type of pump or motor designed with internal gear teeth or motor is the gerotor unit (see Fig. 6–28). This pump consists of two elements, an inner gerotor and an outer gerotor. The inner element always has one less tooth than the outer.

The volume of the "missing tooth" multiplied by the number of driver teeth determines the volume of fluid pumped for each revolution. As the toothed elements, mounted on fixed centers but eccentric to each other, turn, the chamber between the teeth of the inner and outer elements gradually increases in size through about 180 deg of each revolution until it reaches its maximum size—equivalent to the full volume of the "missing tooth."

During the initial half of the cycle, the gradually enlarging chamber is exposed to the inlet port, creating a partial vacuum into which the hydraulic fluid flows. During the next 180 deg of the revolution, the chamber gradually decreases in size as the teeth mesh, and the fluid is forced out the discharge or outlet port and into the system. The Gerotor unit as illustrated in Fig. 6–29 is a popular design used in a number of fields of application as both pumps and motors.

Gerotor-type pumps are available as single, double, or two-stage units. The two-stage pump consists of two separate pump assemblies contained within a single housing and driven by a common shaft. The unit is constructed so that flow from the first stage is directed to the inlet of the second stage and, in turn, to the system. This series connection allows a substantial increase in operating pressure.

OUTLET

DRIVE SHAFT

INLET

RIGHT-HAND ROTATION
INTERNAL DRAIN

Fig. 6-28. Gerotor unit. (*Courtesy of Double A.*)

Example. As both pumping units rotate, fluid enters the inlet port of the first stage from the reservoir. The fluid is then transferred through the outlet port of the first unit and into the inlet of the second stage. It is then transferred to the outlet side of the second stage and into the system.

If the resistance due to the work should build the system's operating pressure to 3000 psi, each unit working together would shares the load. This means that each pumping unit or each stage would have a 1500 psi differential between inlet and outlet. Thus, an arrangement of two pumps in a series can double the pressure energy output without exposing either unit to a greater operating pressure than that for which it is designed (in this case, 1500 psi). Figure 6–30 shows the construction of a two-stage pump.

A dividing valve is used to equalize the pressure load on each pumping unit and to correct minor flow differences. The dividing valve spool *A* has an area ratio of 2:1. The larger area of the piston is connected through chamber *B* to the outlet of the first stage pump. The second stage pump outlet is connected to chamber *C,* where the piston area is one-half as great. Thus, any pressure deviation from the fixed ratio will result in valve spool movement to equalize the pressure load between each unit.

$$PA = PA$$

$$3000 \text{ psi} \times 1 \text{ in.}^2 = 1500 \text{ psi} \times 2 \text{ in.}^2$$

This analysis also applies to making flow corrections between the two pumping units.

Greater flow from the first stage produces an increase in pressure at chamber *B*. The pressure increase causes the dividing valve spool to move upward, connecting a flow passage through the center of the valve spool and into passage *D* back to the inlet of the first stage pump.

If the second stage unit has greater flow in passage *C*, the pressure increase will react on the small area side of the dividing valve spool, thus causing the spool to move downward. When the spool is moved downward, passage *C* is connected through the center of the spool with chamber *B,* and fluid escapes back to the inlet side of the second stage unit.

Pumps of this design are used for presses, injection molding machines, and other applications needing higher pressures.

Another internal gear tooth design uses a gear within a gear principle with a crescent which divides the flow and acts as a seal between the inlet and outlet sides of the pump. This unit operates equally well in either direction.

Fig. 6-29. Gerotor standard.

GEROTOR STANDARD

Fig. 6-30. Two-stage pump. (*Courtesy of Double A.*)

The pumping action is similar to other gear designs, except that both gears operate in the same direction. Gear-within-a-gear-type pumps are used advantageously where a drive shaft must go through the pump, as in automatic transmissions used in automobiles.

Internal gear pumps are produced with capacities up to 200 gpm. In the larger sizes they are generally applied as fluid transfer pumps.

Screw Pumps

There are three types of screw pumps which are commercially available. These three types are the single-screw, the two-screw, and the three-screw pumps. A single-screw pump consists of a spiraled rotor which rotates eccentrically in an internal stator. A two-screw pump consists of two parallel rotors with intermeshing threads rotating in a closely machined housing. These pumps use external or internal timing gears. A three-screw pump consists of a central drive rotor with two meshing idler rotors; the three rotors are surrounded by a closely machined housing.

The flow through a screw pump is axial and in the direction of the power rotor. When the inlet side of the pump is flooded with hydraulic fluid, a certain volume of the liquid that surrounds the rotors is caught as the rotors rotate. This fluid is pushed uniformly with the rotation of the rotors along the axis and is forced out the other end. In operation, when the power rotor is turning clockwise, the idler rotors

are turning counterclockwise. This is true is all types of screw pumps.

The volume delivered by screw pumps is easy to maintain, because the fluid does not rotate but moves linearly. The rotors work like endless pistons which continuously move forward; thus, there are no pulsations even at higher speed. Figure 6–31 shows the construction of the screw type pump. The absence of pulsations and the fact that there is no metal-to-metal contact results in a very quiet operating pump.

Screw pumps are produced for volumetric capacities from 2 to 3400 gpm and for pressures up to 4000 psi. Operating speeds of this design have been as high as 24,000 rpm.

The larger pumps are used as low pressure, large volume prefill pumps on large presses. A typical example of this would be the fast closing phase of a cycle on a large hydraulic press as used in the plywood industry. These presses have wide openings to allow multiple stacking of the plywood sheets without difficulty. The press is then closed at a high rate of speed by using a 500 gpm screw-type pump. When the press is closed on the work, pressure unloads the 500 gpm screw pump and a low volume, high pressure pumping unit comes into operation to finish the pressing action.

Other applications include hydraulic systems on submarines and other uses where noise must be controlled.

Fig. 6-31. Screw pump-ASA symbol. (*Courtesy of Sier Bath.*)

Vane-type Pumps

Vane-type hydraulic pumps may be either fixed delivery or variable volume units.

The fixed delivery design usually has a pressure-balanced rotor which is slotted and contains the flat rectangular vanes. The rotor is splined to the pump shaft and rotates within a cam ring. Wear plates are fitted against both sides of the cam ring, thus making up the pumping element. The wear plates generally contain the fluid passages.

The balanced vane unit shown in Fig. 6–32 is popular for industrial applications requiring continuous operating cycles.

As the rotor is turned, centrifugal force causes the vanes to move out against the hardened and ground contour of the cam-shaped ring. Fluid is

Fig. 6-32. Vane pump–ASA symbol. (*Courtesy of Socony Mobil Oil Co.*)

OUTLET

INLET

DISCHARGE PORT

CAM RING

SUCTION PORT

SUCTION PORT

ROTOR

DISCHARGE PORT

trapped between the rotating vanes when they pass over the inlet port. Because the vanes reciprocate in and out of the rotor as they rotate, while passing over the inlet port the vanes are extended out of the rotor and carry a maximum amount of fluid. As the vanes reach the outlet port, the cam ring contour forces them back into the rotor and the hydraulic fluid must go out the discharge side of the pump.

Balanced vane units have two inlet ports and two discharge ports which are diametrically opposite. This means that both sides of the pump will be pressure-balanced, and equal forces will result on both sides of the rotor.

When resistance builds up in the system and pressure increases, small passages in the wear plates allow fluid under pressure to operate beneath the vanes during their delivery period to keep them held tightly against the cam ring. This helps to increase the efficiency of the pump during the work phase of an operating cycle.

Fixed delivery vane-type pumps are available as double pumps or two-stage pumps. Volumetric capacities of vane units are most popular up to about 60 gpm. Operating pressures are generally held to a maximum of 2000 psi and under.

The unbalanced vane type of hydraulic pump has a rotor which is eccentric to the housing cam ring. As the rotor containing the flat rectangular vanes is turned, hydraulic fluid entering the inlet port is trapped between the housing and vanes. When the vanes rotate to the outlet port, the space between the rotor and the housing cam ring lessens and the vanes are pushed back into the rotor. The fluid must then go out the discharge side of the pump and into the system.

Figure 6–34 shows the slipper-vane type pump which indicates the large number of varied designs that are available.

Most industrial unbalanced vane pumps manufactured are variable volume units. This is easy to accomplish by making the cam ring housing adjustable from maximum eccentricity to a concentric position where the vanes are held equal distant between rotor and the ring through 360 deg.

Most generally, a mechanical spring force is used to hold the housing cam ring eccentric with the rotor, which delivers maximum volume until the system's work load causes a certain pressure at the outlet side of the pump. When this pressure level is reached, forces due to this pressure within the pumping element push the housing cam ring and overcome the force of the spring. This moves the housing cam ring to a center position relative to the rotor, and the

Fig. 6-33. Vane pump–double balanced.
(*Courtesy of Denison Engineering.*)

Fig. 6-34. Unbalanced vane pump; (a) straight vane; (b) slipper vane.

pump delivery becomes nearly zero. Another popular industrial type pump is the variable vane design shown in Fig. 6–35.

Through the application of different spring forces the variable vane-type pump can be adjusted to reach zero delivery or "deadhead" at various pressure levels.

Hooke's law expresses the meaning that within the elastic limits of the spring material the ratio of the stress to the strain produced is a constant:

$$\frac{\text{Stress}}{\text{Strain}} = K$$

The curves in Fig. 6–36 indicate the pump delivery characteristic relative to the type of spring being used.

One of the advantages of the variable-volume-type pump is its low power consumption. The amount of power that is used varies with the work load on the system. This concept is termed "pressure compensated," which means that when the work resistance is the least the volume output is at a maximum. When the work resistance is the greatest, the pump delivery output is the least. It is important to repeat again, "Pressure in a hydraulic system is due to the resistance on the system." Also, "A hydraulic pump pumps the hydraulic fluid, not pressure."

Thus, from these facts, a pressure-compensated pump is one which changes its volume output with changes in system pressure.

Another adaptation of the variable vane design is the slipper vane pump. It features peripheral inlet passages through the cam ring to the center of the slipper-type vane. It also has a special discharge port element supported in the pump housing which absorb system hydraulic shock loads.

121

Fig. 6-35. Variable vane. (*Courtesy of Racine Hydraulics.*)

Fig. 6-36. Pump delivery characteristic relative to type of spring.

Piston Hydraulic Pumps

Piston-type pumps are the oldest form of hydraulic pump. They have been used successfully since about 1900 for naval gun turrets, steering control systems, and other shipboard uses. Piston pumps are now being used in practically all fields.

The piston design itself lends to excellent control of clearances and use of materials in manufacture. At the present time, these units provide the highest degree of sophistication found in the many hydraulic pump designs that are available.

Piston pumps offer the highest volumetric efficiencies (up to 97 per cent), higher pressure ratings (10,000 psi or higher), and operating speeds as high as 12,000 rpm. With their high operating efficiency less power is converted to heat, and better total system economy is provided with less size and weight.

Piston units are available for fixed delivery, variable volume, and in variable reversible flow pumps (over-the-center).

In general, there are three basic designs of piston pumps: radial, axial, and reciprocating plunger.

The radial hydraulic piston pumps operate on the principle of converting the rotary shaft motion to radial reciprocating piston motion (see Fig. 6–38).

Axial hydraulic piston pumps operate on the principle of converting rotary shaft motion to axial reciprocating piston motion (see Fig. 6–39).

Reciprocating plunger pumps may be horizontal or vertical piston pumps and operated by either a rotating or reciprocating power input shaft (see Fig. 6–40). They are generally large piston-type pumps as used in central hydraulic power systems or for extremely large volume installations.

Piston hydraulic pumps may also be classified for their method of porting fluid between inlet and outlet.

There are three common methods of porting fluid: valve plates, check valves, and pintle valve arrangements (see Figs. 6–52, 6–41, and 6–46). As a general rule, units designed with pintle valves or valve plates are adapted to both pumps and fluid motors, but check valve units are available only as pumps.

Radial Piston Pumps

The radial piston pump classification is based on the line of piston motion which is perpendicular to the centerline of the input shaft.

Check Valve Design

An example of the check valve design is shown in Fig. 6–42. When the pump shaft is rotated, fluid from the reservoir enters the inlet side of the pump. Fluid is then ported from the flooded crankcase through the inlet checks and into the piston chamber as the piston travels inward during one-half of one shaft revolution. During the other one-half of one revolution, the eccentric bearing forces the piston out-

TWO PIECE BODY AND CAP CONSTRUCTION PROVIDES
RIGID BEARING SUPPORTS AND FACILITATES PROPER
SHAFT ALIGNMENT--EXTRA CAPACITY ANTIFRICTION
BEARINGS.

PERIPHERAL SUCTION PASSAGES IN RING PERMITS THE
COMPLETE FILLING OF VANE CAVITIES BY FEEDING
FROM THE CENTER.

CARTRIDGE DESIGN** ALL ROTATING PARTS HAVE BEEN
COMBINED INTO ONE SEPARATE UNIT WHICH CAN BE RE-
MOVED AS A PACKAGE. THE VANE, PORTING RING, SIDE
PLATE AND ROTOR ASSEMBLY HAS THE PROPER RUN-
NING CLEARANCES MACHINED IN, ELIMINATING ANY NEED
FOR SHIMMING.
  1. LONG ROTATING SEAL BETWEEN REACTION RING
     AND SIDE PLATES.
  2. LEAKAGE ALONG SHAFT COMPLETELY ELIMINATED
  3. VANE EDGES SWEEPING SIDE PLATES ELIMINATED,
     SHORT VERTICAL VANE MOVEMENT ONLY ON SIDE
     PLATES.

FLOATING PORTING AND REACTION RING (HEART OF A
VARIABLE VOLUME VANE PUMP). ALL HYDRAULIC LOADS
ARE BALANCED OUT EXCEPT FOR A SMALL FORCE NEC-
ESSARY TO KEEP THE RING FIRMLY ON THE GUIDE BAR
AND A COMPONENT NECESSARY TO REACT AGAINST THE
SPRING GOVERNOR, PERMITTING MAXIMUM RESPONSE TO
SYSTEM FLUID DEMANDS.

O-RING SEAL BETWEEN SUCTION AND DRAIN AREAS PRE-
VENTING CAVITATION PROBLEMS RESULTING FROM AIR
BEING DRAWN THROUGH SHAFT SEAL.

SPRING GOVERNOR WITH OVER-RIDE LIMIT STOP FOR
ADJUSTING COMPENSATING PRESSURE AND PREVENTING
THE EXTREMELY RESPONSIVE REACTION RING FROM
OVER SHOOTING ON SUDDEN REDUCTION OF FLOW.

PATENT PENDING

COMBINATION GUIDE BAR AND DISCHARGE PORT ELE-
MENT, RIGIDLY SUPPORTED IN HOUSING, RELIEVES THE
REACTION RING FROM ANY HYDRAULIC LOADS AND PER-
MITS IT TO SLIDE FREELY ON A PRESSURE LUBRICATED
OIL FILM.

Fig. 6-37. Slipper vane pump. (*Courtesy of Hartmann Manufacturing Co.*)

Fig. 6-38. Radial action.

Fig. 6-39. Axial action.

Fig. 6-40. Reciprocating plunger.

VOLUME-
CONTROL
HAND
WHEEL

CONNECTING ROD

UNIVERSAL
JOINT SHAFT

STATIONARY
PINTLE

DRIVE
SHAFT

DRIVING
FLANGE

PUMP YOKE
CYLINDER BLOCK
VALVE PLATE

VIEW A

TO SYSTEM

TO SYSTEM

FROM RESERVOIR

Fig. 6-41. Principle of check valve porting.

PISTON
STROKE

VIEW
B

a

a

PISTON
STROKE

b

b

CYLINDER
BLOCK

DRIVING
FLANGE

VIEW C

ward, causing fluid delivery to the system through the discharge check valve. This design is a fixed delivery pump and is available in a number of different volumetric capacities.

It is also important to observe that each piston is essentially an independent pumping unit, which is generally true with all check valve piston pump de-

Fig. 6-42. Check valve design-symbol. (*Courtesy of Racine Hydraulics.*)

signs. These units can be assembled to supply fluid independently to several circuits simultaneously, if desired. The check valve radial piston pump previously shown can be operated at pressures up to 10,000 psi.

Pintle Valve Arrangement

The radial piston pump shown in Fig. 6–43 uses the pintle valve method of porting fluid. It is available as a fixed delivery, variable volume, or reversible vari-

Fig. 6-43. Radial piston pump. (*Courtesy of Oilgear Co.*)

able volume unit. The volumetric capacity of the fixed stroke type varies with the size of the unit and the rotational speed.

Input power applied to the pump shaft rotates the cylinder, piston, and rotor assembly. Centrifugal force keeps the pistons against the thrust rings to rotate the rotor with the cylinder. During the lower one-half of one revolution, the pistons move outward to allow fluid to enter the piston cavity through the inlet port and pintle arrangement. During the upper one-half of one revolution, the pistons are thrust inward to discharge the fluid from the piston cavity through the pintle and out into the system.

The pumping pistons (see Fig. 6–44) have a rounded head and are fitted in the cylinder at a slight angle. This causes the pistons to roll and reciprocate simultaneously. Both motions are uniformly accelerated and decelerated for smooth action.

The reversible variable unit shown in Fig. 6–45a provides an infinitely variable volume of fluid flow in either direction through the pump. The internal porting arrangement of the pump facilitates flow in either direction as the slide block race is positioned from side to side. The pump shaft rotates in one direction constantly; only the fluid changes direction.

This type of hydraulic pump can be controlled manually, mechanically, by fluid, electrically, or with a servo system to discharge the exact amount of fluid and to operate a cylinder or fluid motor in either direction.

This type of system is called a "closed loop" system.

Figure 6–45b illustrates a closed loop hydrostatic transmission with a replenishing pump. Both the fixed

Fig. 6-44. Piston. (*Courtesy of Oilgear Co.*)

Fig. 6-45(a). Variable reversible radial pump. With a double-rod-end cylinder, which has the same displacement on both ends, the reversing-pump circuit can theoretically consist of nothing but the pump, cylinder, and suitable pipe connections. The oil pumped from one end of the cylinder is delivered to the other. Speed is controlled by adjusting the pump stroke, and no reversing valve is needed. Actually, a small fixed-volume pump—usually a gear pump—and reservoir is also required to make-up main pump leakage losses and keep the entire system under positive pressure so that no air can get in. This "super-charging" gear pump, the make-up check valves, and the relief valve are sometimes built into the main pump. A rotary fluid motor may be used instead of a cylinder as shown. This type of circuit is used for variable-speed transmissions, ship steering equipment, and windlass drives. (*Courtesy of Socony Mobil Oil Co.*)

delivery and the variable-reversible pump are on the same shaft and driven by the electric motor. The small pump charges the system and then during the work phase of the cycle makes up any losses in the system's fluid due to leakage. The replenishing pump delivers fluid through the check valve system to the low pressure side of the circuit. Fluid not needed to replenish the system is discharged over the relief valve at relatively low pressure (usually 40 to 120 psi).

Another version of the pintle-type arrangement is shown in Fig. 6–46. The operating principle is the same as previously described.

Axial Piston Pumps

Axial piston hydraulic pumps constitute many different versions of converting rotary shaft motion to an axial reciprocating motion of a piston.

Both check valve and valve plate methods of porting fluid are used with axial piston pumps.

Check Valve Design

The check valve design illustrated in Fig. 6–47 is rotatable in either direction. The rotating input motion is converted to a reciprocating axial piston motion by the cam plate, which is keyed to and revolves with the shaft. One face of the cam plate is at an angle to the center line of the shaft.

Fig. 6-45 (b). *Principle of Variable-Discharge Radial-Piston Pump.* This type of variable-stroke pump has a cylinder block that rotates about a stationary spindle, or pintle. Several closely fitted pistons are carried around with the block, their outer ends being held in contact with a rotating outer ring by centrifugal force and by oil pressure. The outer ring is supported by bearings not shown. Suction and discharge ports in the pintle communicate with the open ends of the cylinders. When the ring and pintle are concentric, the latter can rotate without any movement of the pistons. However, if the ring is moved off-center, rotation causes the pistons to reciprocate in the cylinders, drawing oil from the suction ports on the outward stroke and delivering it through the discharge ports on the inward stroke. By shifting the position of the ring, it is possible to vary the length of stroke and the volume of oil delivered from zero to a maximum. Moreover, by adjusting the eccentricity of the outer ring either to the right or to the left (as shown), it is possible to reverse the direction of oil flow without stopping the pump or changing its direction of rotation. Any intermediate volume of flow in either direction can be secured by this adjustment. This principle of operation is employed in many makes of pumps; however, the exact manner of carrying out the principle varies widely in different designs. (*Courtesy of Socony Mobil Oil Co.*)

Fig. 6-46. Another pintle type arrangement. (*Courtesy of Socony Mobil Oil Co.*)

INLET PORT
PISTON RETURN SPRING
PISTON
OUTLET PORT
MOUNTING FLANGE
WOODRUFF KEY
DOUBLE OIL SEAL
RADIAL BEARING
THRUST BEARINGS
PISTON CAP
OUTLET CHECK BALL
INLET CHECK BALL

Fig. 6-47. Check valve axial piston pump. (*Courtesy of Dynex.*)

As the input shaft and cam plate assembly are rotated, the *thin section* of the cam slides under the piston cap. The piston return spring forces the piston toward the inlet side, causing a low pressure condition within the tightly sealed piston cavity. This causes the inlet check valve to open and allow fluid to enter the low pressure area of the piston cavity.

When the cam plate rotates 180 deg, *the heavy* section slides under the piston cap, which pushes the piston toward the discharge side. The pressurized fluid forces the discharge valve open, and the fluid goes out into the system. As the thin section again slides under the piston cap, the process is repeated.

Each piston contained in the cylinder barrel of this pump design is an independent pumping unit. This arrangement provides for multiple circuitry from the use of one pump.

The variable pump concept for the check valve axial piston design is shown in Fig. 6–48.

Volume is varied by by-passing a portion of the displaced fluid back to the inlet side on each piston stroke. A cam-operated by-pass valve connects each piston discharge cavity with a passage back to the inlet side. The pumping action is the same as in the fixed delivery pump, except for the by-pass feature.

The cycling phase relationship between the cam plate and the by-pass cam can be varied during pump operation. If both the cam plate and the by-pass cam are phased so that the by-pass valves are open during the full 180 deg of rotation when the pistons are moving toward the discharge side, the pump

output is zero. The fluid merely circulates through the open passages back to the inlet side. However, if the by-pass cam is phased with the full 180 deg of rotation when pistons are being returned by the springs on the inlet stroke, the by-pass valves will be closed on the discharge stroke and all the fluid will be delivered to the system. The by-pass cam can be

Fig. 6-48. Variable volume check valve pump. (*Courtesy of Dynex.*)

128

adjusted for any increment of overlapping to provide infinitely variable volume to the system.

This type of variable volume pump delivers flow in only one direction and is used with "open-loop" systems (see Fig. 6–49). Notice that in this circuit a directional control is used to change the direction of the fluid motor rotation.

The check valve design can also be used for multiple circuitry. It is also possible to install both fixed delivery and variable volume features in one pump (see Fig. 6–50).

Valve Plate Design

Figure 6–51 shows examples of the valve plate porting design of axial piston pumps. These pumps are available as fixed delivery, variable volume, and as variable-reversible units.

In this design a valve plate with two kidney-shaped ports are used, one which allows fluid to enter the piston chamber and the other to direct the fluid out into the system.

As the pump shaft and cylinder barrel rotate, the pistons maintain a fixed pumping action due to the cam angle. During one-half of one revolution the pistons move in a direction away from the kidney port, and fluid enters to fill the cavity of the cylinder bore. On the next one-half of one revolution the pistons are forced toward the kidney port, and fluid is forced out into the system.

The design of valve-plate-type axial piston pump in Fig. 6–52 consists mainly of a two-section housing which establishes a fixed angularity. The rotating elements maintain the fixed angle relationship during operation. This angle determines the distance through which a piston travels in its bore during shaft rotation. Because of the angular relationship between cylinder block and drive shaft, pistons reciprocate in their respective bores as they rotate with the cylinder block. Each piston accepts fluid during one-half of one revolution and forces it out during the other one-half of one revolution. The inlet and outlet ports located in the valve block lead into separate kidney-shaped slots which are mated with corresponding valve plate slots. The valve plate slots are also in phase with port openings in the cylinder barrel bores. Narrow walls separate inlet from outlet.

Units of the valve plate design type are readily available as variable volume units or as reversible-variable volume units.

When the cam angle or angular relationship between cylinder barrel and drive shaft can be adjusted on only one side of center, the pump is a variable

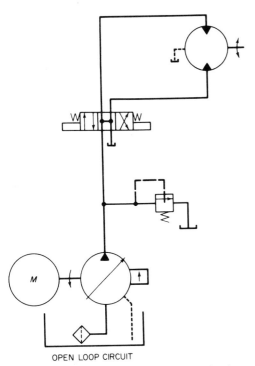

OPEN LOOP CIRCUIT

Fig. 6-49. "Open Loop" circuit.

Fig. 6-50. Multiple circuit from one pump.

Fig. 6-51. Valve plate-fixed delivery axial piston pump. (*Courtesy of Denison Engineering.*)

Fig. 6-52. Bent axis-axial piston pump-fixed delivery. (*Courtesy of Vickers Co.*)

THE BASIC PISTON PUMP IS USED IN BOTH FIXED AND VARIABLE DISPLACEMENT PUMPS. A ROTATING GROUP CONSISTING OF DRIVE SHAFT (A), UNIVERSAL LINKAGE (B), CONNECTING RODS (C), PISTONS (D) AND CYLINDER BLOCK (E) IS DRIVEN THRU SHAFT (A) TO PRODUCE PUMPING ACTION. A NON-ROTATING VALVE PLATE (F) PORTS INLET AND OUTLET FLUID TO AND FROM CYLINDER BLOCK (E).

CYLINDER BLOCK (E) AND DRIVE SHAFT (A) ROTATE IN UNISON WITH ALIGNMENT MAINTAINED BY UNIVERSAL LINKAGE (B). THE ANGLE FORMED BETWEEN THE AXIS OF CYLINDER BLOCK (E) AND DRIVE SHAFT (A) CAUSES RECIPROCATION OF PISTONS (D) WITH RESPECT TO THE CYLINDER BLOCK BORES. FIGURE 1 SHOWS THE RELATIVE PISTON STROKE DURING ONE HALF REVOLUTION WITH A CYLINDER BLOCK ANGLE OF "x". RELATIVE PISTON MOTION DECREASES AS ANGLE "x" APPROACHES ZERO. EACH PISTON THUS FUNCTIONS TO CREATE A VOID WHICH FILLS WITH FLUID DURING ONE HALF REVOLUTION AND DISCHARGES THIS FLUID DURING THE REMAINING HALF. THE AMOUNT OF FLUID DISPLACED PER REVOLUTION IS DEPENDENT ON ANGLE "x".

CYLINDER BLOCK (E) ROTATES AGAINST THE BEARING SURFACE OF VALVE PLATE (F) WITH CONTACT MAINTAINED INITIALLY BY SPRING FORCE. VALVE PLATE (F) IS PINNED TO PREVENT ROTATION AND TO POSITION IT PROPERLY IN RELATION TO THE PLANE OF THE CYLINDER BLOCK ANGLE. THE CYLINDER BLOCK PORTS ARE ALIGNED WITH THE SEMICIRCULAR PORTS OF VALVE PLATE (F). THE CLOSED SECTIONS BETWEEN THE SEMICIRCULAR PORTS OF (F) PROVIDE A SEAL AT CROSS-OVER POINTS WHERE PISTONS (D) REVERSE THEIR STROKE. RELATIVE PISTON STROKE WHILE CROSSING BETWEEN INTAKE AND DISCHARGE PORTS IS PRACTICALLY NIL.

VALVE PLATE AND CYLINDER BLOCK CONSTRUCTION IS SUCH THAT OPERATING PRESSURE, ACTING ON THE DIFFERENCE IN EFFECTIVE AREAS, ASSISTS THE SPRING IN MAINTAINING CONTACT.

PUMP FLOW REVERSES WHEN CYLINDER BLOCK (E) IS MOVED ACROSS CENTER TO A REVERSE ANGLE.

Fig. 6-53 (a). The component parts of an axial piston variable volume pump that is reversible to provide infinitely variable volume in either direction. This unit servo-controlled. *(Courtesy of Socony Mobil Oil Co.)*

Fig. 6-54. Variable volume axial piston reversible unit with valve plate method of displacing fluid. *(Courtesy of Denison Engineering.)*

Fig. 6-53 (b). This view shows the manual version of controlling pump delivery using the same bent axis concept. *(Courtesy of Socony Mobile Oil Co.)*

volume unit. This means that the pump delivers flow out one specific port. However, if the pumping angle can be adjusted on both sides of center (over-the-center), then the pump can deliver variable flow from either port for reversing fluid motors or cylinders (see Figs. 6–53a and b 6–54).

Fixed Delivery
Pump Selection Factors

1. Volumetric capacity (gpm).
2. Operating speed (rpm).
3. Pressure rating (psi).
4. Type of duty (continuous or intermittent).
5. Type of mounting.
6. Shaft size and type (splined or keyed).
7. Cost.
8. Type of fluid.
9. Operating temperature °F.
10. Design type (gear, vane, or piston).
11. Size and weight.
12. Operating conditions (shock, loads, etc.).
13. Environmental factors (temperature, dust, etc.).
14. Type of drive (direct coupled, belt and pulley, etc.).

Questions

1. Name several prime movers which are used to impart energy to hydraulic pumps.
2. Why are electric motors especially popular for supplying energy to a hydraulic system?
3. What is a power take-off?
4. When are gear reductions used to drive hydraulic pumps?
5. What precaution must be taken regarding the side-loading of pumps driven by belts, chains, or other devices?
6. How much hydraulic horsepower would a pump produce when working at 4000 psi and delivering 12 gpm? What size (in horsepower) electric motor would be selected to drive this pump if it were 90 per cent efficient?
7. What are the two main classes of pumps?
8. Name the three popular construction types of nonpositive displacement pumps.
9. What is meant by staging centrifugal pumps?
10. What is a positive displacement pump and

15. Inlet characteristics (atmospheric pressure, etc.).

Variable Volume
Pump Selection Factors

1. Volumetric capacity (0 to max. gpm).
2. Operating speed (rpm).
3. Pressure range (0 to max. psi).
4. Type of volume control (manual, pressure, servo).
5. Type of duty (continuous or intermittent).
6. Type of mounting.
7. Shaft size and type (splined, keyed).
8. Cost.
9. Type of fluid.
10. Operating temperature °F.
11. Design type (vane, piston).
12. Size and weight.
13. Operating conditions (shock, loads, etc.).
14. Environmental factors (temperature, dust, etc.).
15. Type of drive (direct coupled, pulley and belt, etc.).
16. Inlet characteristics (atmospheric pressure, etc.).

how does it differ from a centrifugal type of pump?
11. How is the volumetric efficiency of a positive displacement pump determined?
12. How is the theoretical output of a positive displacement pump determined?
13. What is mechanical efficiency and how is it determined?
14. Explain how atmospheric pressure forces hydraulic oil up into the pump inlet.
15. What is the recommended velocity for good inlet conditions when a pump is operating in normal atmospheric conditions?
16. List several reasons for pressurizing the hydraulic pump inlet.
17. What might happen to the inlet supply to a hydraulic pump if the air breather on a reservoir becomes dirty and clogged?
18. Describe the operation of a fixed delivery hydraulic pump. Describe the operation of a variable volume hydraulic pump.
19. List several construction types of fixed delivery hydraulic pumps.

20. When is it necessary externally to drain a hydraulic pump or motor unit?
21. Name three designs of external gear pumps.
22. Name two designs of internal gear pumps.
23. Describe the operation of a two-stage internal gear pump.
24. Why is the operation of a screw-type pump quiet?
25. Describe the operation of a balanced-vane-type pump.
26. Why is an unbalanced vane pump used when variable volume characteristics are important?
27. What is meant when a variable volume vane pump "deadheads"?
28. Explain the main differences between radial, axial, and reciprocating plunger piston pumps.
29. What are the three common methods of porting fluid from inlet to outlet in piston-type pumps?
30. Which methods of displacing fluid from inlet to outlet in a piston unit allows operation either as a pump or motor?
31. What is meant when a piston pump is said to be variable and reversible?
32. What kind of a pump is used in a closed loop system?
33. What is the purpose of a replenishing pump?
34. What is meant by multiple circuitry using a single pump?
35. List the important considerations in the selection of a pump for any given application.

LOBED ELEMENT   SPUR GEAR   HELICAL GEAR

HERRINGBONE GEAR   TANDEM GEAR   CRESCENT GEAR

GEROTOR TYPE   SINGLE SCREW   2 ROTOR SCREW

3 ROTOR SCREW   UNBALANCED DESIGN   BALANCED VANE

VARIABLE VANE

FIXED DELIVERY
(CHECK VALVE)

VARIABLE DELIVERY
(PINTLE)

FIXED DELIVERY (BENT AXIS)

FIXED DELIVERY (CAM PLATE)

FIXED DELIVERY (CHECK VALVE)

VARIABLE VOLUME (CHECK VALVE)

PINTLE USE

VARIABLE VOLUME

VALVE PLATE

(CHECK VALVE)

IN → → → → OUT

RECIPROCATING PLUNGER
(SINGLE)

RECIPROCATING PLUNGER
(DOUBLE)

CHAPTER 7

THE CONTROL
OF HYDRAULIC
POWER

Fluid power is power transmitted and *controlled* through use of a pressurized fluid.

In the above definition the "controlling" factor in the transmission of fluid power not only is important but provides a real challenge to the ingenuity of system designers. Through slight modification in the designs of standard components, many new techniques in the art of controlling hydraulic power are being discovered.

Primarily, controlling hydraulic power involves the pressure, direction, and volume of fluid flow.

Controlling fluid pressure concerns the maximum system operating pressure, or a pressure-sensing device that controls the direction or volume of the fluid. Because of the flexibility of applying hydraulic controls, they are often named because of what they do rather than what they actually are. Therefore, it is important to know the primary function of the various controls available and how they operate. Knowing the operational characteristics of a control leads to the discovery of new techniques in applying the control to any given hydraulic circuit.

Pressure Controls

Relief Valves

One of the most important pressure controls is the relief valve or safety valve. Its primary function is to limit system pressure. Relief valves usually establish or limit a working pressure. Safety valves operate only when there is a circuit malfunction. A relief valve may function in dual capacity as both relief and safety device.

There are two basic designs, which are the directly operated or inertia type, and the pilot-operated design (see Fig. 7–1).

The direct type of relief valve has two basic working port connections. Additional ports may be provided for convenience in making system connections. System pressure opposes the poppet, which is held on its seat by an adjustable spring. The adjustable spring is set to limit the maximum pressure that can be attained within the system. When pressure exceeds the spring setting, the poppet is forced off its seat and

DIRECTLY OPERATED   PILOT OPERATED

PILOT VALVE

PISTON

MAIN VALVE

FROM PUMP

TO RESERVOIR   TO MACHINE

TO RESERVOIR

FROM PUMP   TO MACHINE

Fig. 7-1. Relief valves-direct and pilot ASA symbols.

excess fluid in the system is by-passed back to the reservoir. When system pressure drops to or below established set value, the valve automatically reseats.

The pilot type of relief valve is a two-stage design sometimes referred to as a compound-type valve. The inline type has three port connections. The main pressure line passes straight through the valve body, and the secondary port is connected to the reservoir. The main spool is fitted with a pistonlike skirt. An orificed passage is provided to direct fluid at a controlled rate from one side to the other. This orifice may be in the piston section of the spool itself. System pressure operates on both sides of the piston, causing it to be pressure-balanced. Because the main spool is pressure-balanced, any increase in system pressure will operate on both sides of the piston, and the spool remains seated with a relatively light bias spring.

From the chamber on top of the piston, observe from the diagram, (Fig. 7–1b) there is a fluid passage that leads up to the small pilot valve. (A pilot valve is applied to operate another valve or control.) The pilot valve is also held on its seat by a spring that is adjustable.

When system pressure exceeds the setting of the pilot valve bias spring, the small pilot valve opens, permitting escape of pressurized pilot fluid from the top side of the main spool piston. This action causes the main valve to open because of the greater force below the piston. The orifice through the spool piston is smaller than the passage to drain, so the relief valve remains open until the system pressure drops and allows the pilot valve to close. The main spool again becomes pressure-balanced and assumes its normally closed position.

A choice between the directly operated and the pilot-type relief valve depends primarily on the application. When the pressure level within a circuit must remain relatively constant, the pilot type is generally used. This design gives less pressure differential during its normal cycle of operation. This cycle involves its cracking point to fully opened and then reseating when the pressure drops to normal.

Pilot-operated relief valves are popular in the machine tool field, where more exact pressure levels are maintained. This design also provides for greater flexibility in control.

Example. A vent line or drain connected to the chamber on top of the main spool piston and fitted with an on-off valve could be used to operate the relief valve remotely.

This technique is used for by-passing pump delivery back to the reservoir through the relief valve at the relatively low bias pressure created by the

Fig. 7-2. Dual pressure control.

spring during the idle time of a machine. As long as the vent line is open to the reservoir, the relief valve is held open at a low pressure, and flow from the pump is by-passed or circulated back to the reservoir, for the relief valve now provides the path of least resistance to the reservoir.

Another technique shown in Fig. 7–2 is dual pressure control.

The directly operated design type is often used as either an adjustable or nonadjustable safety valve. It is used to dampen hydraulic shock and to prevent overpressures in long lines or in branch circuits which may be subject to local intensification.

Relief valves must be properly sized and placed in the hydraulic system to provide maximum protection at all times. Generally, the main relief valve is installed close to the pump. Other relief or safety valves may be placed throughout the system if needed.

Once the relief valve is adjusted for the work cycle of a machine's operation, it should not be tampered with. Generally, the relief valve is set about ten per cent higher than the normal pressure necessary to perform the actual work done by the hydraulic circuit, unless it is used to establish the working pressure level, a case which might be encountered in a press or clamp circuit.

Sequence Valves

A sequence valve's primary function is to direct flow in a predetermined sequence. It is a pressure-actuated valve similar in construction to a relief valve. The sequence valve operates on the principle that when main system pressure overcomes the spring setting, the valve spool moves up, allowing flow from the secondary port (see Fig. 7–3).

A sequence valve may be direct or remote pilot-operated. These valves are used to control the operational cycle of a machine automatically or in response to a manual input signal.

Figure 7–4 shows an automatic sequencing operation using the system's pressure as an actuating signal. When the clamp cylinder moves out and clamps the work piece, resistance is met and pressure in the system increases. When the pressure reaches the spring setting in the sequence valve, it opens, and flow is directed out of the secondary port to operate the work cylinder.

A good rule to remember with almost all pressure control valves is, "When a pressure control valve operates—if the flow from the secondary port performs work or is pressurized—the valve must be externally drained."

It is obvious that trapped fluid in the spring

Fig. 7-3. Sequence valve.

chamber would not allow the valve spool to move upward, because hydraulic fluids are practically incompressible. It is, therefore, necessary to allow clearance flow or internal valve leakage to drain back to the reservoir unrestricted.

Unloading Valves

An unloading valve is used to permit a pump to operate at minimum load (see Fig. 7–5).

The unloading valve operates on the principle that pump delivery is diverted to the secondary port and back to the reservoir when sufficient pilot pres-

Fig. 7-4. Clamp and sequences. (*Courtesy Socony Mobil Oil Co.*)

Fig. 7-5. Unloading valve.

sure is applied to move the spool against the spring force. The valve is held open by pilot pressure until the pump delivery is again needed by the circuit. The pilot fluid applied to move the spool upward becomes a static system. In other words, it merely pushes the spool upward and maintains a static pressure to hold it open. When the pilot pressure is relaxed, the spool is moved down by the spring, and flow is directed through the valve and into the circuit.

The unloading valve is useful in systems having one or more fixed delivery pumps to control the amount of flow at any given time. It is especially popular for use in feed and traverse circuits where rapid approach, feed, and return strokes are needed.

Unloading valves also help to prevent heat buildup in a system, which is caused by fluid being discharged over the relief valve at its pressure setting. A well-designed hydraulic system uses the correct amount of fluid for each phase of a given cycle of the machine operations. When pressure builds up during the feed phase of the cycle, pilot pressure opens the unloading valve, causing the large pump to bypass its 15 gpm back to the reservoir. The check valve isolates the high pressure side from the low.

A typical application of an unloading valve is shown in Fig. 7–6. The total delivery of the two pumps in the figure is 20 gpm. This would provide 20 gpm for rapid advance, 5 gpm for feed, and 20 gpm for rapid return (see Fig. 7–7).

Example Problem. What size electric motor drive would be needed in the "unloading circuit" if the rapid advance operating pressure is 250 psi and the feed operating pressure is 2000 psi? (Rapid return is the same psi as rapid advance; assume pumps to be 100 per cent efficient.)

$$hp = gpm \times psi \times 0.000583$$
$$\text{Rapid advance } hp = 20 \times 250 \times 0.000583$$
$$\text{Rapid advance } hp = 2.91 \ hp$$
$$\text{Feed } hp = 5 \times 2000 \times 0.000583$$
$$\text{Feed } hp = 5.83 \ hp$$

Fig. 7-6. Application of unloading valve.

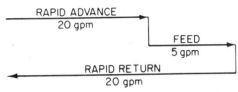

Fig. 7-7. Machining cycle for an unloading circuit.

A standard 5 hp electric motor could be used if the duty cycle were intermittent. Generally, it is good practice to select the next highest hp motor, which would be $7\frac{1}{2}$ hp.

This sample problem shows that using an unloading valve to eliminate the large pump during the high pressure feed portion of the cycle reduces the input horsepower. Without the unloading valve, the horsepower input would be

$$hp = 20 \times 2000 \times 0.000583$$
$$hp = 23.32 \; hp$$

By using the unloading valve, only 5 hp input is needed, about 18 hp being saved. This saves in the cost of the electric motor and prevents hp from turning into heat during the feed portion of the cycle.

A great number of hydraulic systems in all fields of usage are faced with the common problem of fast approach or high volume at low pressure, then low feed strokes during the high pressure work phase of the cycle. It is always the prime target of the circuit designer to conserve horsepower and prevent system heat buildup. If an unloading circuit is not feasible, a more sophisticated approach is necessary, using variable volume pumps.

Counterbalance Valves

A counterbalance valve is used to maintain back pressure to prevent a load from falling. Common applications include vertical presses, loaders, lift trucks, and other machines that must position or hold suspended loads.

The counterbalance valve operates on the principle that fluid is trapped under pressure until pilot pressure, either direct or remote, overcomes the spring force setting in the valve. Fluid is then allowed to escape, letting the load descend under control (see Fig. 7–8).

This same type of pressure control valve is used as a "braking valve" for decelerating heavy inertia loads. It can be applied to either linear or rotary motion.

When the counterbalance valve is used on large vertical presses, it may be important to analyze the source of pilot operating pressure. Figure 7–9 shows

Fig. 7-8. Counterbalance valve.

Fig. 7-9. Comparison between direct pilot and remote pilot operation.

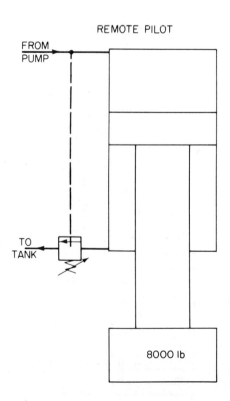

a comparison between direct pilot and remote pilot operation.

Through the application of Pascal's law P (psi) = force/area = 800 psi, a pressure of 800 psi is needed to support the heavy load. If pilot pressure is taken directly, then the counterbalance valve will operate at about 800 psi or slightly higher because of inertia and friction. In the other case, where remote pilot pressure is taken from the pressure line at the top of the cylinder, a choice of the operating pressure can be made for the valve. A counterbalance valve is a normally closed valve and will remain closed until acted upon by the remote pilot pressure source. Therefore, a much lower spring force can be selected to allow the valve to operate at a lesser pilot pressure. It should also be noted that the press member cannot move downward unless flow from the pump is directed into the top of the cylinder; this a normal function of the machine.

A number of the pressure control valves manufactured can be used as relief, sequence, unloading, counterbalance, or braking valves by making some modification in the drains, pilot connections, or other slight changes. Most designs offer unlimited applications because of their designed flexibility.

Pressure-reducing Valves

A pressure-reducing valve is used to limit its outlet pressure. Usually these valves are available in various operating ranges and are adjustable. Their application is important where limited pressure must be controlled for such things as clamping light metal objects, or other light duty uses.

Reducing valves are also used for the operation of branch circuits, where pressure may vary from the main system pressures. The pressure-reducing valve is normally an open-type valve. A free flow passage is provided through the valve from inlet to secondary outlet until a signal from the outlet side tends to throttle the passage through the valve. The valve operates on the principle that pilot pressure from the controlled pressure side opposes an adjustable bias spring normally holding the valve open. When the two forces are equal, the pressure downstream is controlled at the valve setting. Thus, it can be visualized that if the spring has greater force, the valve opens wider; if the controlled pressure has greater force, the valve moves toward the spring and throttles the flow (see Fig. 7–10).

When branch circuits are used for clamping heavy work pieces or for other important uses, the circuit designer must analyze the safety aspects of his de-

IN (HIGH PRESSURE)

DRAIN

OUT (LOW PRESSURE)

IN →

MAIN PRESSURE
SIDE

OUT →

CONTROLLED
PRESSURE

Fig. 7-10. Pressure reducing valve.

sign. There have been cases where pump failure due to the work cycle caused accidental unclamping of heavy pieces, which caused serious damage. Sometimes a separate clamp system, independent of the main hydraulic system, should be considered.

Directional Control Valves

Directional control valves have a primary function to direct or prevent flow through selected passages. They are used to extend, retract, position, or reciprocate hydraulic cylinders and other components for linear motion.

When rotary motion is required, directional controls are used to operate fluid motors in either direction with smooth reversals.

When directional controls are used to operate other controls, they are called pilot directional valves.

Directional control valves range from a simple shutoff valve (Fig. 11a) to a complex, multiple-segment, banked valve package (Fig. 11b) which is used for operating a number of circuits from one hydraulic power source. Generally speaking, each linear or rotary circuit in an open-loop-type system is operated independently by a directional control valve or segment of a multiple control.

Directional control valves as shown in Fig. 12a

and 12b, are used to direct the fluid through selected passages to or from actuators during the work.

Valves may be classified as one-way, two-way, three-way, and four-way valves with manual mechanical, pilot, solenoid, or servo actuators. Many of these valves are standard units with a combination of solenoid, pilot, mechanical, and manual actuators.

There is a valve or combination of valves for every purpose, with special ones available for handling exotic fluids, high temperatures, high pressures, or to solve problems concerning weight or space.

One-way Valves

The simplest directional valves are the one-way valves or check valves (see Fig. 7–13). These units allow flow in one direction through the valve and are usually designed for in-line or right-angle connections. They are used primarily to allow free flow around other controls when actuating cylinders and fluid motors are being reversed (see Fig. 7–14).

Check valves are also built with various spring forces to act as low pressure relief valves, or to create a *pressure drop* between the system and the return line to the reservoir, for use as pilot pressure (see Fig. 7–15).

Two-way Valves (Straightway designs)

Two-way valves, or on-off valves, are used primarily as simple shutoff valves. These valves are of the seating action design which utilizes a gate, disc, ball, or other device to obstruct the flow path. Generally, fluid flow can be in either direction through the valve. However, some two-way valve designs specify the pressure side from the downstream side.

Gate valves are generally restricted to low pressure applications, whereas globe- and ball-type valves are used for high pressures. Gate valves are used as shutoff valves and should be operated either fully opened or fully closed. Otherwise, high flow rates or hydraulic shock may crack or break off the wedge-like gate. Globe and ball designs may be adjusted to provide flow regulation where accuracy is not of prime importance. They are manufactured in almost all standard port configurations and are used to connect or isolate component parts of a hydraulic system.

Two-way valves may also be of the spool or poppet design and are available as normally opened or as normally closed valves. They are also available as solenoid, pilot, mechanical, or manually actuated (see Fig. 7–17).

Fig. 7-11 (a). Shows in the following order: Simple shut-off valve *(Republic Valve.)*; Directional control valve; Directional control valve bank; Dual directional control valve; Dual directional control valve *(All courtesy of Commercial Shearing.)*

Fig. 7-11 (b). Combination solenoid and manual operated directional control assembly. *(Courtesy of Racine Hydraulics.)*

Fig. 7-12. *Constant-Volume Hydraulic System.* Elements common to many hydraulic systems are shown: Reservoir, pump, piping, control valves, and hydraulic cylinder, or "motor." The hydraulic fluid, is the "life blood" of such systems. In view *A*, the four-way directional-control valve is in its forward position so that hydraulic fluid is delivered by the pump to the left side of the piston, causing it to move to the right on its work stroke. Hydraulic fluid from the right side of the piston flows through the four-way valve to the reservoir. A relief valve controls maximum pressure. In view *B*, the four-way valve is in its reserve position, and hydraulic fluid is delivered by the pump to the right side of the piston, causing it to move to the left on its retraction stroke. Hydraulic fluid from the left side of the work cylinder flows through the four-way valve to the reservoir. *(Courtesy Socony Mobil Oil Co.)*

Fig. 7-13. Check valves-ASA. *(Courtesy of Republic Mfg.)*

Fig. 7-14. Check valve application.

Fig. 7-15. Check valve for pilot pressure.

GATE VALVE

A S A SYMBOL

GLOBE VALVE   BALL VALVE

Fig. 7-16. Two-way valves.

SINGLE PILOT CHECK   DOUBLE PILOT CHECK

Fig. 7-18. Double and Single pilot check-ASA. *(Courtesy of Republic Valve.)*

Pilot-operated Check Valve

A pilot-operated check valve is a design of two-way valve which allows free flow in one direction but prevents reverse flow until actuated by pilot pressure or other means. They are available with either single or double checks (see Fig. 7–18).

These valves are applied as a safety device to prevent an accidental occurrence of a cycle, or phase of a cycle, out of sequence. In other words, they may be used as a safety interlocking device. They are also used for controlling the sequence of a machining cycle and to prevent a load from dropping (see Fig. 7–19).

Three-way Valves

A directional control valve whose primary function is alternately to pressurize and exhaust *one* working port is called a three-way valve. Generally, these valves are used to operate cylinders or fluid motors which need power in only one direction (see Fig. 7–20). Single-acting cylinders are used when loads are extended or positioned by fluid and retracted by some other means, such as a spring, or by the weight of the load itself.

Three-way directional valves are available for manual, mechanical, pilot, solenoid, or servo actuation. There are three basic designs, which include the spool, poppet, and shear seal types (see Fig. 7–21).

Three-way valves may be two-position or three-position. Most commonly they have only two posi-

Fig. 7-17. Solenoid, Pilot, Mechanical and Manual spool 2-way and poppets.

SOLENOID   PILOT   MECHANICAL

CAM

MANUAL

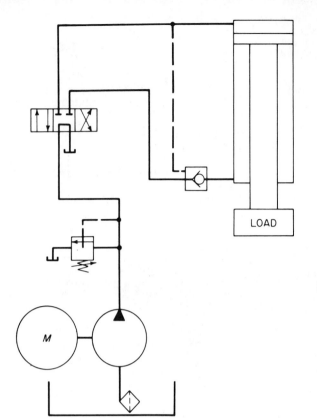

Fig. 7-19. Use of pilot operated check valve.

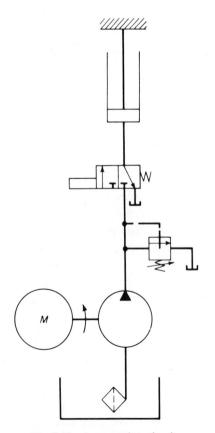

Fig. 7-20. 3-way valve circuit.

SPOOL TYPE   *3 pos 4 way*

SLIDE PLATE

ROTARY PLATE

SHEAR SEAL TYPE

POPPET TYPE

Fig. 7-21. Spool, poppet and shear seal type 3-way valves.

148

Fig. 7-22. Operation of spool type, two-position and 3-way valves.

tions, but in some cases a neutral position may be needed. Figure 7–22 shows the operations of a spool type, two-position, three-way valve.

The three-way-valve working ports are inlet from the pump, outlet to the cylinder, and exhaust to reservoir. These ports are generally identified by P = pressure; C = cylinder; and T or R = tank or reservoir. Sometimes the letters A or B are used to identify the ports connected to an actuating cylinder or fluid motor.

When port identification cannot be readily found, blow low pressure air through the various ports and shift the valve from one position to the other to determine port connections. (Make sure that clean dry air is used.)

A three-way valve may be used to direct pressurized flow from one system to another i.e., a tractor hydraulic system may be used alternately to actuate a front-end loader or a plow control mechanism. The diversion-type three-way valve will direct the pump delivery to the system that is to function at a specific time.

In cases of emergency or other reasons, two three-way valves can be used to reciprocate a double-acting cylinder or for reversing fluid motors (see Fig. 7–23).

Fig. 7-23. Two 3-way valves can operate as a 4-way valve.

ROD END PUMP HEAD END

TO RESERVOIR

WORK STROKE

RETURN STROKE

NEUTRAL

OPEN-CENTER TYPE
VALVE IN NEUTRAL

Fig. 7-24. Operation of 4-way valve.
(Courtesy of Socony Mobil Oil Co.)

Four-way Valves

The type of the valves generally used to operate cylinders and fluid motors in both directions are four-way valves. These valves are manufactured as two-, three-, four-, five-, and even six-position valves in the spool, poppet, or shear seal designs. The primary function of a four-way directional control valve is alternately to pressurize and exhaust two working ports (see Fig. 7–24).

These valves are also available with a choice of actuators, including manual, mechanical, solenoid, servo, pilot, or in any combinations.

Four-way valves can easily be converted to three-way valves by plugging one of the working ports.

Two-position four-way valves are used to reciprocate and hold an actuating cylinder in one position. They are also used on machines where fast reciprocating cycles are needed. Since the valve actuator moves such a short distance to operate the valve from one position to the other, this design is used for punching, stamping, and for other machines needing fast action.

Figure 7–25 shows a two-position four-way directional control valve holding a cylinder in the "up"

position. According to fluid power standards, the connecting lines are shown in the position with the spring. It means that the valve assumes this position unless actuated manually by an operator. This applies for any means of actuation.

When fast actuation is needed, electrical solenoids, pilot fluid, and/or mechanical springs are used to shift the four-way valve from one position to the other (see Fig. 7–26).

When electricity is supplied alternately to the two solenoids, magnetic force causes an iron core to be pulled into the magnetic field. Since the valve actuator is attached to the iron core, it shifts back and forth as the solenoids are energized and de-energized.

It is possible to actuate a four-way valve with both a solenoid and mechanical spring plus a manual override. In Fig. 7–27 the solenoid is energized to position the valve in one direction; then when de-energized the spring force moves it back, thus reciprocating the valve spool. The manual lever can also be used to override the solenoid or spring in an emergency.

Pilot fluid is also an excellent means of positioning control valves. The fluid may be either a liquid

Fig. 7-25. Two-position 4-way directional control valve holding cylinder in up position.

Fig. 7-26. Solenoid-actuated 4-way valve. *(Courtesy of Racine Hydraulics.)*

Fig. 7-27. 4-way valve with solenoid and mechanical spring-plus manual override.

Fig. 7-28. 4-way valve-needle valve to control shifting.

151

or a gas. Air-piloted hydraulic valves provide rapid operation, freedom from fire hazard, and independent power source for the pilot function.

Figure 7–28 shows a four-way valve where the rate of shifting the spool from one side to the other can be controlled by a needle valve. Fluid entering the pilot pressure port on the X end flows through the check valve and operates against the piston. This forces the spool to move toward the opposite position. Fluid in the Y end passed through the adjustable needle valve and is exhausted. The amount of fluid bled through the needle valve controls how fast the valve will shift. Some machines, where heavy work loads are encountered, cannot be reversed too rapidly. Pilot-controlled valves, such as the one described, provide flexibility to adjust machine reversals.

Figure 7–29 shows a typical example of using air as the pilot fluid. In this case, limit switches are used to energize and de-energize the air pilot valve solenoids. Their operation is similar to any off-on electric switch, such as the three-way type used between a house and garage. Either switch can be operated to supply electricity.

Limit switches are generally located on a machine to provide the proper stroke length and are actuated automatically by the movement of the machine.

As solenoids A and B of the pilot valve are energized and de-energized by the limit switches, air

is alternately supplied to either end of the main hydraulic four-way valve to reciprocate the machine table automatically.

In recent years, air controls have become available to replace electrical solenoids, switches, and other devices to increase service life and eliminate potential fire, explosion, and shock hazards.

Figure 7–30 shows a four-way hydraulic control valve being actuated automatically by air components.

Air controls are clean and safe to use, having a minimum of moving parts. Therefore, air controls eliminate much of the maintenance generally found with electrical devices. Burned-out solenoids, fire hazards from switches arcing, and wear of moving parts are common problems with electrical controls.

Directional controls may also be actuated with hydraulic fluid obtained directly from the system or from an auxiliary circuit. In the circuit shown in Fig. 7–31, as the cylinder reciprocates, the pilot directional valve is operated by mechanical means to supply fluid to either end of the main four-way valve. This causes the machine to reciprocate automatically, and it will not stop until the pump is shut off. Automatic cycling of a machine is accomplished with controls that allow the electric motor to operate continuously because of the high starting torque requirements.

Directional control valves with a center position allow the machine to stop without shutting down the entire system.

Fig. 7-29. Air pilot hydraulic circuit.

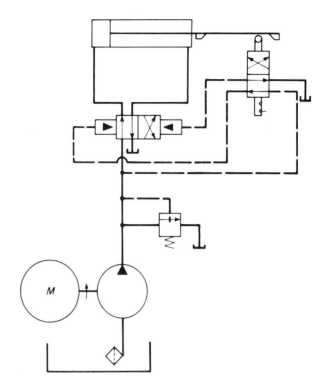

Fig. 7-30. Hydraulic 4-way operated valve with air controls.

Fig. 7-31(b). Circuit diagram.

Fig. 7-31 (a). The method of shifting the directional control valve using a rotary pilot valve which is cam-operated by machine members. (*Courtesy of Socony Mobil Oil Co.*)

WORK STROKE   RETURN STROKE

153

Fig. 7-34. 3-position open-center valve-hydraulic circuit.

Fig. 7-32. Directional valve center chart.

Fig. 7-33. Open-center circuit.

Valves with three or more positions provide flexibility for starting, stopping, reversing, holding, positioning, speed control, stabilizing, and many other operational characteristics.

The common types of standard three-position four-way valves are available in the following center configurations:

Open	Closed	Tandem-P to R
P to A	B to R	P to B
A to R	A and B to R	P to A and B

A variety of center configurations provides great flexibility for system designers.

Open Center Circuit

(All ports are open to each other in the center position.)

When a three-position, open-center-type valve is used in a hydraulic circuit, pump flow is directed to the reservoir when the valve is in the center. The other two positions are used to reciprocate the cylinder (see Fig. 7–34).

Open center valves help to prevent heat buildup in the system by allowing the pump flow to go back to the reservoir at a minimum pressure during the idle time of a machine. However, it should be understood that the hydraulic fluid from the pump always follows the path of least resistance, and no work can be done by any part of the system depending on the pump so long as the valve remains in the center position. This means that pilot valves or other auxiliary devices using pressure energy cannot function. Air pilot or separate hydraulic pilot sources are often used with open center systems.

Closed Center Circuit

(All ports are closed to each other in the center position.)

When a three-position, closed-center-type, four-way valve is used in a hydraulic circuit, pump flow must go over the relief valve when it is not being used to move a cylinder or fluid motor. A variable displacement pressure-compensated pump must automatically reduce delivery rate under the same circumstances.

Closed center versions are used when multiple circuits or functions must be accomplished from one hydraulic power source. There are many techniques used to prevent heat buildup when closed center valves are used. Venting or unloading through a pressure control is one method which was discussed previously.

Fig. 7-35. Closed center circuit.

Fig. 7-36. Tandem center 4-way valves.

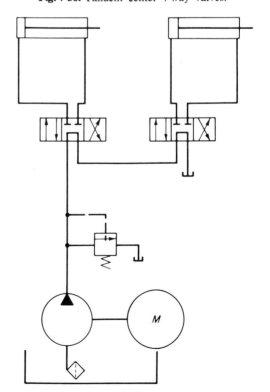

Variable displacement pumps with pressure-actuated controls are another economical method of avoiding heat and horsepower loss.

Tandem center valves direct the pump flow out the reservoir port with the other two working ports closed when in the center position. The application of this design may be to hold a cylinder or fluid motor under load or to permit the pump flow to be connected to a series of valves for multiple circuitry.

In addition to the flexibility in types of centers, directional valves are built with many other special features, such as integral restrictors, load-holding check valves, and relief valves (see Fig. 7–37). They are also available as banked units providing several valves in one package. Banked valves are commonly used on mobile equipment, where several circuits are operated from a central hydraulic power source (see Figs. 7–38 through 7–41).

Three- and four-way directional controls are available in a wide range of sizes and pressures.

Fig. 7-37. ASA symbols of integral restrictor, load holding check valve and relief valve.

Spool-type valves are available for pressures up to 3000 psi with nominal valve sizes up to 3 in.

Poppet and shear seal types are generally available for pressures up to 10,000 psi, but in smaller nominal valve sizes.

The flow capacity of a directional control valve depends on the design type and the size of ports. Most generally, valves are selected on the basis of their port size. However, pressure drop across the working ports of a valve is important. If the valve is too small for the flow, excessive pressure drop causes power losses. If the valve is too large for a given flow, it adds excessive cost and more size and weight to the circuit.

Pressure drop is generally determined by the valve designer through actual bench testing. These data are available from the manufacturer.

A number of the control valve manufacturers use the flow coefficient Cv to express valve capacity and flow characteristics. The Cv coefficient of a directional control valve is defined as the flow of water at 60°F in gallons per minute at a pressure drop of one pound per square inch across the valve.

The following formula based on the Darcy equation is generally useful for determining valve size.

$$Cv = \frac{Q}{\sqrt{\dfrac{\Delta P}{SG}}}$$

$Cv =$ capacity factor (above definition).

$Q =$ gallons per minute.

$\Delta P =$ pressure drop.

$SG =$ specific gravity (water $= 1 = 62.4$ lb/cu ft).

$$\left[SG = \frac{\text{fluid density (lb/cu ft)}}{62.4}\right]$$

The flow in gallons per minute of any liquid having a viscosity close to that of water at 60°F can be determined as follows:

$$Q = Cv \sqrt{\Delta P \frac{(62.4)}{(\rho)}}$$

$\rho =$ density of fluid in lb per cu ft

The pressure drop ΔP can be computed from the same formula changed to

$$\Delta P = \frac{\rho}{62.4} \frac{(Q)^2}{(Cv)^2}$$

There have been many improvements in valve designs which have increased the flow capacity of controls so that smaller-sized pipe ports can be used. The chart below can serve as a guide for typical size and capacity of valves.

Fig. 7-38. Valves banked for multiple circuitry.

Fig. 7-39. Lift truck and circuit showing directional control valve with built-in relief and load handling check valve. (*Courtesy of Socony Mobil Oil Co.*)

Fig. 7-40. Back-hoe and front end loader. *(Courtesy of Formrite Tube.)*

are used to obtain any amount less than maximum speed. This means that fluid energy developed in the hydraulic system that cannot be used for useful work may be converted into heat.

Example. An actuating cylinder must extend 12 in. in the first second and 6 in. in the next second. What is the maximum gpm needed if the cylinder has a 2 in. bore?

Solution: Area = 0.7854 × dia²

Area = 3.14 sq in.

$$Q(\text{gpm}) = \frac{A \times \text{length (in.)} \times 60 \text{ sec}}{231 \times \text{time (sec)}}$$

$$\text{gpm to extend first sec} = \frac{3.14 \,(\text{in.}^2) \times 12 \text{ in.} \times 60 \text{ sec}}{231 \text{ in.} \times 1 \text{ sec}}$$

$$\text{first sec gpm} = 9.7$$

$$\text{gpm for next 1 sec} = \frac{3.14 \text{ in.} \times 6 \text{ in.} \times 60 \text{ sec}}{231 \text{ in.}^3 \times 1 \text{ sec}}$$

$$\text{gpm} = 4.8 \text{ gpm/sec}$$

Thus, it can be seen that a 9.7 gpm pump would be needed for the maximum speed during the 12 in. stroke, whereas only 4.8 gpm would be needed for the 6 in. stroke. The pump would be selected on the basis of the 9.7 gpm. Therefore, at the end of the first second, 4.9 gpm would have to be diverted from the system, or a pressure-compensated pump, if used, would need to reduce output flow accordingly.

The speed of the cylinder for the 6 in. stroke can be controlled in three different ways:

1. Meter the fluid flow into the cylinder.
2. Meter the fluid flow out of the cylinder.
3. Bleed off the excess fluid flow and discharge it back to the reservoir.

If meter out (2) is used, fluid could be discharged through the control with a maximum pressure drop. In other words, the fluid leaving the system via the control valve may have maximum pressure on one side and atmospheric pressure on the reservoir side, depending on load conditions.

Metering the fluid into the cylinder (1) will re-

Fig. 7-41. Banked solenoid valves. *(Courtesy of Racine Hydraulics.)*

Valve (port) pipe size	(Valve capacity) Nominal rating (gpm)
$\frac{1}{8}$ in.	up to 2 gpm
$\frac{1}{4}$ and $\frac{3}{8}$ in.	up to 8 gpm
$\frac{1}{2}$ and $\frac{3}{4}$ in.	up to 20 gpm
$\frac{3}{4}$ and 1 in.	up to 40 gpm
$1\frac{1}{4}$ and $1\frac{1}{2}$ in.	up to 85 gpm
2 in.	up to 180 gpm
3 in.	up to 320 gpm

If in doubt, consult the manufacturer.

Flow Control Valves

A flow control valve has the primary function of controlling the rate of flow. Basically, the amount of fluid delivered to a given actuating cylinder or fluid motor determines the speed in which it will do the work.

When fixed delivery pumps are used for a hydraulic system, the size of the pump is selected to provide a maximum of speed. Flow control devices

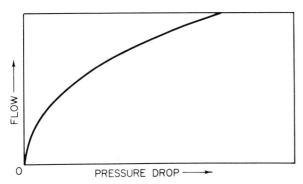

Fig. 7-42. Flow vs. pressure drop across a fixed orifice.

Fig. 7-43. Restricting devices-ASA symbol, orifice plate and check valve.

flect system pressure on the inlet and work load resistance on the outlet.

Flow and pressure drop across a control are related, and as flow in the system increases, pressure drop increases as the square of the flow changes. The reciprocal is also true in that as the pressure drop increases, flow increases in proportion to the square root of the change in pressure.

When flow versus pressure drop is plotted for a fixed orifice, the curve in Fig. 7-42 results.

In spite of the inefficiencies in applying flow control devices, they are useful and important fluid power components. Variable volume pumps can only be used as a flow control means if a pump is limited to one controlled cylinder or motor. If several flow control valves are used in parallel, a compensated variable delivery pump will automatically adjust to the needed volume, if it has sufficient capacity to actuate the multiple units at maximum value. If there are two or more operations being done at the same time, a separate power source is needed for each operation.

In contrast, a fixed delivery pump sized for the maximum speeds could supply several circuits with each actuator speed controlled by a flow device.

Restrictors

Restricting devices may be an orifice plate or a check valve with a small hole drilled through it. These devices are classified as noncompensated flow controls (see Fig. 7-43).

In some cases, simple on-off valves are drilled with a small hole to provide an orifice control (see Fig. 7-44). These are the simplest types of flow control devices. They are used to create a fixed rate of pressure drop to operate pilot valves, to stabilize a system with a slight back pressure, to prevent control overrides, and for decompressing systems under high pressure.

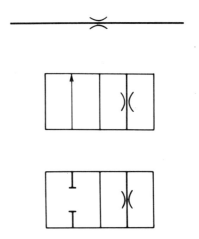

Fig. 7-44. Orifice control.

A needle valve is a variable restrictor device which allows the orifice size to vary by adjustment (see Fig. 7-45). Some designs incorporate a check valve and a needle valve which provides adjustable flow in one direction and free flow in the opposite direction. Flow control devices, such as the needle valves and the combination needle and check valves, are used for controlling cylinder and fluid motor speeds where small adjustments are essential. However, when the work load in a system varies, system pressure also varies with each change in resistance, and precise accuracy in speed control cannot be maintained. This is because flow inceases with pressure drop across the fixed orifice, and each time the system pressure changes the flow also changes.

CONTROLLED FLOW

FREE FLOW

Fig. 7-45. Needle valve (3).

In some applications these designs are used to obtain equal speeds with two or more cylinders or fluid motors.

Noncompensated flow controls are also used on some mobile equipment where higher pressures and low volume are required. Generally, in mobile applications several circuits are operated by a central hydraulic power source. Flow controls are used to help balance the entire system and to counter for uneven loads which normally cause fluid energy to seek the path of least resistance. They are also used to prevent gravity loads from dropping too rapidly.

Pressure-compensated Flow Controls

Because flow through an orifice or valve varies as the square root of the pressure drop across it, any change in pressure at the control inlet or outlet changes flow through the valve. Pressure-compensated flow controls adjust automatically to pressure changes and maintain a constant pressure drop from inlet to outlet, thus providing constant flow.

Principle of Pressure-compensated Flow Control

Assume that the pump capacity in Fig. 7-46 is 5 gpm and that the orifice is sized to pass 3 gpm.

When the load is zero, 3 gpm is going into the system and 2 gpm is going back to the tank over the by-pass valve. Since the by-pass poppet has 1 sq in. of area and the spring has a force of 50, gauge A will read 50 psi. Gauge B will read zero because of no load. There will be a 50 psi pressure drop across the metering orifice.

If the system load increased, causing gauge B to read 100 psi, then this same increase in pressure would be added to the by-pass valve spring through the pilot line. This means that gauge A would now have to read 150 psi to by-pass the extra 2 gpm back to the tank. Since the pressure difference between gauge A and gauge B remains the same (50 psi pressure drop), the flow across the metering orifice remains the same. Thus any increase or decrease in load will either add or subtract to the spring, maintaining constant pressure drop and flow across the metering orifice.

One type of pressure-compensated flow control regulates or restricts fluid flow entering the valve. The other type by-passes the excess fluid back to the reservoir. In both types a constant pressure drop across the metering orifice provides constant flow (see Fig. 7-48).

Fluid entering the control at the inlet port is restricted at the needle orifice. Pressure builds up, causing the compensator spool to move toward the spring force, which is generally about 40 lb. When the spool moves in the direction of the spring, fluid is restricted across the inlet port.

When fluid passes across the needle orifice, back pressure builds up on the spring side of the compensator and the compensator spool is forced to the right, which opens the restriction at the valve inlet.

Fig. 7-46. Principle of pressure compensation.

Thus, the compensating spool is pressure-balanced, regardless of any changes in system pressure. Changes in pressure affect both sides of the compensator because of the equal areas. In this case the spring creates a constant pressure differential from the inlet to the outlet. In other words, the inlet pressure is always slightly higher, only because of the spring force.

The by-pass type of flow control valve maintains constant flow across the metering orifice and by-passes excess fluid back to the reservoir at a pressure only slightly above the system pressure.

$$A_1 = A_2 + A_3$$

In this design type, shown in Fig. 7–48b, the compensator spool is held closed by the spring in A_1. Fluid can still get past the spool and across the

Fig. 7-48. (a) Restrictor inlet type pressure compensated flow control.

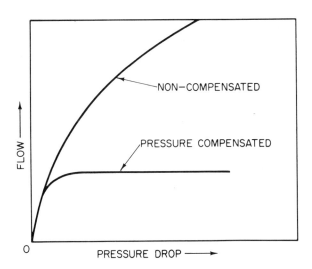

Fig. 7-47. Curves comparing pressure drop.

metering orifice to cause back pressure to build up on the A_1 side. Since the areas are equal on both ends of the compensator spool, it becomes pressure-balanced. However, since the spring force is relatively light and the spool is pressure-balanced, a slightly greater force on the inlet side opens the compensator to by-pass the excess fluid.

With the compensator spool balanced, any change in system pressure does not affect the flow rate across the metering orifice.

If the pressure on the outlet side exceeds the pilot relief valve setting, it opens and causes fluid to bleed off the spring side of the compensator, and with the spool unbalanced it opens wide to by-pass the system fluid directly to the reservoir.

This design has one main advantage over the restricted inlet type. The by-pass design allows excess

fluid to by-pass back to the reservoir at a slightly higher pressure than the system's working pressure. However, the restricted inlet design causes the surplus fluid to by-pass over the main relief valve at its pressure setting. This means that during periods when any

Fig. 7-48. (b) By-pass type of flow control.

Fig. 7-49. Metering-in circuit.

Fig. 7-50. Metering-out circuit.

portion of the pump delivery is restricted it blows across the main relief valve at its maximum pressure setting.

Metering-in Circuits

Metering-in circuits control the flow of fluid before it enters the working cylinder or fluid motor (see Fig. 7–49).

Metering-out Circuits

Metering-out circuits control the flow exhausting from a working cylinder or fluid motor (see Fig. 7–50).

Bleed-off circuits control the fluid by bleeding off the excess not needed by the working cylinder or fluid motor (see Figs. 7–51a and b).

The speed control circuit selected depends on the load characteristics of the cylinder or fluid motor.

Meter-in circuits are generally used when the load characteristics are constant and positive. If the load is

FORMULAS FOR APPLYING FLOW CONTROLS

p = pressure setting, maximum pressure valve, psi.
p' = pressure, cylinder inlet, psi.
p'' = pressure, cylinder outlet, psi.
V = velocity, piston, in./min.
Q = discharge, pump, cu in./min.
q = flow through flow control valve, cu in./min.
q' = discharge out of maximum pressure valve, cu in./min.
A = area, piston, inlet side, sq in.
A' = area, piston, outlet side, sq in.
F = work load, opposing motion, lb.
l = pressure drop @ maximum pressure valve, psi.
l' = pressure drop @ flow control valve, psi.

Meter-in	Meter-out	Bleed-off
$p' = \dfrac{F}{A}$	$p' = p$	$p' = \dfrac{F}{A}$
$p'' = 0$	$p'' = \dfrac{pA - F}{A'}$	$p'' = 0$
$V = \dfrac{q}{A}$	$V = \dfrac{q}{A'}$	$V = \dfrac{Q - q}{A}$
$q' = Q - q$	$q' = Q - \dfrac{Aq}{A'}$	$q' = 0$
$l = p$	$l = p$	no flow
$l' = p - p'$	$l' = p''$	$l' = p'$

Note: Formulas based on 0 pressure drop through piping and open valves. Courtesy of Vickers Inc.

erratic or negative, the cylinder or fluid motor will have jerky motion.

The by-pass type of flow control is used for meter-in circuits and has an integral relief valve which provides protection between the work cylinder or motor and the control. This type of flow control is much more efficient than the inlet restricting type for meter-in, because the by-pass feature allows fluid to

Fig. 7-51. (a) Bleed off for both directions; (b) bleed off for inlet to cylinder or motor.

Fig. 7-52. Intensified back pressure.

be exhausted to the reservoir at just slightly higher pressure than that necessary to do the work.

With the inlet restricting type, pump delivery not used would discharge over the main relief valve at maximum pressure.

Meter-in circuits are used on surface grinders, welders, milling machines, and other machine tools where fine speed control is essential. By-pass meter-in controls are usable only with one pump and one basic actuator because of the integrated relief valve type of compensation.

Meter-out circuits are best where negative loads may occur, because back pressure is maintained on the exhaust side of the actuator, preventing erratic motion. This type of speed control circuit operates satisfactorily for drilling, boring, reaming, or tapping operations.

When cylinders with large piston rods are used at high pressure, intensified back pressures may occur. Relief valve protection should be used for such conditions.

$$P = \frac{F}{A}$$

$$P = \frac{30,000}{5}$$

$$P = 6,000 \text{ psi.}$$

Figure 7–52 shows that an intensification may result in very high pressures on the rod side of cylinders; if it is not protected with a relief valve, a serious accident might occur. If a relief valve is installed ahead of a meter-out flow control valve, it may negate the flow control function. The use of a meter-in flow control is mandatory in many circuits because of this factor.

Bleed-off circuits provide less accuracy in speed

control, because they do not compensate for any change in fluid losses due to pressure change. In other words, pump slip increases with pressure, and so do other internal losses which are not sensed by the compensator of a flow control used with bleed-off circuits.

This method of speed control does have an advantage with regard to efficiency, because system pressure is only high enough to accomplish the work and fluid does not discharge over the main relief valve.

Deceleration Flow Control

A deceleration flow control valve gradually reduces flow rate to provide deceleration of heavy loads (see Fig. 7–53).

In Fig. 7–54, the deceleration valve gradually restricts the flow from the rod end of the cylinder to slow down the heavy moving table as the machine starts the feed phase of the cycle. The angle of the cam on the end of the cylinder piston rod and the designed features of the valve determine the time to decelerate.

There is a tremendous amount of flexibility in

Fig. 7-53. Deceleration valve.

Fig. 7-54. Deceleration circuit.

hydraulic components to control speeds, feeds, positions, pressures, and various motions. The discussion in this chapter has barely scratched the surface.

Hydraulic pumps are designed with flow adjustment mechanisms as integral parts of the pump itself. Directional controls are available with metering orifices built into any given flow passage through the valve. One combination control device might have a directional control, a flow control, and other features all built into one package.

Questions

1. What are the three important things that are controlled in a hydraulic system?
2. What is the function of a relief valve in a hydraulic system?
3. What are the two general types of relief valves?
4. Which type of relief valve is used as a means of unloading a pump during the idle time of machine?
5. What rule generally applies to the pressure setting of a relief valve?
6. What is the function of a sequence valve?
7. What is the general rule for draining pressure control valves?
8. What is the function of an unloading valve?
9. Is an unloading valve directly or remotely pilot-controlled?
10. What is the advantage of using an unloading circuit when feeds and speeds of a machine must be varied?
11. What is the function of a counterbalance valve?
12. When is a pressure reducing valve used in a hydraulic system?
13. Where is pilot oil taken to operate or control a pressure-reducing valve?
14. What are directional control valves called that are used to operate other control valves?
15. What is the simplest type of directional control valve?
16. Name three types of on-off valves.
17. What is the purpose of the pilot-operated check valve?
18. When are three-way valves used in a hydraulic circuit?
19. How are directional control valves actuated?
20. What three general designs of four-way valves are available?
21. What is the difference between a two-position and a three-position four-way valve?
22. Illustrate the various center configurations that are standard with most four-way three-position directional control valves.
23. Explain the difference between a closed center system and an open center system.
24. In general, what are the maximum operating pressures for spool type valves? What are the maximum operating pressures for shear-seal-type valves?
25. How is pressure drop across directional control valves determined?
26. What are the three ways of applying flow control valves?
27. What is meant when a flow control valve is said to be "pressure-compensated"?
28. What is the difference between a by-pass type of flow control valve and a restricted inlet type of flow control valve?
29. How are the by-pass type of flow control valves used in a circuit?
30. What is the purpose of a deceleration valve?

CHAPTER **8**

ACTUATORS PROVIDE FLEXIBILITY IN THE USE OF FLUID POWER

Fluid power can provide either linear or rotary motion through the use of actuators, called *cylinders,* and fluid motors.

A tremendous advantage of using fluid power rather than mechanical systems is the fact that it can be applied directly to the work itself. In recent years fluid power has replaced the old-fashioned factory system of shaft and belt drives, where steam engines provided the input power. This was a dangerous hazard to operators and workmen and caused many accidents with loss of arms and legs. Fluid power is doing many of the jobs formerly done by mechanical linkage, such as mechanical brakes, chain drives, drive shafts, cams, and other devices.

The flexibility of distributing fluid power through conductors and directly to the actuator does not affect the geometry of the machine. This gives excellent freedom for the design of equipment.

Generally, there is no need for heavy, cumbersome mechanical linkage between the power source and the work. Actuating cylinders and fluid motors can be directly attached to the work, and the power can be piped in to meet any force, speed, or positioning requirement.

Fluid power can be transmitted and controlled to provide the correct force and speed for any job from simple to complex.

Actuating Cylinders

A hydraulic cylinder is a device which converts fluid power into linear mechanical force and motion. It usually consists of a movable element, such as a piston and piston rod, plunger or ram, operating within a cylindrical bore.

The operating principle of the piston and piston rod type is that fluid entering one port drives the movable piston and rod assembly in one direction as shown in Fig. 8–1. Fluid from the opposite side is

Fig. 8-1. Function of hydraulic cylinder.

Fig. 8-2. Function of hydraulic cylinder-retract.

exhausted back to the reservoir through a directional control valve.

When the directional control valve is shifted to the opposite position, fluid then enters the cylinder port at the rod end, pushing the piston and rod assembly back again as shown in Fig. 8–2. Fluid exhausting from the large area side is directed back to the reservoir again through the four-way valve.

Cylinder Velocity

The velocity of a cylinder as it extends or retracts depends upon how fast the *oil* is fed into the cylinder and the *area* of the cylinder. With 10 gpm being delivered to each of the cylinders in Fig. 8–3, the illustration shows the length of travel in one minute (231 in.³ = 1 gal). Velocity varies directly with the gal/min (gpm). Velocity varies inversely with the area. Mathematically, this may be expressed as

$$V = \frac{\text{gpm}}{A}$$

Substituting units into the formula, we have

$$V(ft/\text{min}) = \frac{\text{gal/min}}{\text{in.}^2}$$

In order to get the correct units on both sides of the equation, convert gallons to cubic inches (231 in.³/

Fig. 8-3. Length of travel in one minute.

gal), and inches to feet (1 ft/12 in.). Now,

$$V(\text{ft/min}) = \frac{\text{gal/min} \times 231\ \text{in.}^3/\text{gal} \times 1\ \text{ft/12 in.}}{\text{in.}^2}$$

Grouping and simplifying, we have

$$V(\text{ft/min}) = \frac{\text{gal/min}}{\text{in.}^2} \times \frac{231\ \text{in.}^2\ \text{ft}}{12\ \text{gal}} = \frac{\text{gpm}}{A} \times 19.25$$

The factor 19.25 in.² ft/gal is a constant which can be used for finding the velocity of any cylinder in feet per minute. If feet per second is required, the formula is $V(\text{ft/sec}) = \text{gpm} \times 0.3208/A$ and is derived in the same manner as the previous formula.

Example. Ten gpm is being fed into a cylinder whose area is 5 in.² What is the velocity of the cylinder in feet per minute?

$$V(\text{ft/min}) = \frac{\text{gpm} \times 19.25}{A} = \frac{10 \times 19.25}{5} = 38.5\ \frac{\text{ft}}{\text{min}}$$

The velocity in feet per second is

$$V(\text{ft/sec}) = \frac{\text{gpm} \times 0.3208}{A} = \frac{10 \times 0.3208}{5}$$

$$= 0.64\ \text{ft/sec}$$

Being mathematical expressions, these formulas may be expressed in different forms.

Formula:

$$V(\text{ft/min}) = \frac{\text{gpm} \times 19.25}{A}$$

Other forms:

$$\text{gpm} = \frac{V(\text{ft/min})A}{19.25}, \qquad A = \frac{\text{gpm} \times 19.25}{V(\text{ft/min})}$$

Formula:

$$V(\text{ft/sec}) = \frac{\text{gpm} \times 0.3208}{A}$$

Other forms:

$$\text{gpm} = \frac{V(\text{ft/sec})A}{0.3208}, \qquad A = \frac{\text{gpm} \times 0.3208}{V(\text{ft/sec})}$$

Example. A cylinder 12 in. long with a cross-sectional area of 19.25 sq. in. contains 231 cu. in. of hydraulic fluid (see Fig. 8–4).

Volume = area × length
Volume = 19.25 × 12
Volume = 231 cu in.
(231 cu in. = 1 gal)

This indicates that if a hydraulic pump delivers 1 gpm it will take exactly one minute to move the piston assembly 12 in. If the pump delivery is 2 gpm, it would take only 30 sec. A pump delivering 60 gpm would require only 1 sec to move the piston assembly 12 in.

Fig. 8-4. Example problem.

$$\text{Speed in ft/min} = \frac{\text{gpm} \times 19.25}{\text{area}}$$

This formula can be used to calculate cylinder piston speed in feet per minute.

Example. If the rod diameter in the previous cylinder is 2 in., what will the return speed of the piston assembly be if the pump delivers 10 gpm (see Fig. 8–5).

Solution. Find the rod area.

$$\text{Area} = 0.7854 \ D^2$$
$$\text{Area} = 0.7854 \times 4$$
$$\text{Area} = 3.1416$$

Subtract the rod area from the cylinder area.

$$19.25 \text{ sq in.} - 3.14 \text{ sq in.} = 16.11 \text{ sq in.}$$

$$\text{Speed in ft/min} = \frac{\text{gpm} \times 19.25}{A}$$

$$\text{Speed in ft/min} = \frac{10 \times 19.25}{16.11}$$

$$\text{Speed in ft/min} = 12.9 \text{ ft/min}$$

The volume of the cylinder and pump delivery determine the speed of a piston assembly, but the area and pressure determine the amount of force being applied.

Work or resistance against the cylinder rod causes fluid pressure to oppose the pump flow as shown in Fig. 8–6.

Fig. 8-5. Example problem.

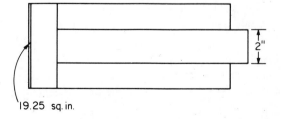

Example. If the resisting force against the piston rod is 10,000 lb, how much pressure will be developed in the system using a 4 in. diameter cylinder?

$$\text{Pressure} = \frac{\text{force}}{\text{area}}$$

$$\text{Area of cylinder} = 0.7854 \ D^2$$
$$= 0.7854 \times 16$$
$$\text{Area} = 12.56 \text{ sq in.}$$

$$\text{Pressure} = \frac{10{,}000 \text{ lb}}{12.56 \text{ in.}^2}$$

$$\text{Pressure} = 796 \text{ psi}$$

The ram- or plunger-type actuating cylinders have a movable element of the same cross-sectional area as the piston rod. Shown in Fig. 8–7.

The ram or plunger type of cylinder is generally powered in only one direction. Auxiliary cylinders

Fig. 8-6. Example problem.

called "kicker cylinders" are sometimes used to retract the ram. This arrangement is shown in Fig. 8–8. They can also be retracted by gravity or weighted machine members if applied in the inverted position.

These designs are available for creating tremendous forces such as those needed in large presses.

Cylinders may also be used for precisely clamping delicate objects with controlled force as shown in Fig. 8–9.

Hydraulic cylinders are manufactured to produce a few ounces of force or thousands of tons. Some metal fabricating presses used for aircraft parts are equipped with large cylinders that produce up to 35,000 tons of force.

In addition to the high force capabilities, cylinders give high power per weight and size, mounting flexibility, and excellent speed control. They are available in a wide range of sizes, strokes, and pressure ranges.

Cylinders may be single-acting, double-acting, or telescoping.

Single-acting cylinders (see Fig. 8–10) use fluid power only in one direction. The return stroke uses a gravity load or a spring retracting force.

Single-acting cylinders with spring return are

Fig. 8-7. Ram or plunger type cylinder.

Fig. 8-8. Press. *(Courtesy of Socony Mobil Oil Co.)*

Fig. 8-9. Force capabilities.

SPRING RETURN

A S A SYMBOL

LOAD

GRAVITY RETURN

Fig. 8-10. Single-acting cylinder.

cause of friction and other factors. Fluid metered out through the flow control valve would allow the cylinder to retract.

Double-acting cylinders use fluid power in both directions and may be of either the single or double rod types (see Fig. 8–13). Double-acting cylinders of the single rod type are by far the most popular design. This type generally uses the large area side to develop the greatest power requirement in a reciprocating circuit.

Example. If the cylinder in Fig. 8–14 has a piston area of 4 in.² and is fitted with a piston rod having 2 in.², what is the difference in force that can be developed between the two sides if the relief valve is set for 2000 psi?

$$\text{Force} = \text{pressure} \times \text{area}$$

$$F = 2000 \times 4$$

Force
extending $F = 8000 \text{ lb}$

$$\text{Force} = \text{pressure} \times \text{area}$$

$$F = 2000 \times 2 \text{ in.}^2$$

$$F = 4000 \text{ lb} \qquad \begin{array}{c}\text{Force}\\ \text{retracting}\end{array}$$

Thus, it is apparent that the piston end without the rod has the most effective area and is capable of twice the force. However, it is also true that the piston and rod assembly will travel twice the speed retracting as it will extending.

When one is selecting tubing and component sizes it is very important that the difference in volumes exhausting from this type of cylinder during extending and retracting be taken into consideration. Cylinders with large piston rods return rapidly, exhausting large quantities of fluid that may cause excessive back pressures and heat buildup. This indicates the importance of a separate drain line for sequence valves or other such components. When cylinders of this type extend, it is also possible to intensify pressure within the rod end if fluid is restricted or metered out of that end.

However, a volume differential cylinder of this type generally conforms to almost all of the important

sometimes used for clamping operations, small presses, and other such applications where little or no load is subjected on the return stroke (see Fig. 8–11).

Gravity return cylinders are used on mobile equipment, farm implements, and other machine applications controlling vertically suspended loads. Fluid power entering the cylinder raises the load, and fluid under control leaving the cylinder lowers the load. Figure 8–12 shows that fluid pumped into the cylinder causes it to rise, and fluid metered out of the cylinder allows the load to drop. A pressure slightly higher than 600 psi would be needed to raise the load be-

Fig. 8-11. Clamp-single-acting (spring return).

WORK
PIECE

Fig. 8-12. Vertical load-single-acting (gravity return).

SINGLE ROD END

CYLINDER

ASA SYMBOL

DOUBLE ROD END

CYLINDER

ASA SYMBOL

Fig. 8-13. Double-acting cylinders.

4 sq. in.
PISTON AREA

2 sq. in.
ROD AREA

Fig. 8-14. Example problem.

Fig. 8-15. Double-acting, double rod type cylinder.

MACHINE FEED TABLE

TELESCOPING CYLINDER

Fig. 8-16. Telescoping cylinder in action. Telescoping cylinder; concrete road runner.

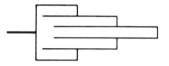

A.S.A. SYMBOL FOR
TELESCOPING CYLINDER

Fig. 8-17. Telescoping cylinder; ASA symbol.

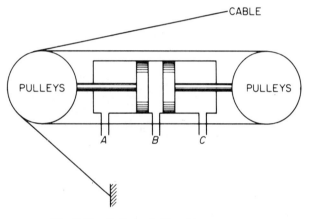

Fig. 8-18. Cylinder with cable.

requirements of any circuit. The cylinder's rod size can be varied to conform to side loads, tensional and compressional stresses, or other factors. The cylinder bore can be varied according to the greatest force requirement. There are essentially no limiting factors, because the blind end is solid metal and can be made with tremendous strength.

A cylinder with a single piston and a piston rod extending from each end is a double-acting, double rod type (see Fig. 8–15). This design of cylinder can provide the same force and speed in both directions. They are generally applied to machines where work is being done in both directions and at equally controlled speeds. This design of cylinder is also used as a metering cylinder where the fluid is directed to another actuator for controlling speed or position.

Telescoping cylinders (see Fig. 8–16) are used where long work strokes are needed.

A telescoping cylinder has nested multiple tubular rod segments which provide a long working stroke in a short retracted envelope. It operates on the principle that the rod with the largest area gives a greater force at the least pressure and moves out first; the next largest moves at a slightly higher pressure; the next at a pressure higher yet; and so on for as many stages as the unit may contain.

Other special design types are shown in Fig. 8–18. Where very long strokes are needed, this cylinder design uses a cable which provides a long pull with a relatively short stroke. Fluid entering *B* extends one or both piston assemblies, which stroke the cable a distance determined by the pulley arrangement.

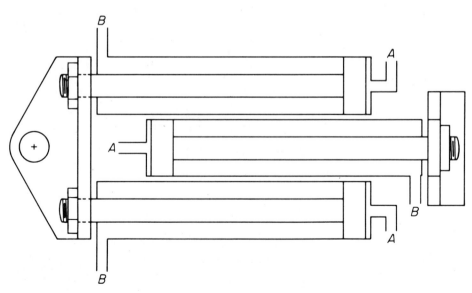

Fig. 8-19. Three cylinders nested together.

Fig. 8-20. Tandem design.

Fig. 8-21. Positioned type cylinder; moving barrel and lower piston; fixed upper piston.

Fig. 8-22. Cylinder body styles; tie rod, mill type, one-piece with retainers, threaded construction.

TIE ROD CONSTRUCTION

ONE PIECE WITH RETAINER

MILL TYPE CONSTRUCTION

THREADED CONSTRUCTION

Fig. 8-22. (Cont.)

Fig. 8-23. Standard cylinder identification code.

Fluid entering *A* and *C* retracts the piston assemblies.

Another cylinder design uses three cylinders nested together for greater strength with a long stroke (see Fig. 8–19). Cylinder ports labeled *A* extend all three piston assemblies, and ports labeled *B* retract piston assemblies.

Positional or tandem designs have two or more cylinders with interconnecting piston assemblies (see Fig. 8–20).

Fluid entering, leaving, or holding at any port or combination of ports can actuate piston assemblies to the various positions. Greater force can be exerted when both piston assemblies are moving in contact with the work.

The positional type cylinder in Fig. 8–21 uses a moving cylinder barrel, one moving piston assembly, and one stationary piston assembly. This design allows a full stroke, which might be used to place the work into a machine, and then a half stroke for repeating a series of work strokes. This technique is sometimes used on welding machines, where a wide stroke is necessary to insert the work piece into the welder. The short stroke is then used to make a number of welds on the work. Figure 8–22 shows popular body styles of cylinders.

Figure 8–23 is a standard cylinder identification coding system which is used by practically all cylinder manufacturers. Specific dimensions can be compared for interchangeability or for ordering replacement cylinders.

Fig. 8-24. Standard bore and rod sizes.

Fig. 8-25. (a) (b) (c). Cylinder mounting charts.

RECOMMENDED BORE AND PISTON ROD SIZE FOR FLUID POWER CYLINDERS

(Nominal)

Bores (in.)		Piston rod diameters (in.)		
1	$4\frac{1}{2}$	$\frac{5}{16}$	$1\frac{1}{4}$	4
$1\frac{1}{8}$	5	$\frac{3}{8}$	$1\frac{3}{8}$	$4\frac{1}{2}$
$1\frac{1}{4}$	6	$\frac{1}{2}$	$1\frac{1}{2}$	5
$1\frac{1}{2}$	7	$\frac{5}{8}$	$1\frac{3}{4}$	$5\frac{1}{2}$
2	8	$\frac{11}{16}$	2	6
$2\frac{1}{2}$	10	$\frac{3}{4}$	$2\frac{1}{4}$	7
3	12	$\frac{7}{8}$	$2\frac{1}{2}$	8
$3\frac{1}{4}$	14	1	$2\frac{3}{4}$	9
$3\frac{1}{2}$	16	$1\frac{1}{8}$	3	10
4	20		$3\frac{1}{2}$	

CYLINDER MOUNTING CHART

A. Fixed centerline mounts	ASA Symbol
Tie rod end	″
Blind end flange	″
Rod end flange	″
Centerline lugs	″
B. Fixed noncenterline mounts	ASA Symbol
End lug mounts	″
Side lug mounts	″
Integral key mounts	″
Flush mounts	″
C. Pivoted centerline mounts	
Rod end trunnion	″
Blind end trunnion	″
Clevis trunnion	″
Center trunnion	″

A. Fixed centerline mounts are used for thrusts that occur linearly or in a centerline with the cylinder. Proper alignment is essential to prevent compound stresses that may cause excessive friction and binding as the piston rod extends. Additional holding strength may be essential with long stroke cylinders.

B. Fixed noncenterline mounts are convenient where exceptionally heavy linear thrusts are encountered. Generally, integral keys or pins are used if excessive hydraulic shock is expected. This helps to relieve shear loads. Since the cylinder has to expand and contract with temperature change, only one end should be keyed or pinned.

C. Pivoted centerline mounts are used to compensate for thrusts occurring in multiple planes or if the attached load travels in a curved path. Ball joints, trunnions, and clevis mounts allow thrusts to be taken up along cylinder centerlines.

(a) FIXED CENTERLINE MOUNTS

TIE ROD END

BLIND END FLANGE

ROD END FLANGE

CENTERLINE LUGS

(b) FIXED NON-CENTERLINE MOUNTS

END LUG MOUNTS

SIDE LUG MOUNTS

INTEGRAL KEY MOUNTS

FLUSH MOUNTS

(c) PIVOTED CENTERLINE MOUNTS

ROD END TRUNNION

BLIND END TRUNNION

CLEVIS

CENTER TRUNNION

COMPRESSION

TENSION

SHEAR

Fig. 8-26. Forces effecting a cylinder.

Figure 8–27 illustrates some of the various methods of applying linear motion using cylinders and how to determine resultant loads.

Seals for Cylinders

Proper seals are important to prevent leakage and loss of hydraulic power. The sealing material must be compatible with the hydraulic fluid and the operating temperatures.

The seals must be applied to prevent seal "break-off" which contaminates the system. They should also be easy to replace.

Sealing devices are used in cylinders to seal piston and piston rod assembly, end caps to barrel, and for the piston rod sealing gland (see Fig. 8–28). Rod scrapers or wipers should be used to prevent abrasives or dirt from damaging the rod gland seal and getting into the system.

Cushioning

For the prevention of shock due to stopping loads at the end of the piston stroke, cushion devices are used. Cushions may be applied at either end or both (see Fig. 8–29). They operate on the principle that as the cylinder piston approaches the end of the stroke, exhaust fluid is forced to go through an adjustable needle valve which is set to control the escaping fluid at a given rate. This allows the deceleration characteristic to be adjusted for different loads.

When the cylinder piston is actuated, fluid enters the cylinder port and flows through the little check valve so that the entire piston area can be utilized to produce force and motion.

Method of Solving for Linear Acceleration and Deceleration

$$F = MA$$
$$S = \tfrac{1}{2} at^2$$
$$V = at$$

Combining the above gives

$$S = \frac{1}{2} \frac{V^2}{a}$$

$$a = \frac{V^2}{2S}$$

$$F = \frac{W}{g} \times \frac{V^2}{2S}$$

Force F acts over distance S to start and stop mass M to or from velocity V.

F = force, lb.

M = slugs.

W = weight, lb.

S = distance, in.

t = time, sec.

a = acceleration in./sec.^{2}

V = velocity, in./sec.

Decelerating distance:

$$d = v_0 t - \tfrac{1}{2} at^2$$

v_0 = initial velocity.

Acceleration and Deceleration Force Determination

The uniform acceleration force factor chart and the accompanying formulae can be used rapidly to determine the forces required to accelerate and decelerate a cylinder load. To determine these forces,

HORIZONTAL CASE

$F = (P \times A) = $ FRICTION
$= \mu N$
$= \mu W$

$\mu = $ COEFFICIENT OF FRICTION
$N = $ NORMAL FORCE ON THE PLANE

INCLINED CASE

$R_d = $ RESULTANT DOWN INCLINE
$R_u = $ RESULTANT UP INCLINE
$\alpha = $ ANGLE OF THE INCLINE

$F = (P \times A) = $ FRICTION $\pm F_g$
$= \mu N \pm F_g$
$= \mu W \cos \alpha \pm W \sin \alpha$
$= \dfrac{W \sin (\theta \pm \alpha)}{\cos \theta}$

$F_g = $ FORCE OF WEIGHT ACTING DOWN THE INCLINE

$\tan \theta = \mu$
OR $\theta = \tan^{-1} \mu$

(a)

VERTICAL CASE

$F = (P \times A) = W$

(b)

JIB CRANE

POINT OF CONCURRENCE

$F = P \times A = \dfrac{W \sin (\alpha + \beta)}{\sin \beta}$

NOTE: AS THE JIB RAISES TO VERTICAL, F APPROACHES W; AS THE ANGLE γ BECOMES LARGER, F APPROACHES ∞

(c)

"LEVER" JIB CRANE

$F = P \times A = \dfrac{l_1}{l_2} W$ SECANT $\gamma = \dfrac{l_1}{l_2} \dfrac{l_3}{l_4} W$

NOTE: AS γ BECOMES SMALLER
F APPROACHES $l_1 / l_2 W$
AS γ BECOMES LARGER
F APPROACHES ∞

(d)

BENT LEVER

$F = P \times A = \dfrac{l_1}{l_2} W$

(e)

FIRST – CLASS LEVER

$F = P \times A = \dfrac{l_1}{l_2} W$

(f)

Fig. 8-27. Various methods of applying linear motion using cylinders.

HORIZONTAL RACK AND PINION

$$F = P \times A = \frac{r_1}{r_2} \mu N = \frac{r_1}{r_2} \mu W$$

(g)

WEDGE

FORCE ACTING THROUGH INCLINE:

$$F_i = P \times A = \text{TAN}\,(\theta \pm \alpha)\,W$$

R_u = RESULTANT UP α = ANGLE OF THE INCLINE
   INCLINE
                          θ = $\tan^{-1} \mu_{\text{INCLINE}}$
R_d = RESULTANT DOWN
   INCLINE μ = COEFFICIENT OF FRICTION

FORCE ACTING THROUGH SLIDING SURFACE:

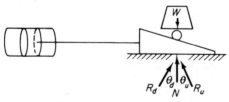

$$F_h = A \times P = \mu \frac{W}{\text{HORIZONTAL}}$$

$$\text{TOTAL } F = P \times A = \left[\tan\,(\theta \pm \alpha) + \mu_h\right] W$$

(h)

BLOCK AND TACKLE

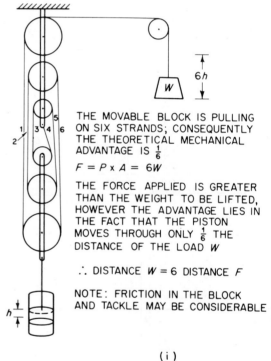

THE MOVABLE BLOCK IS PULLING ON SIX STRANDS; CONSEQUENTLY THE THEORETICAL MECHANICAL ADVANTAGE IS $\frac{1}{6}$

$$F = P \times A = 6W$$

THE FORCE APPLIED IS GREATER THAN THE WEIGHT TO BE LIFTED, HOWEVER THE ADVANTAGE LIES IN THE FACT THAT THE PISTON MOVES THROUGH ONLY $\frac{1}{6}$ THE DISTANCE OF THE LOAD W

$\therefore$ DISTANCE W = 6 DISTANCE F

NOTE: FRICTION IN THE BLOCK AND TACKLE MAY BE CONSIDERABLE

(i)

Fig. 8-27.   (Cont.)

the following factors must be known: total weight to be moved, maximum piston speed, distance available to start or stop the weight (load), direction of movement, i.e., horizontal or vertical, and load friction. By use of the known factors and the *g* factor from the chart, the force necessary to accelerate or decelerate a cylinder load may be found by solving the formula applicable to a given set of conditions. (See Graph 8–1.)

Nomenclature

V = velocity in feet per minute.

S = distance in inches.

F = force in pounds.

W = weight of load in pounds.

g = force factor.

f = friction of load on machine ways in pounds.

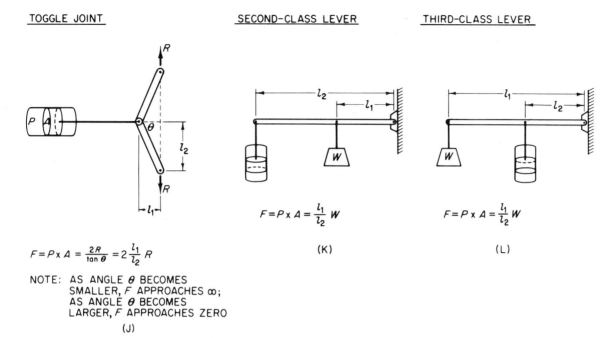

TOGGLE JOINT

$$F = P \times A = \frac{2R}{\tan \theta} = 2\frac{l_1}{l_2} R$$

NOTE: AS ANGLE θ BECOMES
SMALLER, F APPROACHES ∞;
AS ANGLE θ BECOMES
LARGER, F APPROACHES ZERO

(J)

SECOND-CLASS LEVER

$$F = P \times A = \frac{l_1}{l_2} W$$

(K)

THIRD-CLASS LEVER

$$F = P \times A = \frac{l_1}{l_2} W$$

(L)

Fig. 8-27. (Cont.)

Fig. 8-28. Depicts: Cross section of cylinder rod gland; cutaway of a hydraulic cylinder; and cross section showing seals. (*Courtesy of Socony Mobil Oil Co.*)

To determine the force factor g from the chart, locate the intersection of the maximum piston velocity line and the line representing the available distance. Project downward to locate g on the horizontal axis.

Example. Horizontal motion of a freely moving 6000 lb load is required with a distance of $\frac{1}{2}$ in. to a maximum speed of 120 ft per minute. Formula (1) $F = Wg$ should be used.

$$F = 6000\,\text{lb} \times 1.50\,(\text{from chart}) = 9000\,\text{lb}$$

Assuming a maximum available pump pressure of 1000 psi, a 4 in. bore cylinder should be selected, operating on push stroke at approximately 750 psi pressure at the cylinder to allow for pressure losses from the pump to the cylinder.

Assume the same load to be sliding on ways with a coefficient of friction of 0.15. The resultant friction

METALLIC ROD WIPER

Fig. 8-29. Cushioning.

load would be $6000 \times 0.15 = 900$ lb. Formula (2) $F = Wg + f$ should be used.

$$F = 6000 \text{ lb} \times 1.5 \text{ (from chart)} + 900 = 9900 \text{ lb}$$

Again allowing 750 psi pressure at the cylinder, we find that a 5 in. bore cylinder is indicated.

Example. Horizontal deceleration of a 6000 lb

load is required by using a 1 in. long cushion in a 5 in. bore cylinder having a 2 in. diameter piston rod. Cylinder bore area (19.64 sq in.) minus the rod area (3.14 sq in.) results in a minor area of 16.5 sq in. at the head end of cylinder. A 1000 psi pump delivering 750 psi at the cylinder is being used to push the load at 120 ft per minute. Friction coefficient is 0.15, or 900 lb.

In this example, the total deceleration force is the sum of the force needed to decelerate the 6000 lb load, and the force required to counteract the thrust produced by the pump.

W = load in pounds = 6000.

S = deceleration distance in inches = 1 in.

V = maximum piston speed in feet per minute = 120.

g = 0.74 (from chart).

f = 900 lb.

Use formula (3) $F = Wg - f$.

$$(F = Wg - f) = (F = 6000 \times 0.74 - 900)$$
$$= 3540 \text{ lb}$$

Fig. 8-30. Rotary actuators: 1-vane; 2-vane; opposed piston; piston chain type.

The pump is delivering 750 psi acting on the 19.64 sq in. piston area, producing a force (F_2) of 14,730 lb. This force must be included in our calculations. Thus $F + F_2 = 3540 + 14,730 = 18,270$ lb total force to be decelerated.

The total deceleration force is developed by the fluid trapped between the piston and the head. The fluid pressure is equal to the force (18,270 lb) divided by the minor area (16.5 sq in.) equals 1107 psi. This pressure should not exceed the nonshock rating of the cylinder.

Rotary Actuators

Rotary actuators are devices for producing limited reciprocating rotary force and motion. They are used for lifting, lowering, opening, closing, indexing, and transferring movements (see Fig. 8–30).

When the fluid enters on one side it causes the movable element to rotate the shaft in one direction. The rotation is limited, depending upon the specific type. Fluid entering the opposite port will reverse the actuator shaft.

The single vane rotates through an arc of 280 deg. Power in the two-vane design is doubled, but the turning arc is reduced to 100 deg.

The opposed piston type produces a high turning force at low pressure. As the pistons extend or retract, they rotate the pinion gear, which is in mesh with each rack. It produces the same turning force in both directions.

TABLE 8-1

THEORETICAL PUSH AND PULL FORCES FOR HYDRAULIC CYLINDERS

Push Force and Displacement

Cyl. bore size (inches)	Piston area (sq in.)	Cylinder push stroke force in pounds at various pressures										Displacement per inch of stroke (gallons)
		50	80	100	250	500	750	1000	1500	2000	2500	
1	.785	39	65	79	196	392	588	785	1177	1570	1962	.00339
1½	1.767	88	142	177	443	885	1325	1770	2651	3540	4425	.00765
2	3.14	157	251	314	785	1570	2357	3140	4713	6280	7850	.0136
2½	4.91	245	393	491	1228	2455	3682	4910	7364	9820	12275	.0213
3¼	8.30	415	664	830	2075	4150	6225	8300	12450	16600	20750	.0359
4	12.57	628	1006	1257	3143	6285	9428	12570	18856	25140	31425	.0544
5	19.64	982	1571	1964	4910	9820	14730	19640	29460	39280	49100	.0850
6	28.27	1414	2262	2827	7068	14135	21203	28270	42406	56540	70675	.1224
8	50.27	2513	4022	5027	12568	25135	37703	50270	75406	100540	125675	.2176

Deductions for Pull Force and Displacement

Piston rod dia. (inches)	Piston rod area (sq in.)	Piston rod diameter force in pounds at various pressures										Displacement per inch of stroke (gallons)
		To determine cylinder pull force or displacement, deduct the following force or displacement corresponding to rod size, from selected push stroke force or displacement corresponding to bore size in table above										
		50	80	100	250	500	750	1000	1500	2000	2500	
½	.196	10	16	20	49	48	147	196	294	392	490	.00085
⅝	.307	15	25	31	77	154	230	307	461	614	768	.0013
1	.785	39	65	79	196	392	588	785	1177	1570	1962	.0034
1⅜	1.49	75	119	149	373	745	1118	1490	2235	2980	3725	.0065
1¾	2.41	121	193	241	603	1205	1808	2410	3615	4820	6025	.0104
2	3.14	157	251	314	785	1570	2357	3140	4713	6280	7850	.0136
2½	4.91	245	393	491	1228	2455	3682	4910	7364	9820	12275	.0213
3	7.07	354	566	707	1767	3535	3502	7070	10604	14140	17675	.0306
3½	9.62	481	770	962	2405	4810	7215	9620	14430	19240	24050	.0416
4	12.57	628	1006	1257	3143	6285	9428	12570	18856	25140	31425	.0544
5	19.64	982	1571	1964	4910	9820	14730	19640	29460	39280	49100	.0850
5½	23.76	1188	1901	2376	5940	11880	17820	23760	35640	47520	59400	.1028

The piston chain type has two cylinders in parallel. The large piston is used for powering the chain and drive sprocket, while the smaller cylinder provides a seal. Fluid entering one end operates against both pistons, but the large piston creates the greater force,

so it moves, causing rotation of the output shaft. Flow directed to the other port reverses the operation.

Rotary actuators are available for operating pressures up to about 5000 psi. Shaft sizes, type of bearings, and cycle frequency are important factors to consider when rotary actuators are selected. They are generally mounted by foot, flange, and end mounts. Cushioning devices are also available in most designs.

Fluid Motors

A fluid motor is a device which converts fluid power into mechanical force and motion. It usually provides rotary mechanical motion.

Fluid motors and pumps from the same manufacturer use interchangeable parts and look alike in many respects, but the motors work in a manner just opposite to the way in which the pumps work. They use the fluid delivered by the pump to provide a rotating force and motion.

There are two general classes of fluid motors, *fixed displacement* and *variable displacement*. In the fixed displacement class, gear, vane, and piston designs are used. In the variable displacement class, piston designs are used.

As the term implies, fixed displacement fluid motors displace a specific amount of fluid for each revolution. Therefore, the speed of any fixed type of

Fig. 8-31. Open-loop hydrostatic transmission.

TABLE 8-2
FLOW VELOCITY AND PRESSURE DATA FOR HYDRAULIC SYSTEMS

The chart below may be used to calculate pressure loss in connecting lines at various flow velocities. The data is useful when one is determining hydraulic cylinder size and port size for applications where cylinder force and speed requirements are known.

S = standard (schedule 40) pipe
H = extra strong (schedule 80) pipe
EH = double extra strong pipe

Tabulations based on a hydraulic oil having a viscosity of 155 SSU at 100° F.—specific gravity of 0.87.

To determine tubing or hose losses, use I.D. closest to tubing or hose I.D. Pressure drop does not vary with operating pressure. Avoid high pressure losses in low pressure systems. Use largest pipe size practical. Avoid flow velocities greater than 15 ft/sec to reduce hydraulic line shock.

Nominal size (Inches)	O.D. (Inches)		Wall thickness (Inch)	I.D. (Inches)	I.D. area (sq. in.)	5 Gal./min.	5 Loss	7 Gal./min.	7 Loss	10 Gal./min.	10 Loss	15 Gal./min.	15 Loss	20 Gal./min.	20 Loss	25 Gal./min.	25 Loss	30 Gal./min.	30 Loss	Tee	Tee	Elbow std.	Elbow sq.	45°
1/8	0.405	S	0.068	0.269	0.057	0.89	1.25	1.24	1.79	1.75	2.60	2.67	3.16	3.56	5.47	4.45	6.20	5.34	7.07	—	—	—	—	—
		H	0.095	0.215	0.036	0.56	1.89	0.78	3.05	1.12	4.26	1.68	5.20	2.24	8.38	2.80	1.10	3.36	12.65	—	—	—	—	—
1/4	0.540	S	0.088	0.364	0.104	1.62	0.67	2.27	1.05	3.24	1.64	4.96	1.92	6.48	2.97	8.10	2.73	9.72	8.73	—	—	—	—	—
		H	0.119	0.302	0.072	1.12	1.11	1.57	1.49	2.24	2.11	3.36	2.84	4.48	4.15	5.60	5.08	6.72	6.30	—	—	—	—	—
3/8	0.675	S	0.091	0.493	0.191	2.98	0.39	4.18	0.57	5.96	0.86	8.94	1.05	11.92	1.69	14.90	4.27	16.88	5.78	2.7	2.7	2.7	1.2	0.6
		H	0.126	0.423	0.140	2.18	0.54	3.06	0.74	4.36	1.10	6.54	1.34	8.72	1.97	10.90	5.19	13.08	7.20	3.5	3.5	3.5	1.5	0.75
1/2	0.840	S	0.109	0.622	0.304	4.74	0.24	6.65	0.37	9.48	0.49	14.22	0.68	18.96	0.68	23.70	3.38	28.44	4.28	2.9	2.9	2.9	1.4	0.68
		H	0.147	0.546	0.234	3.65	0.30	5.12	0.45	7.30	0.71	10.95	0.78	14.60	2.47	18.15	3.61	21.90	5.00	4.5	4.5	4.5	2.1	1.0
		EH	0.294	0.252	0.050	0.78	1.54	1.09	2.19	1.56	3.08	2.34	3.65	3.12	6.13	3.90	7.48	4.68	9.55	4.0	4.0	4.0	1.6	0.8
3/4	1.050	S	0.113	0.824	0.533	8.32	0.14	11.65	0.22	16.64	0.27	24.96	0.78	33.28	1.47	41.60	2.19	49.92	3.00	5.7	5.7	5.7	2.6	1.2
		H	0.154	0.742	0.432	6.74	0.16	9.45	0.26	13.48	0.37	20.22	0.87	26.96	1.71	33.70	2.48	40.44	3.52	5.2	5.2	5.2	2.5	1.1
		EH	0.308	0.434	0.148	2.31	0.53	3.24	0.67	4.62	1.05	6.93	1.31	9.24	1.94	11.55	5.06	13.86	7.02	3.0	3.0	3.0	1.5	0.75
1	1.315	S	0.133	1.049	0.863	13.45	0.10	18.85	0.13	26.90	0.34	40.35	0.57	53.80	1.42	67.25	1.64	80.70	2.24	7.5	7.5	7.5	3.7	1.6
		H	0.179	0.957	0.719	11.21	0.11	15.70	0.15	22.42	0.24	33.63	0.62	44.84	1.23	56.05	1.84	67.26	2.93	7.0	7.0	7.0	3.5	1.5
		EH	0.358	0.599	0.282	4.39	0.26	6.16	0.37	8.78	0.53	13.17	0.67	17.56	2.25	21.95	3.29	26.34	3.30	3.0	3.0	3.0	1.5	0.75
1¼	1.660	S	0.140	1.380	1.496	23.35	0.05	31.68	0.08	46.70	0.25	70.05	0.39	93.40	0.78	116.75	1.18	140.10	1.47	7.5	7.5	7.5	3.7	2.4
		H	0.191	1.278	1.280	19.95	0.07	28.06	0.09	39.90	0.26	58.85	0.44	79.80	0.85	99.75	1.27	119.70	1.80	7.0	7.0	7.0	3.5	2.1
		EH	0.382	0.896	0.630	9.83	0.13	13.75	0.16	19.66	0.24	29.49	0.71	39.32	1.35	49.15	2.01	58.98	2.76	4.9	4.9	4.9	2.3	1.05
1½	1.900	S	0.145	1.610	2.036	31.75	0.04	44.49	0.11	63.50	0.19	95.25	0.33	127.00	0.64	158.75	0.96	190.50	1.26	9.0	9.0	9.0	4.3	2.0
		H	0.200	1.500	1.767	27.55	0.04	38.62	0.08	55.10	0.21	82.65	0.36	110.20	0.71	137.75	1.06	145.30	1.36	8.2	8.2	8.2	4.0	1.8
		EH	0.400	1.100	0.950	14.81	0.09	20.75	0.09	29.62	0.32	44.43	0.51	59.24	1.05	74.05	1.51	88.86	2.14	6.5	6.5	6.5	3.0	1.4
2	2.375	S	0.154	2.067	3.355	52.30	0.04	73.45	0.08	104.60	0.14	159.20	0.24	209.20	0.48	261.50	0.69	313.80	0.85	11.0	11.0	11.0	5.5	2.5
		H	0.218	1.939	2.953	46.00	0.03	64.60	0.09	92.00	0.15	138.00	0.26	184.00	0.52	230.00	0.73	276.00	0.98	10.8	10.8	10.8	5.0	2.4
		EH	0.436	1.503	1.773	27.65	0.04	38.78	0.12	55.30	0.21	82.95	0.36	110.60	0.72	138.25	1.34	165.90	1.36	8.2	8.2	8.2	4.0	1.8
2½	2.875	S	0.203	2.469	4.788	74.75	0.03	104.80	0.07	149.50	0.11	224.25	0.20	299.00	0.37	373.75	0.53	448.50	0.72	14.0	14.0	14.0	6.5	3.0
		H	0.276	2.323	4.238	66.11	0.04	92.60	0.07	132.22	0.12	198.33	0.21	264.44	0.39	330.55	0.57	396.66	0.87	13.0	13.0	13.0	6.1	2.9
		EH	0.552	1.771	2.464	38.45	0.03	53.40	0.10	76.90	0.17	115.35	0.30	153.80	0.59	192.25	0.79	230.70	1.15	10.3	10.3	10.3	4.8	2.2

*Consult valve manufacturer for pressure drops in a particular type of valve and port-to-port flow pattern.

Courtesy of Parker-Hannifin Corp.

183

fluid motor depends on the displacement per revolution and the amount of fluid supplied to it by the pump.

A variable displacement piston-type fluid motor is built with a device that can adjust the displacement per revolution. The speed of these designs can be adjusted over a range from zero to the maximum speed for the size of the unit being used.

Fluid motors are becoming more and more popular for use with modern machinery. Fluid motors can be applied directly to the work, and they provide excellent control for acceleration, operating speed, deceleration, smooth reversals, and positioning. They provide flexibility in design and eliminate much of the bulk and weight of mechanical and electrical power transmissions.

The application of fluid motors in their various combinations with pumping units are termed *hydrostatic transmissions*. There are two classes of applying hydrostatic transmissions, the *open loop* and the *closed loop*.

The open loop shown in Fig. 8–31 uses the conventional reservoir and a directional control valve for reversing the direction of motor rotation. The fluid motor speed depends on the capacity of the pump and the displacement of the motor.

The closed loop hydrostatic transmission shown in Fig. 8–32 uses a separate replenishing pump to keep the system charged at all times. Only a small reservoir is needed, because most of the fluid is retained and circulates within the system.

The variable reversible (over-the-center) type of pump is controlled to vary the speed and direction of the fluid motor.

Regardless of the transmission type, fluid motors provide rotational power to do work. The turning

FLUID AT 600 psi

GEAR MOTOR

ASA SYMBOL

Fig. 8-33. Example problem.

force is a measure of torque which is equal to the force multiplied by the radius arm. This means that torque produced by a fluid motor depends on its pressure capabilities and its geometric configuration.

Fig. 8-32. Closed-loop hydrostatic transmission.

Torque produced by a fluid motor is expressed by the following equations:

Pounds inch of torque

$$= \frac{\underset{\text{psi}}{\text{pressure}} \times \underset{(\text{in.}^3)}{\text{displacement/rev.}}}{2\,\pi}$$

Pounds foot of torque

$$= \frac{\underset{\text{psi}}{\text{pressure}} \times \underset{(\text{in.}^3)}{\text{displacement/rev.}}}{24\,\pi}$$

Sample Problem. How many pounds inch of torque will a gear-type motor produce at 600 psi if its displacement is 1.25 cu in. of oil per revolution (see Fig. 8–33).

$$\text{Torque (lb in.)} = \frac{600 \text{ psi} \times 1.25 \text{ in.}^3}{6.28}$$

$$\text{Torque (lb in.)} = 114.6$$

Fig. 8-34. Typical performance curves; hp, speed and torque.

The relationship between the output horsepower, torque, and speed of a fluid motor is shown by the following:

$$\text{Horsepower (hp)} = \frac{\text{torque} \times \text{rpm} \times 2\pi}{33,000 \text{ ft/lb min}}$$

or

$$\text{hp} = \frac{T \times \text{rpm}}{63,025}$$

(torque in pounds inch)

or

$$\text{hp} = \frac{T \times \text{rpm}}{5252}$$

(torque in pounds foot)

Example. In the previous problem, how much horsepower would the gear motor produce if the pump delivers 10 gpm?

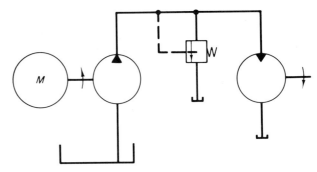

Fig. 8-35. Constant hp, torque and speed.

Solution:

$$\text{hp} = \frac{T \times \text{rpm}}{63,025}$$

First,

$$\text{Fluid motor speed} = \frac{\text{gpm} \times 231 \text{ in.}^3}{\text{disp/rev in.}^3}$$

$$\text{Speed (rpm)} = \frac{10 \text{ gpm} \times 231 \text{ in.}^3}{1.2 \text{ in.}^3}$$

$$\text{Speed (rpm)} = \frac{2310}{1.2}$$

$$\text{rpm} = 1924$$

Then,

$$\text{hp} = \frac{T \times \text{rpm}}{63,025}$$

$$\text{hp} = \frac{114.6 \times 1924}{63,025}$$

$$\text{hp} = 3.4$$

There are various combinations of pumps and motors used for hydrostatic transmission drives. The type of hydrostatic transmission drive shown in Fig. 8–35 provides fixed ratios of horsepower, torque, and speed.

The pump capacity is selected to provide the maximum output speed needed for a motor whose displacement and operating pressure meet the torque requirements.

Fig. 8-36. Variable hp and speed, constant torque.

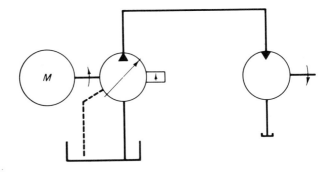

Flow controls are generally applied to control speeds less than maximum. The setting of the relief valve in the system controls torque level.

Constant torque will result when the type of transmission drive shown in Fig. 8–36 is used. The pump delivery is adjusted to vary the motor speed, but the torque remains practically constant. This combination is excellent for systems that have constant loads.

In the transmission drive shown in Fig. 8–37, the input horsepower remains constant, but the output will vary by adjusting the fluid motor displacement. Speed will depend on the maximum delivery of the pump and the maximum displacement adjustment in the fluid motor. The speed is generally limited to a ratio of 4:1 in this type of transmission.

Fig. 8-37. Constant hp, variable speed and torque.

Fig. 8-38. Variable hp, speed and torque.

Fig. 8-39. External gear type design: single-unit; tandem-units.

SINGLE UNIT   TANDEM UNIT

HELICAL GEAR   TANDEM GEAR

The torque developed during operation depends on the displacement and relief valve settings.

The combination of hydrostatic transmission shown in Fig. 8–38 provides for a wide range of varying loads. The maximum horsepower, torque, and speed depend on the size of the units selected. Infinitely variable speed and torque between zero and maximum provide the most flexibility available in transmission today.

New fields of application are being discovered constantly for hydrostatic transmissions. Farm implements, road machinery, materials handling equipment, numerically controlled machines, high performance aircraft, military uses, and special machinery are only a few of the new fields expanding through the use of fluid power transmissions. Some farm tractors are completely hydraulic, as well as combines and other mobile machinery. In Europe, automobiles, railway locomotives, and buses are using hydrostatic transmissions with excellent success.

Gear-type Fluid Motors

Gear-type fluid motors are often used as either pumps or motors. However, since fluid motors are supplied with fluid from the pump and are under pressure while doing work, bearings, type of seals, and drains are sometimes affected.

The type of bearings used is important because of the high operating speeds that are sometimes encountered. It is not uncommon for gear-type motors to operate up to 5000 rpm.

Seals are affected by both operating speed and pressure. Since most fluid motors are reversible, it means that both ports are alternately under pressure or exhaust. Shaft seals are generally limited to a specific operating pressure. Also, proper lubrication and low pressure drains are essential.

Most reversible types of fluid motors must be externally drained to prevent shaft seals from blowing and to provide adequate lubrication and circulation of fluid for keeping bearings cool.

Gear-type motors are either internal or external gear designs.

The external gear design shown in Fig. 8–39 is economical and is available as single or tandem units. A set of matched gears is fitted into a closely machined housing. The gears rotate together as fluid enters the space between the major and minor diameters. Some designs use pressure plates on either side of the rotating gears to prevent leakage as the pressure increases with resistance due to the load. System pressure is directed through internal passages to the out-

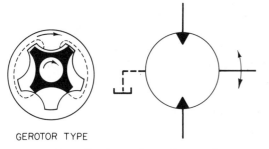

GEROTOR TYPE

Fig. 8-40. Internal gear-type motor.

BALANCED VANE

Fig. 8-41. Balance-vane motor.

side of the pressure plates or wear plates, holding them closely against the gears as the motor operates. This technique has helped to increase the efficiency of gear units.

The internal gear design shown in Fig. 8–40 consists of a pair of rotating gears, one inside the other. With fluid entering one port and the other open to exhaust, the gear elements rotate. As the teeth disengage, the space between the two elements increases on the pressure side and then decreases on the exhaust side as they gradually engage again. During one-half of one revolution, as the space increases, fluid enters from the pump; during the other one-half of one revolution, as the space decreases, fluid is exhausted from the motor.

The internal gear design has high starting torque characteristics and operates at relatively high speeds.

Vane-type Fluid Motors

Vane-type fluid motors are of the same general design as vane pumps, except that the vanes must be held tightly against the contour of the retaining ring to provide torque. This is generally done with springs applied underneath the vanes to hold them against the intersurface of the retaining ring. Seals, bearings, and drains are important considerations, as they are with all other designs of fluid motors.

Vane motors of the design shown in Fig. 8–41

are balance-type units. Fluid enters from the pump at two points 180 deg apart. Fluid also exhausts from two points 180 deg apart. This means that the rotating element of the motor is pressure-balanced, preventing uneven loads on the bearings.

Fluid forced into the vane motor from the pump pushes against the vanes and causes the rotor and shaft assembly to turn. Flow directed to the opposite port reverses the rotor and shaft assembly rotation.

The porting arrangements can be located in the retaining ring or in the side plates.

Another type of vane motor uses rolling vanes

Fig. 8-42. Roll-vane motor.

Fig. 8-43. Axial piston motor. (*Courtesy Socony Mobil Oil Co.*)

FIXED DISPLACEMENT

VARIABLE DISPLACEMENT

(see Fig. 8–42). The vanes are attached to a gearing arrangement that operates with the rotating shaft. This maintains the timing of the rolling vanes as they roll in and out of the rotor. Close tolerances are maintained between the vanes and the retaining ring to provide high efficiencies.

Piston-type Fluid Motors

Piston designs may be axial piston units or radial piston-type fluid motors. Both designs can be either fixed or variable motors. Their construction features are nearly the same as the pump units:

The axial piston motors use the valve plate method of displacing fluid, and the radial design uses the pintle arrangement of porting.

Piston-type fluid motors provide excellent operating characteristics and low leakage. They have high starting torque and can operate at much higher pressures than other types. Output speeds normally range between 0 and 3000 rpm. Operating pressures range from 0 to 5000 psi. In addition to high efficiency, piston units give long life expectancy.

Axial Piston Fluid Motors

Axial piston motors as shown in Fig. 8–43 operate on the principle that fluid entering a port pushes against the pistons during one-half of one revolution, which causes the cylinder barrel and shaft assembly to turn. During the next one-half of one revolution the same pistons exhaust, while other pistons following in sequences provide continuous operation.

As shown in Fig. 8–44, fluid entering the A port from the pump passes through the A kidney port in the valve plate and pushes against the piston and shoe

Fig. 8-44. Axial piston motor in operation.

Fig. 8-45. Example of a torque curve for a piston motor.

FIXED DISPL.   VARIABLE DISPL.

Fig. 8-47. Radial motor cutaway showing fixed and variable displacement.

assembly, causing it to seek the lower side of the sloping face of the cam plate. As the slope decreases, the piston moves outward in relation to the rotating cylinder barrel. Each sequential piston and shoe assembly imparts a tangential force to the cylinder barrel, causing it to rotate.

When the piston and shoe assembly reaches the other one-half of the revolution, the cam pushes the piston back into the cylinder barrel, exhausting fluid through the *B* kidney port and back to the pump. Each sequential piston and shoe assembly reacts to the cam plate in the same manner to exhaust their fluid.

Piston-type motors usually have an odd number of pistons—five, seven, nine, eleven, or more, depending upon the size of the unit. Generally, the greater number of pistons means more power and a smoother operation. (An odd number of pistons is used for positive starting torque.) Torque characteristics of a typical fluid motor are shown in Fig. 8–45.

Fig. 8-46. Variable displacement axial piston motor. (*Courtesy of Denison Engineering.*)

In the fixed unit the cam plate is stationary, whereas in the variable unit the cam angle is adjustable as shown in Fig. 8–46.

Radial Piston Motors

The radial piston design fluid motor has a cylinder barrel with an attached output shaft. The cylinder barrel contains a number of radial bores fitted with precision-ground pistons. The cylinder barrel and shaft assembly rotate within a thrust ring. If the fluid motor is a variable displacement unit, the thrust ring is adjustable. If the unit is fixed displacement, the thrust ring is stationary.

Fluid entering the motor from the pump passes through the pintle arrangement and into the cylinder chamber beneath the pistons. The pistons are forced against the thrust ring, which imparts a tangential force to the cylinder barrel and shaft assembly, causing it to rotate.

Mass rotating forces are low with the design, provides excellent performance characteristics for slow speeds with high efficiency.

With new metal alloys and improved techniques in manufacturing, both pumps and motors have improved tremendously, giving higher efficiency than ever before.

This high efficiency, coupled with flexibility in design and control, offers many new possibilities for the machines of the future using hydrostatic transmission drives.

Questions

1. What is one of the main advantages in applying actuating cylinders or fluid motors over mechanical devices?
2. What is the function of a hydraulic cylinder in a hydraulic system?
3. What kind of a valve is necessary to operate a double-acting cylinder?
4. How many feet per minute will a cylinder's piston travel if the pump is rated for 12 gpm and the diameter of the cylinder is $2\frac{1}{2}$ in.
5. How many seconds does it take for a cylinder's piston to stroke 8 in. if the area is 7 sq in. and it is operated by a 6 gpm pump?

$P = \frac{F}{A}$

6. What operating pressure is necessary to produce a force of 12,000 lb if the area of the cylinder is 3.14 sq in.?

$A = \frac{F}{P}$

7. What is the area of a cylinder if the force is 10,500 lb and the operating pressure is 1500 psi?

$F = AP$

8. What force will a cylinder produce if it has a bore diameter of $3\frac{1}{2}$ in. and is operating at a pressure of 2200 psi?

170

9. What precautions are necessary in the selection of conductors when cylinders operate faster on the return stroke than they do on the extending stroke?

10. When is a telescoping cylinder used?
11. Explain the operation of a tandem-type cylinder.
12. Why are wiper rings used on cylinder rods?
13. Explain the function of cushioning in cylinders.
14. Explain the operation of a rotary actuator.
15. What is the main difference between a rotary actuator and a fluid motor?
16. What are the two general classes of fluid motors?
17. How many rpm would a fluid motor make if the displacement of the motor per revolution is 0.85 cu in. and it is operated by a 7.5 gpm pump?
18. How much torque would this same motor produce if it is operated at 3000 lb per sq in.? What horsepower would this same motor produce?
19. What are the four general combinations of hydrostatic transmissions? Explain their output characteristics.
20. How are most reversible types of fluid motors drained? *186*
21. Name the construction types of fluid motors available.

fixed pump – fixed motor

fixed m

VP TOVM

CHAPTER 9

SYSTEM COMPONENTS AND CIRCUITS

The material presented in previous chapters has dealt with basic fundamentals and system components. In addition to the material already covered, there are numerous component parts of a hydraulic system that can be more simply explained in a discussion of circuitry.

Designing circuits is an intriguing responsibility. The circuit designer is limited only by the boundaries of his own ingenuity and knowledge of the components available

A circuit is an arrangement of components interconnected to provide a desired form of fluid power. The circuit is generally a part of the hydraulic system, whereas the system refers to the complete assembly of component parts that transmits and controls fluid power. A system may be composed of several circuits.

There are thousands of different components available for designing hydraulic circuits. Many of these components have multiple uses with merely the change of a cover, the use of an adapter, different ways of connecting, the addition of a port, or some other simple procedure.

Never before has a technology been so simple or so complex. It extends from an ordinary hydraulic jack to a space ship control system. Fluid power overlaps practically every branch of engineering, especially electrical, electronic, and mechanical engineering. All of these branches of engineering sciences are teaming up to use the best of each for any given design problem.

With all of this flexibility to feed the designer's imagination and ingenuity, he has a big responsibility to perform. Among these factors of responsibility there are three important considerations:

1. The system must be designed to operate safely. Safe operations of a system may include such things as:
   a. Pressure ratings of system components.
   b. Temperature ratings of system components.
   c. Operating speeds where pertinent.
   d. Compatibility between system components.
   e. Environmental conditions.
   f. Interlocks for sequential operations.
   g. Emergency shutdown features.
   h. Power failure locks or safety pins.
   i. Fire hazards.
   j. Protective devices for operators and servicemen.

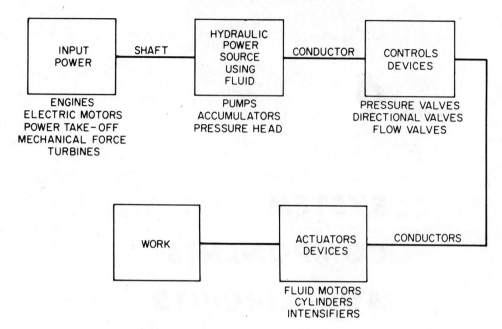

Fig. 9A-1. Block diagram showing system.

Fig. 9-1. Linear circuit-simple reciprocating circuit.

2. The system must be functional. This may include such things as the following:
   a. System components must meet the required performance specifications.
   b. System components should be selected according to the duty cycle for which they are designed.
   c. The life expectancy of the component parts should be essentially the same as the life of the machine.
   d. When the system involves electrical or mechanical hookups with fluid power, compatibility for fabrication and control is essential.
   e. System components should facilitate good maintenance practices.
   f. System components should be selected with a margin of safety to withstand operational hazards, such as hydraulic shock.
   g. Accumulaters should be discharged when being stored or in a rest condition to avoid danger to personnel or equipment.
3. The system must be efficient. The following are some of the factors to consider:
   a. Keep the system as simple as possible, but safe and functional.
   b. Use good fabrication techniques.
   c. Standardize component parts as much as possible.
   d. Maintain clean design with access to impor-

tant parts that may need repair or adjustment.

e. Consider the availability of replacement parts.

f. Design to keep operational costs to a minimum.

g. Design the system to prevent and remove contamination.

h. Supply good drawings.

The factors mentioned concerning safety, function, and efficiency of a system are only a few of the many important considerations in designing hydraulic systems.

One of the most important tools for anyone working in fluid power is a knowledge of the components available and how they operate in a circuit.

The discussion that follows will cover components and circuitry that are representative of the present technology. The various circuits presented will cover component selection factors, methods of calculating specification data, operating information, and graphical diagramming.

Linear Circuits

Figure 9–1 shows a simple reciprocating circuit for bending a piece of sheet steel. The system has the following components:

1. Reservoir.
2. Strainer.
3. Pump inlet line.
4. Pump.
5. Flexible coupling.
6. Electric motor.
7. Discharge conductor and connectors.
8. Relief valve.
9. Directional four-way valve (manually controlled).
10. Double-acting single rod end cylinder.

How is a circuit problem of this nature approached?

1. What are the specifications for the job?
   a. Force requirement to bend the steel = 1800 lb.
   b. Length of work stroke = 6 in.
   c. Speed of piston and rod assembly extending = 0.50 sec.
2. What size cylinder is needed? Since the force required to do the work is 1800 lb, cylinder area and operating pressure must be selected.

$$F = PA \qquad A = \frac{F}{P} \qquad P = \frac{F}{A}$$

Fig. 9-2. Regenerative circuit.

Referring to standard bore and rod sizes for cylinders, we make a selection by reasoning that a larger diameter cylinder operates at a lower pressure but requires a bigger pump to give the required cylinder speed. We also reason that a small diameter cylinder must operate at a higher pressure, but a smaller pump is capable of the cylinder speed.

Cylinder selected = 2 in. bore.
Operating pressure:

$$P = \frac{F}{A} \qquad P = \frac{1800}{3.14} \qquad P = 573 \text{ psi}$$

In considering the types of pumps available, both vane and gear pumps are relatively low in cost and give excellent life at this pressure level.

3. What capacity pump is needed? The pump delivery is selected to provide the maximum cylinder speed required by the circuit.

$$\text{gpm} = \frac{\text{area of cyl.} \times \text{length of stroke} \times 60 \text{ sec}}{231 \text{ in.}^3 \times \text{time in sec}}$$

$$\text{gpm} = \frac{3.14 \text{ in.}^2 \times 6 \text{ in.} \times 60}{231 \text{ in.}^3 \times 0.50}$$

$$\text{gpm} = 9.8$$

The above calculations indicate that 9.8 gpm are needed to extend the 2 in. diameter bore cylinder 6 in. in 0.5 sec. It must deliver this amount at a pressure which will reach 573 psi.

The next step is to determine the pump to be used by considering the selection factors as described in the chapter on pumps. It is generally good practice to select a standard pump that provides the nearest specifications to those desired. It is also important to remember that the pump selected usually dictates the choice of other system components. However, if there are unique design characteristics about a system, such as being mobile or stationary, directly or remotely controlled, having limitations in size and weight, or operating at extreme temperatures, then *complete system compatibility is a must*.

4. What size electric motor is needed to drive the pump?

$$\text{hp} = \text{gpm} \times \text{psi} \times 0.000583$$
$$\text{hp} = 9.8 \times 573 \times 0.000583$$
$$\text{hp} = 3.27$$

This indicates that if the pump and circuit are 100 per cent efficient, then 3.27 hp is needed. A gear or vane type pump is generally about 85 per cent efficient, and this means that the power losses in the pump would be about 15 per cent. The pump capacity would have to be greater than 9.8 gpm to make up the loss and still satisfy the cylinder speed. It is possible that other losses may also occur, such as fluid friction, mechanical friction, and others.

Electric motors generally have an overload factor, but this should not be used under continuous duty operation. A good selection in this case would be a standard 5 hp electric motor of the enclosed type. (Enclosed designs prevent a fire hazard due to arcing.) Shaft size and type, direction of rotation, and mounting standards should also be indicated. The type of flexible coupling between the pump and electric motor is merely a matter of choice. The mating shafts should be compatible.

5. What size reservoir should be used? The rule of thumb is $2\frac{1}{2}$ to 3 times the capacity of the pump. In this case, a standard 30 gpm reservoir would be adequate. If there were a number of large cylinders in the circuit or if fluid were continuously blowing over the relief valve, causing heat, then the size must be calculated.

6. What size pump inlet would be needed? Inlet flow velocities normally must be maintained between 2 and 5 ft/sec. The total pump capacity, including 9.8 gpm for the cylinder speed plus about 1.5 gpm to make up for pump slip, equals 11.3 gpm.

The formula for determining area size of conductor is

$$\text{Area} = \frac{\text{gpm} \times 0.3208}{v}$$

$$\underset{\text{(sq in.)}}{\text{Area}} = \frac{11.3 \text{ gpm} \times 0.3208}{3 \text{ ft/sec}}$$

$$\text{Area} = 1.2 \text{ in.}^2$$

The pipe tables indicate that by using standard schedule 40 pipe the inlet size would be $1\frac{1}{4}$ in. pipe. If the system had to use fire resistant fluid with a high specific gravity or if the circuit operated at a high altitude with less atmosphere, then a larger size might be used.

7. What size discharge conductors are needed? The same procedure is used to determine the size of conductors and connectors on the discharge side of the pump. The velocity of flow for the discharge side should be maintained between 7 and 15 ft/sec.

Since the operating pressure is relatively low, either pipe or tubing can be selected. Tubing offers fewer fittings and is generally easier to install. In this case, standard $\frac{3}{4}$ in. pipe or SAE 1010 dead soft steel tubing would be adequate. Trends indicate that tubing is usually preferred because of less potential leakage.

The inlet strainer would be selected on the basis of the inlet pipe size.

8. How is the relief valve selected? A number of component manufacturers make only a few standard sizes of valves. A standard relief valve having $\frac{3}{4}$ in. NPT ports would be a logical choice. The system capacity determines the control valve sizes, which are compatible with standard conductor sizes.

A simple relief valve fitted with a spring having a range setting between 0 and 1000 psi would be selected. Usually, the relief valve is set from 10 to 12 per cent higher than necessary to perform the work.

9. What determines the type of directional four-way valve? A manual operator was specified

for valve actuation; however, this might have been electrical, mechanical, pilot, or other. The type of valve actuation depends on the application.

The graphical diagram indicates that an open center four-way valve is used. Open center systems allow the pump delivery to by-pass back to the reservoir when not being used to extend or retract the cylinder.

If a closed center valve had been used, the pump delivery would blow across the relief valve and heat would result. The size of the four-way directional control would be the same as the relief valve and have the same port configuration. The procedure used for selecting components for this simple linear circuit generally applies to any circuit. Selecting components for this simple linear circuit generally applies to any circuit. Selecting components for any hydraulic system is usually a compromise, with cost as the major factor.

A Regenerative Circuit

A regenerative circuit is one in which pressurized fluid discharged from a component is returned to the system to reduce power input requirements (see Fig. 9-2).

On single rod end cylinders the discharge from the rod end is often directed to the opposite end to increase rod extension speed. If equal speeds in both directions are required, a 2:1 area ratio is used.

To simplify the example, a cylinder with a 2 in. area bore and a 1 in. area piston rod is shown in Fig. 9-3. The pump has a volumetric capacity of 10 gpm, and the cylinder stroke is exactly 12 in.

1. What speed will the piston and rod assembly travel in each direction? (Calculations to be made in ft/min.)

Using the formula $\text{ft/min} = \dfrac{\text{gpm} \times 19.25}{\text{area in.}^2}$

we have

Forward stroke $\quad \text{ft/min} = \dfrac{10 \times 19.25}{2} = 96.25$

Reverse stroke $\quad \text{ft/min} = \dfrac{10 \times 19.25}{1} = 192.5$

Thus it is shown that a speed ratio of 2:1 is given when one is using a pump with a constant displacement of 10 gpm. Because the cylinder area or volume is a ratio of 2:1, the piston and rod assembly travels twice as fast, retracting as it does on the forward stroke. By connecting the fluid from the rod end to that of the large area end, equal speeds would result in both directions.

This principle is used to provide equal work speeds in both directions, which is desired in many applications, such as presses and bulldozer scraper blades.

It is important to remember that only one-half of the area is effective for producing force as the cylinder's piston and rod assembly extends during regeneration of the fluid. A regenerative circuit does not reduce power input requirements; it trades force for speed.

The press circuit shown in Fig. 9-4 uses a by-pass valve to release fluid on the rod side of the piston during the pressing action. This facilitates the use of the entire area on the large piston end to produce the required force.

With the pump in operation and the manually actuated four-way valve positioned for press down, 20 gpm is delivered through port C-1 and to the large area end of the cylinder. The sequence valve is set to open when the pressure reaches 300 psi. This control also acts as a counterbalance valve to prevent the press from dropping due to gravity.

If friction is neglected, the force relationship may be expressed by using Pascal's principle:

Force acting to push the press down equals

$$F = P \times A = 300 \times 20 = 6000 \text{ lb}$$

plus 1000 lb of weight for press member, which gives a total of 7000# force.

Force resisting the downward motion of the press equals

Pressure on the rod side due to fluid

Pressure on top $= \dfrac{F}{A} = \dfrac{300 \times 20}{10} = 600 \text{ psi}$

Fig. 9-3. Cylinder with 2 in.² area bore and 1 in.² area piston rod.

Fig. 9-4. Press circuit.

Fig. 9-5. Design types of hydropneumatic accumulators.

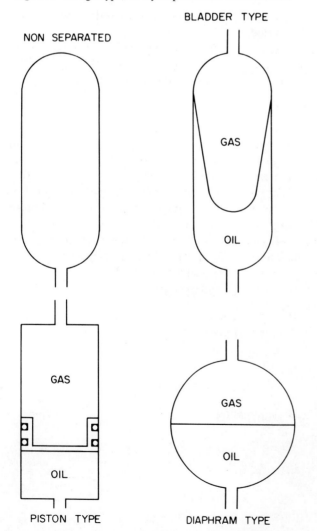

NON SEPARATED

BLADDER TYPE

GAS

OIL

PISTON TYPE

GAS

OIL

DIAPHRAM TYPE

GAS

OIL

Pressure due to weight $= \dfrac{1000}{10} = 100$ psi

Therefore,

Total resisting force $= P \times A = 700$ psi

$\times\ 10$ in.$^2 = 7000$ lb force

This proves that the press is balanced on both sides, which means it would take a slightly higher pressure than 300 psi on top to move the press downward.

As the press moves downward, the fluid in the rod end of the cylinder is forced out and passes over the sequence valve to join the fluid coming from the pump.

$$\text{Time extending} \atop \text{(seconds)} = \dfrac{\underset{(\text{in.}^2)}{\text{area}} \times \underset{(\text{in.})}{\text{stroke}} \times \underset{(\text{sec})}{60}}{\underset{(\text{cu in.})}{231 \times \text{gpm}}}$$

$$\text{Time extending} = \dfrac{20\ \text{in.}^2 \times 10\ \text{in.} \times 60\ \text{sec}}{231 \times 40\ \text{gpm}} = \dfrac{300}{231}$$

$$\text{Time extending} = \dfrac{300}{231} = 1.29\ \text{sec}$$

As soon as the press comes in contact with the work and resistance causes the pressure to build up to the setting of the by-pass valve (500 psi), it opens, relieving the back pressure beneath the piston. Check valve A isolates the high pressure side from the low pressure side of the system.

With the relief valve set for 3000 psi, the maximum force available for the pressing action would be

$F = P \times A = 3000 \text{ psi} \times 20 \text{ in.}^2 + 1000 \text{ lb}$

Total force $= 61,000 \text{ lb}$

When the pressing action is completed, the four-way valve is positioned to raise the press. Fluid from the pump is directed through port C-2 and check valve B into the rod end of the cylinder.

$$\text{Time retracting} = \frac{\text{area} \times \text{stroke} \times 60}{231 \times \text{gpm}}$$

$$\text{Time retracting} = \frac{10 \text{ in.}^2 \times 10 \text{ in.} \times 60 \text{ sec}}{231 \text{ cu in.} \times 20 \text{ gpm}}$$

$$\text{Time retracting} = \frac{300}{231} = 1.29 \text{ sec}$$

Thus, a 2:1 area ratio cylinder provides equal speeds in both directions.

The pilot check valve is used in this case to divide the flow coming out of the large side of the cylinder.

Remember, on the press downstroke 20 gpm from the pump plus 20 gpm from the rod side of the cylinder was used to bring the press down in 1.29 sec. This means that with the equal speeds in both directions 40 gpm must be taken out of the large area side on the up stroke.

If a larger conductor is used between the large area end of the cylinder and the pilot check valve, it allows part of the fluid to pass through a separate line out of the check valve to the reservoir. The rest of the fluid passes through the four-way valve and back to the reservoir. This arrangement allows the use of smaller components in the system, which helps to reduce costs.

Accumulator Circuits

An accumulator is a container in which fluid is stored under pressure as a source of fluid power.

There are two general types of accumulators: the hydropneumatic and the mechanical designs. The hydropneumatic design (see Fig. 9–5) uses a compressed gas which applies force to the stored liquid. Mechanical designs use a weighted member or spring which applies force to the stored liquid (see Fig. 9–6).

The hydropneumatic designs are most commonly used for hydraulic systems and use either air or nitrogen for precharge. This type operates by the principle of Boyle's law:

$$P_1V_1 = P_2V_2$$

Accumulators are used for a number of different functions. Some of these uses are discussed as follows:

1. *Auxiliary power source* to supplement the pump where the cycle time will allow a charge to be stored for peak requirements (see Fig. 9–7). In an intermittent-duty-type system an

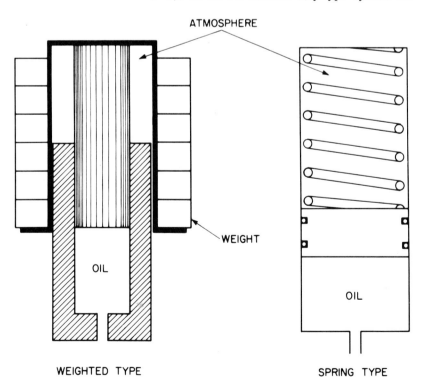

Fig. 9-6. Mechanical designs of accumulators.

ATMOSPHERE

WEIGHT

OIL

OIL

WEIGHTED TYPE

SPRING TYPE

Fig. 9-7. Auxiliary power source.

Fig. 9-8. Emergency power source.

accumulator is used as a secondary power supply. This helps to reduce the input horsepower by storing energy during idle times of the machine. The pilot-operated relief or unloading valve opens when the accumulator is fully charged and allows the pump delivery to return to the reservoir at very low pressure. The accumulator maintains its charge, since it is isolated by the check valve. The solenoid actuated on-off type valve is shown as a safety device to unload the accumulator when the pump is shut down.

2. *Emergency power source* in case of power failure to operate critical circuit functions (see Fig. 9–8). The closed-center-type directional four-way valve allows the accumulator to charge during idle times, and the pump unloads at full charge through the pilot-controlled relief valve.

If emergency fluid power is needed, the on-off valve is shifted and fluid enters the system to extend or retract the cylinder through the four-way valve.

3. *Hydraulic shock absorber* for circuits where sudden impact loads, quick stops, or reversals

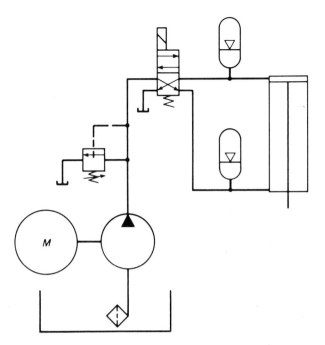

Fig. 9-9. Hydraulic shock absorber.

reduced. The accumulator should be installed as close to the shock source as possible.

The formula for sizing the accumulator to absorb hydraulic shock is as follows:

$$\text{Size of acc.} = \frac{(0.004)(\text{rate of flow}(0.005 \text{ length of in gpm}) \text{ conductor in ft}) \text{ upstream of valve} -\text{valve closing time in sec}}{\text{max. surge psi–psi for max. flow at valve}}$$

(Peak surge pressures can be determined by using a transducer and oscilloscope or hydrauliscope.)

4. *Leakage compensation* in circuits necessary to hold loads for long periods of time. The pressure switch may be used as shown in Fig. 9–10 to start and stop the hydraulic pump for maintaining the system pressure level for adequate holding force.

5. *Pressure transfer barrier* operations generally use accumulators to apply a static test or transfer fluid energy to a secondary fluid system. In the example shown in Fig. 9–11, the fire extinguisher tank is pressure-tested to 2000 psi. City water prefills the tank, then the on-off valve A is closed. Hydraulic fluid then enters

with heavy loads are a characteristic of the system (see Fig. 9–9). Hydraulic shock loads may be reduced considerably if the deceleration time of the flowing fluid mass can be

Fig. 9-10. Leakage compensation.

Fig. 9-11.   Pressure transfer barrier.

Fig. 9-12.   ASA symbol-single and double acting intensifier.

the lower side of the accumulator, acting against the movable barrier and creating a pressure controlled by the relief valve. The operator would hold the test pressure for the required time and then open valve A and close valve B. This would unload the hydraulic pump back to the reservoir and allow a new fire extinguisher to be installed for test.

Charging and Sizing Accumulators

Precharge pressures are selected so that almost all of the hydraulic fluid charge is used to do useful work. In most cases, the precharge pressure equals the minimum pressure that will perform the work.

The operating principle of the accumulator is based on Boyle's law for gases (see Section II, Chapter 1).

Accumulators are preloaded with air or an inert gas, such as nitrogen. Nitrogen is generally used because it is relatively inexpensive and helps to prevent corrosion within the component. As hydraulic fluid is supplied to charge the accumulator, it causes the movable element to compress the gas into a smaller volume to make room for the hydraulic fluid which is being stored. The ASA Standard requires that accumulators working above 200 psi must be charged with dry nitrogen.

Standard sizes of accumulators generally range from $1\frac{1}{2}$ cu in. to 10 gallon capacity and operate at pressures up to 6000 psi. The nonseparator type is available in larger sizes, and specially built units for aerospace use operate at 60,000 psi.

In order to determine the size of an accumulator required, the amount of hydraulic fluid needed must be known. This is determined by the cylinder or fluid motor size and the work speed. Total volume of the accumulator is the sum of the hydraulic fluid stored and the volume of compressed gas necessary to supply the fluid at its working pressure.

The rate of charging or discharging determines whether the process is isothermal or adiabatic. Isothermal expansion occurs when the process is slow enough to allow heat transfer out of the accumulator, which is generally taken as 3 minutes or longer.

The equation for isothermal expansion and compression is

$$P_1V_1 = P_2V_2$$

Adiabatic expansion or compression occurs when the process is fast enough to retain all of the heat.

The equation for adiabatic expansion and compression is

$$P_1V_1^{1.4} = P_2V_2^{1.4}$$

Example. What size accumulator is necessary to supply 2.5 gpm with an allowable pressure drop from 3000 to 2000 psia? The precharge is 1500 psia. (2.5 gpm $\times$ 231 cu in. = 577.5 cu in.) (*Note.* Absolute pressure must be used when one is working with gases.)

Isothermal

$$P_1V_1 = P_2V_2 = P_3V_3$$

$P_1 = 1500$ psia $\quad V_1 = ?$

$P_2 = 2000$ psia $\quad V_2 = V_3 + 577.5$

$P_3 = 3000$ psia $\quad V_3 = V_2 - 577.5$

$$P_2V_2 = P_3V_3$$

$$2000\,(V_3 + 577.5) = 3000\,V_3$$

$$V_3 = 1155 \text{ cu in.}$$

$$V_2 = 1732.5 \text{ cu in.}$$

$$P_1V_1 = P_2V_2$$

$$1500\,V_1 = 2000\,(1732.5)$$

$$V_1 = 2306.6$$

$$V_1 = \frac{2306.6}{231}$$

$$V_1 = 9.9 \text{ gal}$$

Adiabatic

$$P_1V_1^{1.4} = P_2V_2^{1.4} = P_3V_3^{1.4}$$

$P_1 = 1500$ psia $\quad V_1 = ?$

$P_2 = 2000$ psia $\quad V_2 = V_3 + 577.5$

$P_3 = 3000$ psia $\quad V_3 = V_2 - 577.5$

$$P_1V_1^{1.4} = P_2V_2^{1.4} = P_3V_3^{1.4}$$

$$P_2V_2^{1.4} = P_3V_3^{1.4}$$

$$V_2 = \left[\frac{P_3}{P_2}\right]^{0.714} \times V_3$$

$$V_2 = \left[\frac{3000}{2000}\right]^{0.714} (V_2 - 577.5)$$

$$V_2 = 2295.5 \text{ cu in.}$$

$$P_1V_1^{1.4} = P_2V_2^{1.4}$$

$$V_1 = 2295 \left[\frac{2000}{1500}\right]^{0.714}$$

$$V_1 = 2820 \text{cu in.}^3$$

$$V_1 = \frac{2820}{231} = 12.2 \text{ gal.}$$

$$P_1V_1^{1.4} = P_2V_2^{1.4}$$

$$V_2 = V_1\left(\frac{P_1}{P_2}\right)^{1/1.4}$$

$P_1 = 40$ psig $= 54.7$ psia

$P_2 = 90$ psig $= 104.7$ psia

$V_1 = 32$ cu ft

$V_2 = ?$

Take logs

$$\text{Log } V_2 = \log V_1 + \frac{1}{1.4}(\log P_1 - \log P_2)$$

$$\frac{1}{1.4}(\log 54.7 - \log 104.7) \rightarrow \log 54.7 =$$

$$\qquad\qquad\qquad\qquad 1.73799$$

Adjust for subtraction ⑩ $\quad + 10.00000 - 10.0000$

$$\text{Log } 54.7 = 11.73799 - 10$$

$$- \text{ Log } 104.7 = 2.01995$$

Log (0.52245)

$\quad = (\log 54.7 - \log 104.7) = 9.71804 - 10$

$$\frac{1}{1.4}[\log (0.52245)] = \frac{9.71804 - 10}{1.4} =$$

$$\qquad\qquad\qquad 6.94146 - \quad 7.14286$$

Adjust $\qquad\qquad\qquad 2.85714 \qquad 2.85714$

$$\frac{1}{1.4}(\log 0.52245) = 9.79860 - 10.00000$$

Now add

Log $\qquad\qquad\qquad 32 = 1.50515$

Adjust $\qquad\qquad\qquad\qquad 11.30375 - 10$

$$\text{Log } V_2 = 1.30375$$

$$V_2 = 20.125 \text{ cu ft}$$

Check

$$\left(\frac{20.125}{32}\right)^{1.4} = \frac{54.7}{104.7} = 0.52245$$

$$(0.6289)^{1.4} \rightarrow \log 0.6289 = 9.79858 - \quad 10$$

$$\qquad\qquad\qquad\qquad\qquad\qquad\qquad 1.4$$

$$\qquad\qquad\qquad\qquad\qquad 13.71801 - 1.4$$

$$\qquad\qquad\qquad\qquad\qquad - \quad 4$$

$$\qquad\qquad\qquad\qquad\qquad 9.71801 \qquad \equiv 0.5225$$

$$P_1 V_1^{1.4} = P_2 V_2^{1.4}$$

$$\left(\begin{matrix}2000 \\ \text{psig}\end{matrix}\right)\left(\frac{X}{\text{gpm}}\right)^{1.4} = \left(\begin{matrix}3000 \\ \text{psig}\end{matrix}\right)\left(\frac{X - 2.5}{\text{gpm}}\right)^{1.4}$$

$$P_1 = 2000 + 14.7 = 2014.7 \text{ psia}$$

$$P_2 = 3000 + 14.7 = 3014.7 \text{ psia}$$

$$\left(\frac{X}{X - 2.5}\right)^{1.4} = \frac{3014.7}{2014.7} = 1.4963518 + \frac{2854}{20,147} R$$

Machine division

$$\qquad\qquad\qquad 1.4963518$$

Correct for ⑤ digits $+ \ .00005$

$$\qquad\qquad\qquad 1.4964$$

Now

$$\frac{X}{X - 2.5} = (1.4964)^{1/1.4}$$

Solve

$$[(1.4964)^{1/1.4}] \equiv \log 1.4964 = 0.17405$$

$$(\quad) = \left(\frac{1}{1.4}\log 1.4964\right) = \frac{0.17405}{1.4}$$

$$(\quad) = \left(\frac{1}{1.4}\log 1.4964\right) = 0.12432$$

$$(\quad) = \text{antilog of } 0.12432 = 1.3313$$

Now

$$X = (X - 2.5)(1.3313)$$

$$X = 1.3313X - 3.32825$$

$$0.3313X = 3.328$$

$$X = \frac{3.328}{0.3313} = 10.045 \text{ gpm}$$

Let us call $X = 10$ and check.

$$(2000)(10)^{1.4} = 3000(7.5)^{1.4}$$

$$10^{1.4} = 1.5 \times 7.5^{1.4}$$

or

$$1.5 = \left(\frac{10}{7.5}\right)^{1.4}$$

$$= (1.333)^{1.4}$$

$$\text{Log } = 0.12483$$

$$\times \ 1.4$$

$$0.17476$$

Intensifier Circuits

An intensifier is a device which converts low pressure fluid power into higher pressure fluid power.

Intensifiers (also called boosters) are used to multiply forces when a great force is needed through a relatively short distance. Hydraulic presses, riveting machines, and spot welders are typical applications.

There are two general types of intensifiers, the air-to-hydraulic and the all hydraulic. The single-plunger, piston-to-ram intensifier is most commonly found. An automatically reciprocating double-acting design is available for applications needing longer work strokes or for maintaining high pressures for a long period of time (see Fig. 9–12).

The air-over-hydraulic intensifier operates on a principle of converting shop air pressure into higher hydraulic pressure (see Fig. 9–13).

A reciprocating intensifier consists of a large double rod end cylinder with the rod ends simulating the operation of two single-acting pumping pistons. Shop air automatically reciprocates the piston assembly through the use of mechanically actuated

Fig. 9-13. Air-over-hydraulic intensifier.

air pilot valves. The pilot valves supply air to the ends of the directional control valves to alternately shift positions for reciprocation.

As the piston assembly reciprocates by using air, the rod ends act like pumps with an inlet and outlet check valve for each pumping piston. When the intensifier moves toward the *left* side, it causes the inlet check valve to close and the discharge check valve to open and supply high pressure fluid for the system. At the same time, air pressure over hydraulic fluid in the A tank causes fluid to enter the piston chamber at the right end to replace the moving piston.

At the end of the stroke to the left, a mechanical stem is actuated by the large air piston, which allows

the pilot valve to shift the main directional control valve to reverse the intensifier. The procedure repeats until the maximum pressure is reached.

Tank A uses hydraulic fluid pressurized by plant air to give the work cylinder a fast approach when the load is light. Then the intensifier comes into action to do the heavy work. After the work stroke is completed, Tank B supplies hydraulic fluid under pressure to return the work cylinder.

Intensifier ratio is determined by

$$\text{Int. ratio} = \frac{\text{intensifier output pressure (psi)}}{\text{intensifier input pressure (psi)}}$$

Intensifiers are available in a great number of ratios, with output pressures to 10,000 psi and even higher.

Because hydraulic fluid exposed to the atmosphere contains entrained air, it compresses. This may cause a delay in the system response between the input flow and the output motion.

The compressibility factor of the hydraulic fluid becomes important when higher working pressures are used. Compressibility of a liquid is the change in volume due to pressure changes. The recommended

Fig. 9-14. Intensifier ratio (example problem).

RATIO 50:1

Fig. 9-15. Force multiplication.

compression value for intensifiers is about 2 per cent of the total work cylinder volume. When conductors are large or long, all the fluid under compression should be calculated for accuracy.

Sample Problem.

Required work force = 24,000 lb

Total work cylinder stroke = 12 in.

High pressure work stroke = 4 in.

Available air pressure = 100 psi

What stroke intensifier or booster is required?

1. Determine the size of work cylinder needed. (A standard 2.5 in. bore diameter cylinder is selected for this application.) This means that a pressure of nearly 5000 psi will be needed for the work portion of the stroke.

$$\text{Pressure} = \frac{\text{force}}{\text{area}} = \frac{24{,}000\ \text{lb}}{4.9\ \text{sq in.}} = 4890\ \text{psi}$$

2. Determine the fluid needed for the high pressure phase of the work cycle.

Volume in.^{3} = area in.^{2} × stroke (in.)

= 4.9 in.^{2} × 4 in.

Volume = 19.6 in.^{3}

3. Calculate the compressibility factor.

Total cylinder volume = 4.9 in.^{2} × 12 in.

Volume = 58.8 cu in. of hydraulic fluid under 5000 psi.

In this case approximately 2 per cent of the total volume is selected for the compressibility factor of the fluid under pressure.

58.8 cu in.

$\dfrac{0.02\%}{1.176}$ cu in. lost through compressibility of the fluid in the cylinder only.

4. Use the following equation for determining the stroke of intensifier or booster:

Intensifier stroke =

$$\frac{\substack{\text{fluid for high pressure} \\ \text{phase of work stroke} \\ \text{(cu in.)}} + \substack{\text{volume due} \\ \text{to} \\ \text{compress.} \\ \text{(cu in.)}}}{\text{area of ram piston (in.}^2\text{)}} + \substack{\text{Pretravel} \\ \text{(in.)}}$$

The pretravel is that portion of the intensifier stroke that is not effective for motion of the work cylinder and will vary with design. (Assume 1.5 in. pretravel.)

Since nearly 25,000 lb of force is needed and 100

psi plant air is available, a 50 to 1 ratio intensifier is selected (see Fig. 9–14).

$$\text{Ratio } 50{:}1$$

Intensified hyd. pressure $= \dfrac{100 \text{ psi air} \times 50 \text{ in.}^2}{1 \text{ in.}^2}$

Intensified hyd. pressure $= 5000$ psi

Stroke equation:

Intensifier stroke $= \dfrac{19.6 \text{ in.}^3 + 1.17 \text{ in.}^3}{1 \text{ in.}^2} + 1.5 \text{ in.}$

$$\text{Stroke} = 22.27 \text{ in.}$$

Consulting a manufacturer's catalog can determine the exact intensifier or booster that will provide the proper stroke.

The system shown in Fig. 15 used on a riveting machine shows the tremendous flexibility in applying hydraulics where force multiplication is advantageous.

Neutral. With both pumps operating, hydraulic fluid flows through the open centered directional control valve and back to the reservoir.

Prefill. Shifting the directional four-way to the A position directs the flow from both pumps through lines 1 and 2 through the prefill passage and into line 3 for extending the work cylinder for a fast approach.

High pressure stroke. When the machine reaches the rivet and resistance due to work causes a pressure increase great enough to open the sequence valve, fluid is directed to the large end of the intensifier through line 4. At the same time, pilot pressure unloads the large pump. The piston assembly of the intensifier advances covering the prefill port, then furnishes intensified fluid pressure to upset the rivet.

Return stroke. When the rivet is set, the operator shifts the four-way directional valve to the B position. This causes a drop in pressure, and the unloading valve closes to direct the flow from both pumps through line 5, returning the work cylinder and the intensifier ram piston assembly. The check valve allows free flow of fluid from the large end of the intensifier back to the reservoir.

Hydrostatic Transmissions

Hydraulic power transmissions are of two general types, hydrokinetic and hydrostatic. The hydrokinetic type uses the kinetic energy of a high velocity flow of fluid, whereas the hydrostatic power transmission uses high pressures and relatively low velocities.

Detailed Comparison of Hydrodynamic/Hydrostatic Transmissions

There has been confusion in the minds of many engineers over the distinction between a hydrodynamic and hydrostatic transmission. A brief résumé of the points of difference will be given in tabular form below to explain this difference.

Hydrodynamic transmissions
  1. A change in fluid *velocity* yields output energy.
  2. Nonpositive displacement units.
  3. No output torque available at low input speeds.
  4. Approximately 6–1 speed ratio input; output maximum practical.
  5. Tend to creep.
  6. Nonreversing.
  7. Input and output directly coupled.
  8. Torque varies over wide range.
  9. Practically no internal braking.
 10. Speed drops off with load.
 11. Only limited control.

Hydrostatic transmissions
  1. A change in fluid *pressure* yields output energy.
  2. Positive displacement units.
  3. Output torque available at all speeds.
  4. Approximately 60–1 speed ratio is practical with hydrostatic (50–3000 rpm range).
  5. No creeping tendency.
  6. Fully reversible.
  7. Input and output can be remote or directly coupled.
  8. Substantially constant torque available over full speed range of hydraulic motor.
  9. Direct braking available.
 10. Substantially constant speed with load variations.
 11. Positive control available.

The hydrokinetic type is doing an excellent job on such applications as automatic transmissions for automobiles. However, in the field of fluid power the hydrostatic type provides greater flexibility of application and more positive control. The term "hydrostatic transmission" refers to the use of hydraulic pumps and motors for converting fluid power into mechanical rotary motion.

Both fixed and variable piston pumps and fluid motors have been used on hydraulic drives for many years. Applications that require critical control and reliability, such as steering systems for ships, naval gun turrets, and other uses, have found that hydraulic drives are unmatched for response characteristics and power performance.

During the last two decades, improvements in design and operational performances have opened new fields of usage which hold unlimited possibilities for machines and equipment of the future. Hydraulic drives are now available up to 1000 hp, and the trends indicate even higher horsepower with the advent of higher pressures, which are possible through the availability of better metals and materials.

Because hydrostatic transmissions utilize positive displacement pumps and motors, they offer a number of advantages over other systems:

1. Constant speed can be maintained with varying load requirements.
2. Constant torque can be maintained with varying rotational speeds.
3. A wide range of operating speeds is available up to 10,000 rpm, with some aerospace designs operating much higher.
4. Systems are available for rotation in one direction or for reversible operations.
5. Rotational speeds can be precisely controlled over the full torque range of the motor.
6. Direct braking can be obtained by applying back pressure on the rotating motor.
7. Compact and lightweight units are capable of high power.

8. Both "open loop" and "closed loop" systems are available.
9. High over-all efficiencies are available.
10. Hydrostatic transmissions are relatively simple in operation and easy to maintain.

Figure 9–16 shows a typical application which uses a hydrostatic transmission drive.

Sample Problem. Select the pump and motor for a hydrostatic transmission that has the following requirements:

a. The fluid motor must have a speed range between 0 and 2000 rpm.
b. The maximum torque requirement is 680 lb in.
c. Fluid motor must operate in both directions.

In Fig. 9–17 a typical closed loop system is shown in graphical symbols indicating the components necessary for replenishing, filtering, and braking.

a. First determine the fluid motor displacement necessary for the torque requirement. (Select an operating pressure of 1500 psi.)

$$T = \frac{\text{psi} \times \text{dis/in.}^3/\text{rev}}{2\pi}$$

$$680 = \frac{1500 \times \text{dis/in.}^3/\text{rev}}{6.28}$$

$$\text{dis/in.}^3/\text{rev} = \frac{6.28 \times 680}{1500}$$

$$\text{dis/in.}^3/\text{rev} = 2.85 \text{ cu in. per revolution}$$

Fig. 9-16. Hydrostatic transmission application. *(Courtesy of Dynex.)*

Fig. 9-17. Closed loop.

b. Find the maximum pump delivery.

Maximum pump delivery

$$= \frac{\text{maximum motor speed} \times \text{displacement}}{231}$$

$$\text{Pump delivery} = \frac{2000 \times 2.85}{231}$$

$$\text{Pump delivery} = 24.6 \text{ gpm}$$

If the pump and motor together have a volumetric efficiency of 90 per cent, then 10 per cent, or 2.46 gpm, extra fluid would be needed for making up slip and leakage.

$$\text{Total pump capacity} = 27.06 \text{ gpm}$$

c. How much horsepower input is needed?

$$\text{hp} = \text{gpm} \times \text{psi} \times 0.000583$$

$$\text{hp} = 27.06 \times 1500 \times 0.000583$$

$$\text{hp} = 23.66$$

There are many other important factors of consideration which would depend on the specific end use of the system.

1. The duty cycle for the equipment operation may require a heat exchanger.
2. The type of pump control, such as manual, electrical, mechanical, pilot, or servo, may be important for functional machine control.

3. The environmental conditions, such as heat, altitude, contamination, and other physical aspects, may dictate such things as special seals, additional components, or fire resistant fluid.
4. Cost is a big factor which may suggest a less expensive method of doing the job.

In general, the same method of approach for selecting component types, sizes, ports, and performance data is used for both linear and rotary circuits. The previous circuit selected would be a sophisticated solution to the problem. Reversible or over-the-center pumps are used with the closed loop approach. However, an open loop system could also resolve the problem with a little less sophistication.

The main difference in the two systems is that in the closed loop the reversible pump controls the speed in both directions. In the open loop a directional valve controls the direction of motor rotation. Fluid in the closed loop circulates in a closed loop with only the leakage replenished. The pump in the open loop uses fluid from the reservoir for the inlet and then exhausts it back to the reservoir through the four-way valve.

Replenishing during the braking period for a heavy load may present a problem in the open loop circuit. Sometimes, heavy loads keep the fluid motor rotating after the four-way valve is shifted to the center position. This means that fluid must be sup-

REPLENISHING
THROUGH
CHECK VALVES

Fig. 9-18. Open loop transmission.

plied to the motor because, essentially, it is now a pump and could be damaged by lack of fluid.

The method shown in Fig. 9–18 may be a possible solution. Replenishing fluid to the low pressure side of the circuit can be accomplished by connecting the reservoir fluid to both sides of the motor through a check valve arrangement.

Braking heavy inertia loads might be done as shown, with a counterbalance valve put into operation by an on-off type of valve which directs the flow to the reservoir or through the braking type counterbalance valve.

It is important to remember that shifting fourway valves from one position to the other may give an open center condition or a closed center condition in between the various positions.

A heavy load rotating could feasibly rupture a line if the fluid has no place to go, even for a fraction of a second; therefore, an open center condition is necessary.

It is also important to know how much back pressure a four-way valve can stand if the exhaust fluid going back to the reservoir is restricted. This also makes it imperative to connect drains to a separate line going back to the reservoir when components are specified to be externally drained.

The circuit calculations would be made in the same manner as before. Either method used would still require the same horsepower as before. However, the pressure could be lowered if the displacement of the motor were increased proportionally, which would still provide the same torque as before. This would help resolve problems of cost by perhaps changing to vane equipment instead of piston pump and motor. Figure 9–19 shows one type of hydrostatic trans-

mission using a variable volume reversible pump and a fixed displacement fluid motor.

A third solution might be the use of a fixed delivery type of pump. This would necessitate the use of a flow control valve to vary the speed of the fluid motor.

The foregoing discussion indicates the great amount of flexibility available in fluid power components which can help to solve a problem with any degree of sophistication that may be required.

Hydrostatic transmissions hold a tremendous potential for future applications in all fields of usage.

Electrically Controlled Circuits

Electrical control devices play a big role in the application of fluid power equipment.

Switches, timers, relays, solenoids, and other such devices help to control the stopping, starting, sequencing, speed, positioning, timing, and reversing of actuating cylinders and fluid motors.

Automatic machines such as those used in the

Fig. 9-19. Dyna-power system. (*Courtesy of N.Y. Air Brake.*)

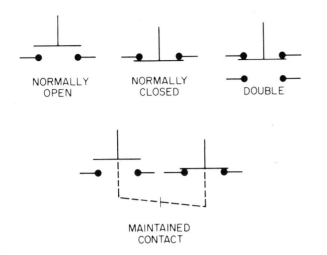

Fig. 9-20. Push button switches.

chine are selected according to the operation that must be performed.

Push button switches (see Fig. 9–20) are used mainly for starting, stopping, or for providing manual override in case of emergency.

Starters are used to protect the electric motor against under-voltage when one is starting hydraulic pumps and bringing them up to operating speed (see Fig. 9–21).

Standard color coding has been adopted by the industry, which requires AC and DC line and load wires to be black or gray; AC control wires should be red, and DC control wires should be blue.

Solenoids (see Fig. 9–22) are used for a push or pull force and consist of a coil and a movable iron core which pushes or pulls due to a magnetic force when the coil is energized. Solenoids are available for forces of a few ounces up to 50 lb. Stroke lengths range from zero to 1 in. Coils capable of producing forces greater than 50 lb are rather large and heavy.

Relays (see Fig. 9–23) are multiple switches which open or close when a coil is energized. A relay may have 15 or 20 contact switches, some being normally closed while others are normally open. In other words, if the coil is energized, the contacts that that normally closed will open and the contacts that are normally open will close. They are generally spring loaded to assume their normal position when the coil is de-energized.

Through an arrangement of the contacts in a

machine tool industry rely mainly on the electrical components to control the hydraulic muscles for doing the work. The aircraft and mobile equipment fields have also found that electricity and fluid power work very well together, especially where remote or dual control is essential.

Through the use of a simple push button switch, an operator can put complex machinery into operation that may complete hundreds of machining operations toward the completion of a product such as an automobile or farm implement.

The wiring, starters, relays, and other devices used in an electrical controller for a hydraulic ma-

Fig. 9-21. Starter circuit: 3-phase wiring diagram for starting and electric motor to drive a hydraulic pump.

Fig. 9-22. Solenoids. *(Courtesy of Decco.)*

Fig. 9-22. Solenoid symbol.

NORMALLY
CLOSED   NORMALLY
OPEN   OVER LOAD

Fig. 9-23. General contacts for starters, relays, etc.

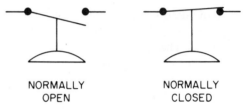

NORMALLY
OPEN   NORMALLY
CLOSED

Fig. 9-24. Pressure switch.

relay, interlocking can be designed into the controls of a machine to prevent accidental operation of a machine out of the proper work cycle sequence. Limit switches are switches that energize or de-energize coils used for positioning relay contact switches. Solenoids used on hydraulic control valves should be actuated through relay contacts, because they require a high current and need protection. Use of relays permits interlock to avoid accidental actuation of the two solenoids in opposite ends of a valve spool, which would caused burnout of one or both of the solenoids.

Pressure switches utilize the hydraulic system pressure to open or close contact switches. A pressure switch may have a high pressure setting and a low pressure setting. For instance, if it were necessary to start and stop a hydraulic power source to maintain a given pressure, the low side would start the pump and the high side would stop it. They are generally used to initiate some phase of a work cycle, or they may be applied as a safety device.

Time delay devices are used to adjust the time phase of a working cycle which may provide a dwell or feed condition. A dwell is sometimes needed in

COIL ENERGIZED

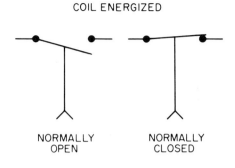

NORMALLY   NORMALLY
OPEN   CLOSED

COIL DE-ENERGIZED

NORMALLY   NORMALLY
OPEN   CLOSED

Fig. 9-25. Timer contacts.

such machine operations as drilling, where the drill pauses at the end of the stroke to clean out the hole. Most timers have an adjustment which gives a precise control for specific feed rates or dwells for machining synchronization. Timers as shown are used to provide measured time sequences of a machine's operation.

The fluid dashpot timing relays are essentially solenoid-operated switches, with the action of the solenoid being delayed by the metering of fluid through an adjustable orifice.

Timers of this type are used where the sequence of operation requires a delay with reasonable accuracy. Because the hydraulic fluid's viscosity varies with temperature change, precise accuracy is not always possible.

Pneumatic timing relays are useful for applications requiring close accuracy. This type is extremely flexible and uses a bellows for controlling the tripping time. They are used for both "on-delay" and "off-delay" operations.

The circuit in Fig. 9–26 shows a three-phase wiring diagram for starting an electric motor to drive a hydraulic pump. When the operator pushes the start button, it completes the circuit from L–1 to L–2, which energizes relay coil M. The relay contacts M–1, M–2, M–3, and M–4 close, completing the circuit to start the electric motor. The contact M–4 around the starter switch provides a holding circuit so the operator can let go of the start button after the motor starts. The quick-disconnect line switch must be closed before the motor can start, and should never be opened with the motor running under load. Higher voltage circuits will arc across the switch contacts if opened under load and cause a severe burn or other accident to operating personnel and equipment.

Generally, a transformer is used to reduce the high voltage of the motor circuit to a lower voltage for the control and solenoid circuits. It is also common practice to go through fuses to provide adequate protection for each circuit. The thermal overloads will open if low voltage occurs during startups or if the hydraulic pump is overloaded.

The stop button switch can be operated at any time to stop the electric motor and pump. It is located to have priority control over other devices in the circuit.

Fig. 9-26.

Fig. 9-27. Combination pump unit operated by an electric motor.

Fig. 9-28. Circuit.

Typical Application Circuits

The hydraulic circuit in Fig. 9–27 shows a combination pump unit operated by an electric motor. As the operator pushes the start button, the electric motor starts and the pump rotates. Fluid delivered by both pumps passes through the open-centered four-way valve and back to the reservoir.

When the cycle start button is pushed, solenoid A is energized, and the flow is directed to the cylinder for the rapid advance portion of the cycle with both pumps. Limit switch (LS-1) is a normally closed, held open switch when the cylinder is retracted; hence, it closes at the start of the rapid advance.

As the cam on the cylinder actuates LS-2, solenoid C is energized, closing the two-way valve. This causes the fluid from the rod end of the cylinder to be metered out through the flow control valve. The machine is now in the feed portion of the cycle, with pilot pressure holding the unloading valve open to unload PF_1 freely back to the reservoir.

At the end of the feed stroke LS-3 is actuated, which must de-energize solenoids A and C, as well as energize solenoid B. As this happens, both the four-way and two-way valves shift to the opposite position. The flow from both pumps is now directed to the rod end of the cylinder, and the machine is in the rapid return stroke.

When the cylinder is fully retracted, the cam actuates LS-1 to open, which de-energizes solenoid B, and the four-way valve is centered by spring force to allow the flow of both pumps to pass back to the reservoir until the cycle is initiated again.

Electrical Circuit

When the operator pushes the motor start button, the hydraulic pumps operate with the flow going back to the reservoir through the four-way directional valve. Figure 9–28 shows the electrical diagram for this system.

Fig. 9-29. Complete servo circuit. *(Courtesy of Vickers.)*

Pushing the cycle start button energizes CR-1 relay, which closes contacts CR1-1, CR1-2, and CR1-3, energizing solenoid A to provide the rapid advance. As the machine moves forward, LS-1 closes. At the end of the rapid advance LS-2 is actuated to energize relay CR-2, closing contacts CR2-1, CR2-2, and CR2-3, which energizes solenoid C to start the feed stroke.

At the end of the feed stroke LS-3 is actuated, energizing CR-3, closing contact CR3-1; however, because CR3-2 and CR3-3 are normally closed, they will open. Then, as CR3-2 and CR3-3 open, contacts CR3-4 and CR3-5 close. This entire action of the relay causes solenoids A and C to de-energize and solenoid B to energize. The machine rapidly returns, and as LS-1 is held open again, it de-energizes relay CR-3, which de-energizes solenoid B as the contacts

in the relay open (CR3-1, CR3-4, and CR3-5) or close (CR3-2 and CR3-3).

The machine is now in the neutral or center position.

The normally opened or closed contacts, as indicated in relay CR-3, provide an interlocking safety feature to provide proper machine sequencing and to prevent two solenoids, on the same valve, from being energized at the same time.

The forward inch button and the return inch button are used to provide a manual override for machine setups or tool breakage.

Electrohydraulic Servo Valves

Electrohydraulic servo valves are used for precision control of acceleration, velocity, and position in the hydraulic control of machine tools, spacecraft, radar, submarines, nuclear reactors, and other applications.

A servo valve (see Fig. 9–30) is a control actuated by a feedback system which compares the output with the reference signal and makes corrections to reduce the difference. Its function is that of a transducer which transforms an electrical signal into fluid power. Servo valves may be used to control flow to and from the hydraulic actuator or to control the volume of a variable pump.

The input electrical signal to the servo valve must be converted into mechanical motion to actuate the control. Generally, an electromagnetic torque motor is used to stroke the mechanical actuator of the valve.

Fig. 9-30. Complete servo valve. *(Courtesy of Pegasus Laboratories.)*

The power level of the electrical input signal to the torque motor is only a few watts, but the power level of the output, measured in hydraulic fluid displacement, may be many thousands of times that of the input. The servo valve provides an output directly proportional to its input and is a very powerful hydraulic amplifier when interconnected with its associated components in a circuit.

Torque Motor

In Fig. 9–31, the armature and valve actuator drive arm are suspended on the flexure tube, which acts as a pivot point.

If an electrical input signal is applied to coil *A,* magnetic flux will be produced in air gap *a,* which tends to draw the armature toward that pole. The armature will move until the centering spring force

AIR GAP a
AIR GAP b
CENTERING ADJUSTMENT
PERMANENT MAGNET
FLUX PATTERN (REF)
STROKE LIMITING SCREW
ARMATURE
COIL B
"C" SECTION
COIL A
FLEXURE TUBE
OUTPUT DISPLACEMENT
COUNTER-WEIGHT
DRIVE ARM

Fig. 9-31. Torque motor. (*Courtesy of Pegasus Laboratories.*)

equals that of the magnetic pull. Thus, the greater the input current signal to the coil, the greater the magnetic pull and the further the stroke.

Since the flexure tube acts as a pivot, the drive arm moves in an opposite direction to actuate the hydraulic valve.

If equal current is applied to both coils *A* and *B* simultaneously, the magnetic pull will be the same in both directions, and the drive arm will remain on center. If one coil receives a weaker signal, the armature will stroke an amount proportional to the difference in the two currents. This is called the differential current input to the torque motor.

The permanent magnet produces magnetic flux, which has the effect of a negative spring applied to the armature. This results in the armature moving toward the C section, and the total reluctance of the magnetic circuit on that side becomes less. Therefore, it increases the magnetic flux, which, in turn, creates more pull. In opposite manner, the pull on the other

side with the wider gap becomes less. Thus, the closer the armature comes to either pole, the stronger the net magnetic force tending to pull the armature to that pole. This negative spring subtracts from the flexure tube spring rate, and, therefore, more stroke is produced for a given electrical input. Also, the permanent magnet increases the gain (output/input) of the torque motor, and the gain can be controlled by changing the permanent magnet charge. This calibration feature is used to build valves with a wide range of input electrical power.

An adjustment of centering springs allows a slight change in the armature neutral position to center the drive arm and valve actuator.

Boost System

The valve boost system is a hydraulic amplifier which amplifies the drive arm motion into a high force level to position the four-way spool valve. In the position feedback valve, it is desirable to have the spool follow the position of the torque motor drive arm.

Figure 9–32 shows the symmetrical arrangement of the design with both sides of the valve equal in all respects. Therefore, with the input signal equal on both coils of the torque motor, the drive arm is centered and the main spool is at neutral.

If the input current signal increases on coil *A,* the armature moves to actuate the drive arm against nozzle *A.* As the flapper arrangement of the drive arm tends to resist the flow out of nozzle *A,* pressure increases in chamber *a.* Force due to the increase in

Fig. 9-32. Boost system-center position. (*Courtesy of Pegasus Laboratories.*)

FORCE MOTOR
SPOOL
a
B A
b
NOZZLE B
C_2 EX C_1
NOZZLE A
ORIFICE
ORIFICE
PS
SCREEN

Fig. 9-33. Force motor. *(Courtesy of Pegasus Laboratories.)*

Fig. 9-34. Force motor. *(Courtesy of Pegasus Laboratories.)*

Fig. 9-35. 4-way spool valve action-center position. *(Courtesy of Pegasus Laboratories.)*

Fig. 9-36. 4-way spool valve action-cylinder. *(Courtesy of Pegasus Laboratories.)*

Fig. 9-37. 4-way spool valve action-cylinder retracting. *(Courtesy of Pegasus Laboratories.)*

pressure causes the main spool to move in direction X as shown in Fig. 9–33. It will move toward the lower pressure end b until the flapper is again in the center between the two nozzles which is shown in Fig. 9–34.

The boost system causes the four-way spool to follow the position output of the torque motor in response to changes in electrical input signals to the torque motor.

Four-way Spool Valve Action

Figures 9–35 through 9–37 show the corresponding motion that would result from a change in elec-

trical input to the torque motor, as previously described.

The spool is designed with four metering edges, L-1, L-2, L-3, and L-4, which are approximately on line when the spool is centered.

The strength of the electrical input signal determines the amount of flow metered to the cylinder and its velocity.

In the design shown in Fig. 9–38 the torque motor armature causes a push rod to actuate the pilot valve spool in a direct proportion to the command current signal. The direction depends upon which coil receives the signal, with a maximum of ± 0.020 deflection possible from the centered position.

Fig. 9-38. Servo valve. (*Courtesy of Vickers.*)

If it is assumed that the pilot spool moves to the left, control pressure can now operate on area $2A$ as well as A. Because the force is greater on the $2A$ end of the main spool, it moves right, allowing supply fluid to flow through port C-1 to the actuating cylinder or fluid motor. The actuator moves in direct proportion to the command current input signal.

If the command signal was applied to the opposite coil, the torque motor armature causes the push rod to move the pilot spool right. This allows the control pressure trapped in area $2A$ to drain to the tank, which, in turn, allows the main spool to move left, reversing the actuator.

The mechanical feedback linkage between the pilot and main spool moves the pilot stage sleeve in order to close off the control pressure at a point which satisfies the input current command.

Fig. 9-39. Servo valve. (*Courtesy of Oilgear Co.*)

The fulcrum position is variable, and it would be ideal to adjust for each application.

The design shown in Fig. 9–39 uses a swing plate arrangement which is attached to the armature of the torque motor by a push rod. A differential current signal causes the push rod to move in a direction depending on the coil receiving the most current. The swing plate is mounted on flat springs which allow a ± 0.015 in. movement from the centered position. The swing plate's underside is contoured and machined flat to match ports. As the push rod moves the plate, hydraulic fluid from the center port will enter the proper actuator port, depending on the direction of the armature. The exhaust or drain connection to the reservoir is provided in the top section of the valve body, which creates a flooded enclosure.

Figure 9–40 illustrates a single stage type of servo valve, manifold mounted with all ports in the bottom plate. It operates on the same principle as the pre-

Fig. 9-40. Servo valve. (*Courtesy of Minneapolis Honeywell.*)

viously described design with a permanent magnet torque motor which serves to produce the linear force to actuate the spool.

However, this design contains no centering spring devices; it relies upon torsional friction in the torque motor armature pivot to provide centering. The spool travel is ± 0.015 in. for zero to full flow.

An important feature of this valve is the linear variable differential transformer (LVDT) and core attached to the end of the spool. It operates on the principle that primary and secondary windings around the core give an output voltage that depends on the position of the spool end. A phase shift occurs each side of zero or center, which provides feedback to

1. Electro hydraulic servo valve assembly
2. Torque motor cover
3. Torque motor cap (permanent magnets – potted)
4. Torque motor coils (potted)
5. Armature, diaphragm, flapper, feedback spring assembly
6. Nozzle body
7. End caps
8. Main body
9. Valve sleeve
10. Primary orifices
11. Internal filter (pilot stage)
12. Valve spool

Fig. 9-41. Disassembly of servo valve. *(Courtesy of Chandler-Evans.)*

correlate the position of the spool. The voltage output indicates how far the spool has moved to either side of center.

The excellent feedback loop helps to provide high dynamic response characteristics in a servo-controlled system.

The design of servo valve shown in Fig. 9–41 is a two-stage, flapper-nozzle type with mechanical feedback. The spool is pressure-positioned, and its displacement is proportional to the differential current applied to the torque motor. In the null or centered position, the flapper drive arm imposes equal restriction to flow from both nozzles, and the spool is stationary.

When the torque motor receives an input signal, the spool moves until the counteracting force transmitted through the feedback spring equals the force developed by the torque motor. At this point, the flapper has returned to neutral, and the pressure differential across the spool is zero. The spool remains in this position until another signal is received by the torque motor.

Servo Control of Variable Volume Pump

Pump flow, in the system shown in Fig. 9–42, is determined by the position of the slide block race. With the pump rotating, the slide block is positioned eccentric to the cylinder. Pistons move outward on the bottom and inward on top. They receive fluid from the bottom side of the pintle valve and force the fluid into the top side of the pintle valve.

In the view in Fig. 9–43, the slide block is positioned concentric with the cylinder and pump flow to the hydraulic system is zero.

Figure 9–44 shows the slide block positioned with eccentricity opposite Fig. 9–43, and flow to the system is reversed.

When the swing-plate type servo valve is deflected downward in Fig. 9–42, fluid from the gear pump enters stroking piston P-1. At the same time, fluid stroking piston P-2 opens to drain, and the slide block moves right.

If the swing plate is deflected upward, the slide

217

Fig. 9-44. Slide block-flow reversed. *(Courtesy of Oilgear Co.)*

Fig. 9-42. Servo control variable volume pump. *(Courtesy of Oilgear Co.)*

Fig. 9-43. Slide block concentric with cylinder-pump flow = zero. *(Courtesy of Oilgear Co.)*

block moves to the left. Centering the swing plate stops and holds the slide block.

The torque motor armature deflects the swing plate proportionally to current received from the amplifier.

The LVDT (linear variable differential transformer) transmits an electrical voltage proportional to the slide block's position. The slide block receives a remote voltage command from a potentiometer (instrument that controls electrical potential) wired to the amplifier in series with the pump LVDT. The slide block moves until LVDT voltage equals potentiometer voltage. When the electrical differential is zero, the swing plate centers.

A reference transformer provides voltages for exactly full pump flow in each direction. The reference transformer then supplies excitation voltage for command potentiometers, calibrating them for zero to full pump flow over zero to full potentiometer rotation. Figure 9–45 is an excellent example of servo-application as shown in the Oilgear steering system for ships.

Checking Safety in Designing Circuits

There is no compromise for safety when fluid power circuits are being designed. However, if errors slip through the design stage of a circuit, trouble may occur, causing additional expense or an unsafe condition.

The following check list of design questions may help to provide safe, efficient, and dependable fluid-powered equipment.

Question 1. *Is the input horsepower adequate?*

Remarks. Prime movers, such as electric motors, can be overloaded for only short intervals during peak loads. However, if peak pressures are to be held for long periods of time, the motor must be large enough to supply the proper horsepower.

Overload periods also occur when relief valves delay in opening or when pressure-compensated pumps fail to stroke immediately. Power input must be sufficient to meet these conditions.

The retraction of high speed cylinders and rams may cause high flow rates at back pressures much too high for the input horsepower to handle.

The duty cycle of fluid power equipment must be considered to match horsepower requirements with conditions of safety, efficiency, and dependability.

Question 2. *Does the system need a heat exchanger?*

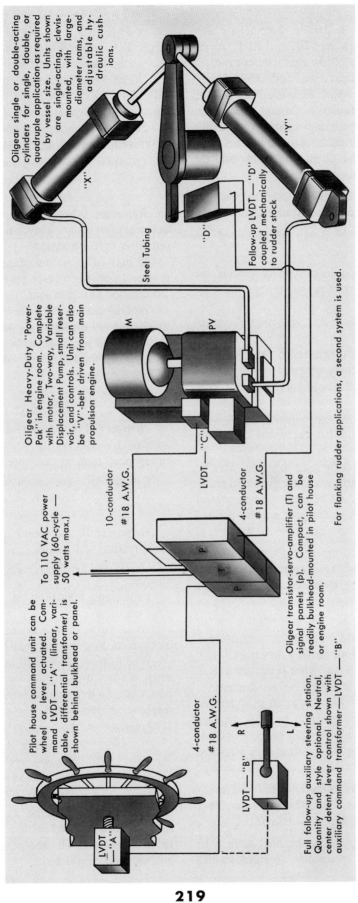

Oilgear single or double-acting cylinders for single, double, or quadruple application as required by vessel size. Units shown are single-acting, clevis-mounted, with large-diameter rams, and adjustable hydraulic cushions.

"X"

"Y"

"D"

Follow-up LVDT —"D" coupled mechanically to rudder stock

Steel Tubing

M

PV

Oilgear Heavy-Duty "Power-Pak" in engine room. Complete with motor, Two-way, Variable Displacement Pump, small reservoir, and controls. Unit can also be "V"-belt driven from main propulsion engine.

LVDT —"C"

10-conductor #18 A.W.G.

4-conductor #18 A.W.G.

To 110 VAC power supply (60-cycle — 50 watts max.)

Pilot house command unit can be wheel or lever actuated. Command LVDT —"A" (linear, variable, differential transformer) is shown behind bulkhead or panel.

P   T   P

Oilgear transistor-servo-amplifier (T) and signal panels (p). Compact, can be readily bulkhead-mounted in pilot house or engine room.

LVDT —"B"

For flanking rudder applications, a second system is used.

R

L

LVDT —"B"

LVDT —"A"

4-conductor #18 A.W.G.

Full follow-up auxiliary steering station. Quantity and style optional. Neutral, center detent, lever control shown with auxiliary command transformer—LVDT —"B".

Fig. 9-45. Servo steering control. (*Courtesy of Oilgear Co.*)

Remarks. Generally, hydraulic systems are recommended to operate between 120° to 140°F. Ambient operating temperatures, flow characteristics within a system, the cycle of operation, and other factors should be recognized to assure adequate heat control.

To ensure proper circulation through a heat exchanger, a separate circulating pump is generally recommended. The electrical hookup should be wired to operate both the cooler circulating pump and the main hydraulic pump simultaneously.

Circulating pumps may also be used as a source of pilot pressure or for filtration systems.

Question 3. Should the pump be unloaded between cycles?

Remarks. In the open loop type of circuit the pump can be unloaded at close to zero pressure by the following methods:

  a. An open center four-way directional control valve.
  b. Venting a pilot-operated relief valve with an on-off type of solenoid-actuated valve.
  c. Pressure-operated unloading valve or two-way directional valve.

Closed loop circuits may give problems of local overheating if the same fluid is circulated constantly for long periods of time.

A bleed-off and replenishing arrangement may be required. Generally, the heat exchanger works directly from the reservoir and would be ineffective to correct local heat problems. Heat exchangers may be applied elsewhere in the system to take care of local heat problems.

Question 4. Is it possible to start hydraulic pumps regardless of the position of actuators or fluid motors?

Remarks. When pumps are unloaded with all solenoids de-energized and controls in neutral, the pump can be started without moving actuators.

If manual, mechanical, or pilot-operated valves are used, they should be checked before startups.

It is also important to check out all cylinders or motors that may be of the constant pressure type or may have trapped pressure energy on one side. (Constant pressure actuators move by creating a pressure drop on one side or the other.)

Question 5. What checks are essential with vertical presses and other such machines that may drift downward?

Remarks. Limit switches may be held open with the ram or cylinder in the up position. A check should

be made to insure that the proper solenoids are automatically energized to pressurize the system to return the actuators to the up position.

When variable pumps are used, they should be adjusted and controlled to hold the actuator in the up position when operating.

Safety locks and pilot check valves are also used to prevent free falling of vertically suspended loads.

Question 6. Is independent movement of any actuator possible regardless of the position of other actuators?

Remarks. When by-pass valves are used to move actuators for setups, they must be checked before the circuit is put into service.

It may also be important to use an interlocking system to protect personnel from injury and prevent damage to the equipment.

Inching actuators in forward or reverse is sometimes essential for adjusting fixtures and clamps. When accumulators are used as a power source when the pump is off, fluid can be used directly through needle adjusting valves.

Question 7. Should all actuators stop immediately with the stop button?

Remarks. For safety reasons, it is good policy to arrange all controls to prevent motion in case of emergency, such as tool breakage or power failure.

Question 8. Are all conductors and connectors properly sized for operating conditions throughout the entire cycle?

Remarks. If the lines are too small, excessive backpressure will result, causing friction and heat. Any controls operated by pilot pressure may also be affected.

Leaks and vibrations may occur from excessive fluid velocity.

Some parts of the system may be subjected to mechanical binding and hydraulic unbalance.

When larger lines and valves than necessary are used, the result is higher component costs. In addition, more fluid is needed, which may affect the circuit response characteristics.

Question 9. When large double-acting cylinders are being used, are the lines and components connected to the large area side adequate to handle exhausting fluid?

Remarks. Because the return speed is much faster than the down speed with a given fluid delivery, a greater flow rate will exhaust from the large area side and must be considered.

Back pressure is sometimes utilized to control fluid delivery from the pumps to the small area side, thus controlling the speed of the actuator.

When the volume exhausting from the large end exceeds approximately four to five times the discharge volume of the pump, a separate unloading or dump valve is recommended.

Question 10. Have drain lines been installed properly on pilot-operated valves?

Remarks. Spring chambers exposed to the entry of fluid must be properly drained when the control valve operates.

Sequence, pressure-reducing valves and other devices, such as pressure switches, must have low pressure drains because trapped fluid will not allow parts to move.

Question 11. Are all parts of the system protected against local intensification?

Remarks. Hydraulic fluid trapped in the small area side of a double-acting cylinder may be subject to intensification when fluid is being applied to the large end.

The delayed action of control valves is also a factor.

Safety fuses should be applied where intensification is likely to occur.

Question 12. Have all ports been indicated properly for each component of the system?

Remarks. Check valves installed backwards and other wrong connections may create serious problems. Factory specialists have been known to travel halfway around the world to turn a check valve around or change a port connection.

Question 13. Are vent connections installed for bleeding off trapped air from vulnerable spots in the hydraulic system?

Remarks. Air trapped in large cylinders can have an accumulator effect which may cause actuators to move without the pump. When air cools in large cylinders during a shutdown, it can cause a negative pressure, which may also cause a machine to jump when a fluid lock is released or the pump is started.

Systems using the pump and reservoir below the actuators require frequent venting to help air out of the system.

Question 14. Is the pump protected against over-pressures at all times?

Remarks. The pump is the source of fluid power and should have priority protection in its discharge side with a relief valve or remote unloading control.

Question 15. Will the fluid under pressure create a decompression problem that may cause shock and damage?

Remarks. Large actuators containing fluid at relatively high pressures may need a valve for decompression. Hydraulic fluids compress at the rate of about one-half of one per cent for every 1000 psi.

Question 16. Are cycle sequences positive?

Remarks. Sequencing is accomplished with pressure controls, timers or position devices.

Sequencing with a pressure control or pressure switch may be subjected to hydraulic shock loads that would create erratic machine operation.

Electrically and mechanically actuated directional controls provide a more positive sequencing, but may be more costly.

Question 17. What visual check devices, such as pressure gauges, are needed to assure proper operation or assist in trouble shooting?

Remarks. Properly calibrated gauges should be installed to indicate operating pressure levels or inlet conditions of the pump.

Temperature gauges and flow meters may also be important in some applications.

All such devices should be protected against hydraulic shock loads with snubbers or check valves.

Question 18. What provisions have been made to facilitate inspection and repair?

Remarks. Working space is important to inspect or remove components for repair.

Temperature gauges, pressure gauges, sight glasses on reservoirs, indicators on filters, and other inspection devices should be visible during the equipment operation without getting too close to the moving parts.

Question 19. Under what conditions must the system operate?

Remarks. This question covers all the factors, such as ambient temperature, contamination in the atmosphere, fire hazards, and such other conditions that may affect circuit design.

When fire resistant fluids are essential, this affects the components' seals that are used, and may have a direct bearing on the life of the equipment selected.

Question 20. If accumulators are used in the circuit, have means been provided to unload pressure energy during shutoffs or for maintenance?

Remarks. When the pump has been shut off, the circuit can still operate with line accumulators.

It is the responsibility of the circuit designer to provide an unloading device to discharge line accumulators before any machine is inspected or worked on. If this is impossible, then accumulators must be isolated automatically from the rest of the system.

Question 21. Does the nature of the work create excessive vibration during operation?

Remarks. Flexible hoses and swivel joints should be used where necessary to connect two parts of a system where motion of one occurs relative to the other.

Rigid conductors should be clamped down, and inline components must be self-supported where vibration occurs.

Questions

1. What are the three main design considerations that are the responsibility of the system designer?
2. What is the most important tool for anyone working in fluid power?
3. Draw a simple reciprocating circuit and name the necessary components of the system.
4. Explain the operating principle of a regenerative-type circuit.
5. Name five different applications for accumulators.
6. Why are accumulators generally preloaded with an inert gas?
7. Illustrate a simple intensifier and explain how it multiplies force.
8. Explain the effect of compressibility when higher pressures are encountered in a hydraulic system.
9. What are the differences between a hydrokinetic type of transmission and a hydrostatic transmission?
10. Draw the graphical diagram for a typical closed loop system. Draw the graphical diagram of a typical open loop system. Compare the important differences between the closed loop and the open loop.
11. Name the common electrical devices that are used to control hydraulic systems.
12. Why are separate circuits used for the electric controls and for solenoids?
13. What is a servo valve, and what is its function in a hydraulic system?
14. Explain the operation of a torque motor.
15. What is the purpose of a feedback in a servo system? *null condition*
16. What is the difference between a single-stage and a two-stage servo valve?
17. What are some common applications of servo valves?
18. Name several safety checks necessary when one is designing circuits, and explain why they are important.

CHAPTER **10**

FLUID POWER
MAINTENANCE
AND SAFETY

General Maintenance Requirements

The first requirement in a completely satisfactory maintenance program for fluid power is good equipment properly installed.

The second requirement is properly trained personnel with a thorough knowledge of the equipment's operations, and the ability to inspect, trouble-shoot, and remedy problems.

The third requisite of a good maintenance program is the establishment of preventive maintenance. This includes keeping good records, continuous inspection of equipment, and heading off difficulties through system cleanliness and making minor repairs before they become major repairs with costly down time.

The Tools of Good Maintenance

The most important tools of the fluid power mechanic are equipment manuals, blueprints, data sheets, parts lists, and other such information.

Another important requirement is an adequate testing facility to bench-test pumps, controls, and actuators that have undergone repair. It may be

dangerous to test repaired components directly on high-production-type machinery. Highly qualified personnel would be needed for maintenance of this procedure.

A third requirement is adequate tools, such as accurate pressure gauges, flow meters, clean working space, hand tools, hoists, tote carts, and other such devices essential to a mechanic.

Another important factor to prevent delay and down time is a good inventory of spare parts, such as seals, bearings, repair kits, extra hydraulic fluid, filter cartridges, and, where important, spare components.

Records

Good records concerning hydraulic machinery are a must for preventive maintenance. These records may include the following:

1. The equipment record. This is generally a card providing model number, ratings, manufacturer or source, location, etc.
2. A repair card which tells repair history, running cost, and other such data.
3. An inspection check list. This is a listing of

points or things to be checked on fluid power machinery, and when the checks should be made, etc.

4. A maintenance schedule of inspections is a day-by-day listing of duties such as changing fluid or filters, making test samples, greasing, cleaning, etc.

5. An inventory control. This is often combined with the repair history and cost. It is important to keep track of spare parts, extra fluid, seals, repair kits, or spare units where essential.

General Safety Rules for the Operator of Fluid-powered Machines

1. Understand the function and operating principle of all the component parts of the machine.

2. Understand the function and operation of the machine's control system.

3. Report any change in operating characteristics, such as abnormal gauge readings, unusual sounds, faulty or erratic performance, and leakage from components.

4. Wear proper clothing that will not catch in rotating or moving parts.

5. Wear safety glasses to protect eyes against flying objects or fluid spray from ruptured lines.

6. Observe the operating rules for the machine, such as safety guards, two-hand control, automatic pull-aways, and other important devices.

7. Know how to shut down the machine and to prepare it for safe inspection or maintenance.

8. Know how to check out controls for restarts after an emergency shutdown.

9. Do not make hasty emergency repairs that may endanger life or damage the equipment.

10. Do not operate controls in a reckless manner that may cause hydraulic shock and damage costly hydraulic components.

11. If the machine is equipped with an accumulator, be sure its pressure energy is released when preparing for shutdowns.

12. Know which valves or controls move the machine parts forward and reverse, in case of an accident.

13. Log any occurrence or observation that may be a clue that preventive maintenance is needed.

14. Contribute information to the safety and well-being of others.

General Safety Rules for the Fluid Power Mechanic

1. Before working on a machine circuit, check the drawings to determine all implications with other parts of the system or other machine operations that may be interconnected.

2. Open and lock out any electrical circuits to the electric motor, controls, and related equipment.

3. For central hydraulic systems, isolate the circuit by closing, wiring, and tagging any connecting valves.

4. When engines are used for input power, disconnect the pump from the drive shaft or use other adequate means to prevent accidental starting of the pump.

5. Shore up or block cylinders and machine parts that may drop because of gravity.

6. Bleed fluid to relieve any pressure from the system by cracking fittings with a rag over the joints to prevent velocity spray of fluid until the pressure is zero.

7. Plug all tube ends and ports of components exposed to the atmosphere with sealing caps or plugs to prevent contamination.

8. Before removing heavy components from a circuit, be sure of the weight and whether or not a hoist is required.

9. Make sure all component parts are marked or tagged during disassembly to prevent such items as valve spools from being installed backwards. (A spool installed wrong may allow a press accidentally to fall during a startup.)

10. When disassembly of a component is made, be careful not to unload a spring force that may cause parts to fly.

11. Do not use harmful or toxic cleaning fluids. A good solvent and some of the machine's fluid is generally recommended for washing parts to assemble components.

12. Be sure all replaced parts, such as seals and other items, are of the proper materials for compatibility with the fluid in the system.

13. Do not over-torque housing bolts or other fasteners when important adjustments for clearances may affect internal moving parts.

14. When applying tube or pipe fittings with NPT threads, be careful not to crack the housing of the component by over-torquing the fitting into the port.

15. Maintain cleanliness during maintenance, and refill all components and lines with clean sys-

tem fluid before replacing them in the circuit to keep out air and contamination.

16. Check and reset all pressure controls with an accurately calibrated pressure gauge.

17. Observe extreme caution when starting up equipment for the first time or after a repair outing.

18. Always use the proper tools for the job.

19. Never use damaging blows or forces when disassembling or assembling fluid power equipment.

20. If conductors or connectors are replaced, be sure they have the proper dimension, pressure rating, threads, ports, and safety factor at operating temperature.

How to Trouble-shoot Hydraulic Circuits

Fluid power systems depend upon flow from the hydraulic pump to provide linear or rotary motion against resistance, which results from the work.

Thus, two major measurements are important for trouble-shooting hydraulic circuits; these are flow and pressure. Temperature is also a factor that affects flow and must be considered.

Flow, pressure, and temperature are affected by external and internal leakages. Pressure drops due to any leakage will cause friction and heat.

External leakage is easy to see, and a good main-tenance program can eliminate such leaks, which are generally due to loose connections, bad seals, pin holes in castings, or worn parts.

Internal leakage can also be detected through the use of pressure gauges, thermometers, and flow meters.

The operating pressure of a system can provide a good indication if leakage problems occur within a circuit. Less load can be accomplished, and if the leak is too bad the actuator will stall under loaded conditions. Adjusting the relief valve will not remedy the problem.

Gauges used to check the pressure in a hydraulic circuit are generally of the bourdon tube or spring-loaded piston types.

The bourdon tube type shown in Fig. 10–1 consists mainly of a socket, tube, and tip. Fluid is communicated through the fitting in the socket to the bourdon tube. As pressure builds up, the internal forces tend to cause the tube to straighten or uncoil. The tip is linked to a movable element which translates the distance moved to the gauge pointer which indicates pressure. A specific ratio between the distance moved by the tip and the rotating distance of the pointer is calibrated for accuracy.

The spring-loaded piston design uses a piston which is opposed by spring force on one side and fluid pressure on the other (see Fig. 10–2). As fluid

Fig. 10-1. Bourdon tube gauge. (*Courtesy of Helicoid Gauge Division—American Chain and Cable Co.*)

Fig. 10-2. Pressure gauge. *(Courtesy of Schrader.)*

Fig. 10-3. Vacuum gauge. *(Courtesy of Marsh Instrument.)*

Fig. 10-4. Compound gauge. (*Courtesy of Marsh Instrument.*)

moves the tightly sealed piston to compress the spring, an indicator rod moves with the piston. The indicator rod motion is translated to a calibrated scale on the gauge body, which gives a pressure reading.

Vacuum and compound gauges are used for determining lower pressures or absolute pressures (see Fig. 10–3).

Vacuum gauges are generally of the bourdon tube type and are calibrated into inches of mercury.

Absolute pressure (in. of Hg)	U.S. gauge pressure (lb per sq in.)
1.0	0.48
2.04	1.00
4.1	2.0
6.1	3.0
8.2	4.0
10.2	5.0
12.2	6.0
14.3	7.0
16.3	8.0
18.4	9.0
20.4	10.0
22.4	11.0
24.5	12.0
26.5	13.0
28.6	14.0
29.92	14.7

Compound gauges are used to measure in both absolute and gauge pressure (see Fig. 10–4).

Hydraulic system components should be equipped with gauge connections so that pressure checks can be made during a complete cycle of operations. Results can be compared to those when the machine functioned properly.

Flow meters are useful for checking flow rates from the pump, to a control, from a control, to an actuator, from an actuator, and to the reservoir.

Complete "hydraulic testers" (see Fig. 10–5) are available for trouble-shooting flow rate, pressure, and temperature. All the components are built into a convenient-to-carry box, which provides portability. Testers of this kind can be used to trouble-shoot a circuit at any point of the circuit by connecting it between desired components and cycling the machine.

When one is trouble-shooting individual components, a test stand can be used which is equipped with all the necessary flow-raters, gauges, and other instruments (see Fig. 10–6).

In the field, a stop watch may be useful to check the cycling time of cylinders. Since the stroke length and bore of a cylinder are known, the cycling time

Fig. 10-5. General test stand set-up for controls.

Fig. 10-6.

Fig. 10-7 Leakage Tests.

provides information to determine the flow rate to the cylinder. A check of the pump capacity gives comparison data to determine losses.

$$\text{Time in sec} = \frac{\text{area of cy.} \times \text{stroke in in.} \times 60}{231 \times \text{gpm}}$$

A hand tachometer is useful in determining the speed of a fluid motor (see Fig. 10–8).

If the pump capacity and the displacement of the fluid motor are known, losses can be analyzed by obtaining the fluid motor speed.

$$\text{Motor speed} = \frac{\text{gpm} \times 231}{\text{displacement/rev}}.$$

Fluid motors make excellent flow meters, and a spare motor may be used in place of a hydraulic tester with the following components as shown in Fig. 10–9: (Temperature can be obtained by using a glass-tube-type thermometer)

This is intended to be a general statement following the diagram.

Care should be taken to avoid exposing low pressure seals to high pressure fluid. External drains may have to be calculated separately by collecting drain fluid into a graduated container for a period of time.

Pump test stands (see Fig. 10–10) are generally equipped with a variable speed drive to operate the test pump over a range of speeds (rpm). It should also be equipped with an accurately calibrated flowrator, various pressure gauges, and a loading device.

Test circuit for pressure control	*Test circuit for directional control*	*Test circuit for flow control*
Set pressure control for desired operating pressure and flow. Connect to test stand flow rator. Gradually increase the test stand pressure until the control opens. Leakage below the setting of the control will be indicated by the test stand flow rator.	Leakage from a directional control can be found by hooking test stand supply to the valve's pressure port. The valve "out" port is hooked to the exhaust, and the two cylinder ports are blocked. Operate valve and observe leakage.	Connect flow control to test stand supply manifold and exhaust; then operate control between zero and full range to determine its operation.

Fig. 10-9. Flow meter using fluid motor.

Fig. 10-8. Hand tachometer.

Fig. 10-10. Pump test stand. (*Courtesy of Kenosha Technical Institute.*)

Fig. 10-11. Pump test stand with variable speed drive, flow rator, etc.

A pump testing station of the design shown in Fig. 10–11 provides for a test over normal operating speeds (rpm) and at any desired pressure level within the range of the test unit. Operating curves can be computed and then compared to those when the pump was factory-tested.

Another easy-to-build test unit is a hydraulic dynamometer arrangement as shown in Fig. 10–12. A standard power unit being used, fluid is directed to operate a fluid motor which, in turn, drives a test pump. The test pump inlet uses a length of non-collapsible hose which is put into a container of oil. The discharge from the pump is connected through a "hydraulic tester" and then exhausts back to the same container.

The load valve, pressure gauge, and flow-rater

Fig. 10-12. Dynamometer. *(Courtesy of Kenosha Technical Institute.)*

can be used to measure the output of the pump. The speed (rpm) of the fluid motor driving the test pump can be determined by a tachometer.

This arrangement could also be used to test the fluid motor that drives the pump. The pump, in this case, would be loaded through the hydraulic tester to determine the torque or horsepower capabilities of the fluid motor.

One of the most important factors to remember in the trouble-shooting of hydraulic systems is that pressure is a result of resistance, and if there is no resistance the pressure will be near zero.

Trouble-shooting Check List

Lack of full pressure in the system:

1. Air in the system.
2. Pump starved by clogged inlet strainer.
3. Loose inlet connections to pump.
4. Relief valve not properly seated.
5. Lack across directional control valve ports.
6. Inadequate input power.
7. Leak in conductor or connector.
8. Defective seals on cylinder piston assembly or rod gland.
9. Loose or cracked piston rod gland seal.
10. Defective pressure gauge.
11. Cylinder crack or scored.
12. Check valve not seated.
13. Defective or worn pump.
14. Defective or worn fluid motor.

15. Scored valve plate in pump or motor.
16. Wrong hydraulic fluid.

When no pressure develops in the system:

1. Input power not coupled properly.
2. Pump not turning in the right direction.
3. Pump housing bolts not properly torqued.
4. Not enough hydraulic fluid in the reservoir.
5. Pressure gauge snubber valve closed, or defective gauge.
6. Ruptured conductor.
7. No resistance in the system.
8. Air breather in the reservoir clogged.
9. Defective relief valve.
10. Inlet to pump blocked.

Actuating cylinder or fluid motor fails to develop full speed:

1. Air in system.
2. Wrong viscosity fluid.
3. Loss of one pump in a dual pump system caused by unloading valve being set too low.
4. Pump operating too slowly.
5. A check valve not seating.
6. Relief valve defective or set too low.
7. Directional valve not shifting far enough, or defective.
8. Defective seals in cylinder piston assembly.
9. Defective fluid motor.
10. Fluid too hot and thin, causing excessive internal leakage.

11. Trapped fluid under back pressure or intensification.
12. Mechanical binding in linkage.
13. Inlet to pump blocked or too small.
14. Excessive friction in conductors, or hose kinked.
15. Air breather in reservoir blocked.

Erratic motion of an actuating cylinder:

1. Air in system.
2. Not enough fluid in the system.
3. Cylinder rod gland not adjusted properly.
4. Mechanical bind.
5. Pump capacity too small.
6. Defective seals in cylinder piston assembly or fluid motor.
7. Unloading or relief valves defective or out of adjustment.
8. Air breather in reservoir blocked.
9. Excessive back pressure on externally drained control valves.
10. Pump inlet leaks.
11. Defective pump.
12. A defective check valve.
13. Flow control valve defective or compensator sticking.
14. Slight back pressure needed to stabilize actuating cylinder or fluid motor due to the nature of the work.
15. Broken spring in a control valve.

Actuating cylinder or fluid motor fails to move:

1. Directional control valve fails to shift.
2. Sequence valve not externally drained.
3. Solenoid burned out, or limit switch not actuated.
4. Relay contacts dirty.
5. Broken spring in control valve.
6. Control sticks open or closed.
7. Pump control defective.
8. System hooked up wrong.
9. Check valve in backwards.
10. Hydraulic lock prevents control from operating.
11. Mechanical obstruction.
12. Insufficient power.

Excessive noise in operation:

1. Air entering the pump inlet line.
2. Entrapped air.
3. Misalignment of pump and drive unit.
4. Conductors vibrating.
5. Sudden release of fluid under pressure.

6. Turbulent flow characteristics.
7. Vibration of system components.
8. Relief valve chattering when set too close to load pressure.
9. Defective pump or fluid motor.

Electric motor driving pump keeps kicking out:

1. Too small to carry load.
2. Carrying too high a system pressure.
3. Electric motor hooked up wrong or operating on low voltage.
4. Misalignment of electric motor and pump.
5. Lack of circulating air to cool electric motor.

Excessive heat, causing high temperature of hydraulic fluid:

1. Components or conductors and connectors too small for the fluid being delivered to the system.
2. Wrong grade of fluid, or dirty fluid.
3. Not enough fluid in the system, or reservoir too small.
4. Pump operating too fast.
5. Small high pressure leaks within components.
6. Dirt holding relief valve partially opened.
7. Defective pump.
8. System overloaded.
9. Cooling system defective or inadequate.
10. Air breather too small or blocked.
11. Entrained air in the hydraulic fluid.

Air in system:

1. Defective seals.
2. Leaks at joints.
3. Loose inlet to pump.
4. Not enough fluid in the system.
5. Loose pump casing.
6. Inlet to pump too long.
7. Replenishing pump defective.
8. Improper maintenance procedures.
9. Permeation through accumulator bladders or diaphragms.

Analysis and Trouble-shooting Circuits

Problem 1. Speed Control Circuit (see Figs. 10–13 and 10–14).

Directions.
1. Trace the flow path of the fluid for each phase of the machining cycle and color lines to indicate pressure, drain, etc. (use standard color code).

Fig. 10-13. Circuit problem #1.

Fig. 10-14. Circuit problem #2.

Fig. 10-15.

RAPID ADVANCE ———————┐ ①

                          └─ FEED ————┐ ②

RAPID RETURN ◄——————————————┘ ③

Fig. 10-16. Circuit problem #2 (regenerative circuit).

Fig. 10-17. Circuit problem #3 (accumulator circuit).

2. List the name of each component in the system and describe its function.
3. Draw the graphical diagram shown in the old symbols, using the new ASA symbols.

Problem 2. Regenerative Circuit (see Fig. 10–16).

Directions.
1. Explain the principle of a regenerative circuit.
2. Draw this circuit in ASA symbols.
3. Color in all lines, using standard color code.

Problem 3. Accumulator Circuit (see Fig. 10–17).

Directions.
1. Explain the function of the accumulator circuit.
2. Color in all lines, using standard color code.
3. Redraw the circuit in ASA symbols and include some safety means of unloading the accumulator when the pump is stopped.

Problem 4. Intensifier Circuit (with error) (see Fig. 10–18).

Directions.
1. Trace the flow path of the fluid through the circuit, and color in all flow lines with standard code.
2. Explain the function of an intensifier.
3. "Trouble-shoot" the sequencing phase of the cycle and describe the change needed.
4. Draw ASA graphical diagram.

Problem 5. Sequencing Circuit (with error) (see Fig. 10–19).

Fig. 10-18. Circuit problem #4 (intensifier circuit with error).

Fig. 10-19. Circuit problem #5 (Sequencing circuit with error).

Fig. 10-20. Circuit problem #6 (press circuit).

Fig. 10-21. Circuit problem #7 (closed-loop circuit).

Fig. 10-22. Circuit problem #8 (Combination air and hydraulic circuit).

Directions.

1. Trace the flow path of the circuit to provide the following work sequence:
   a. Top cylinder extends ⟶
   b. Bottom cylinder extends ⟶
   c. Bottom cylinder retracts ⟵
   d. Top cylinder retracts ⟵

2. Redraw the circuit, eliminating any components not needed and making all necessary corrections for proper circuit operation.

Problem 6. Press Circuit (see Fig. 10–20).

Directions.

1. Trace the flow path of the fluid through the system and color in all lines, using standard colors.
2. Describe the function of a prefill valve.
3. Describe the function of kicker cylinders.
4. Draw an ASA graphical diagram for the circuit.

Problem 7. Closed Loop Circuit (see Fig. 10–21).

Directions.

1. Identify and list each of the component parts of the system.

2. Draw the ASA graphical diagram of the complete system.

3. Color code all active pressure, exhaust, inlet, and drain lines of the circuit.

4. Describe the power and torque output characteristics of the system.

Problem 8. Combination Air and Hydraulic Circuit (see Fig. 10–22).

Questions

1. What records should be maintained on hydraulic machinery for a good maintenance program?

2. Read and discuss the general safety rules for the operator of fluid-powered machines.

3. Read and discuss the general safety rules for the fluid power mechanic.

4. Explain the importance of the three following statements when one is trouble-shooting hydraulic circuits:

a. Pressure in a hydraulic system is due to resistance.

b. A pump pumps hydraulic fluid and not pressure.

Directions.

1. Name and describe the function of each component in the system.

2. Redraw the circuit, using the latest ASA graphical symbols.

3. Name the advantages of using air.

c. Atmospheric pressure forces the hydraulic fluid into the inlet side of the pump.

5. Which is more difficult to find and eliminate: external leakage or internal leakage in a system?

6. What is a compound pressure gauge?

7. What measurements can be found with a hydraulic tester?

8. Explain how a fluid motor can be used as a flow rater.

9. Explain how to build a simple test hydraulic dynamometer and how it can be used as a maintenance tool.

10. Read and discuss the trouble-shooter's check list.

National Fluid Power Association
GLOSSARY OF TERMS FOR FLUID POWER

SECTION 1

PRIMARY TERMS

FLUID POWER: Energy transmitted and controlled through use of a pressurized fluid.

FLUID POWER SYSTEM: A system that transmits and controls power through use of a pressurized fluid within an enclosed circuit.

HYDRAULICS: Engineering science pertaining to liquid pressure and flow.

HYDRODYNAMICS: Engineering science pertaining to the energy of liquid flow and pressure.

HYDROKINETICS: Engineering science pertaining to the energy of liquids in motion.

HYDROPNEUMATICS: Pertaining to the combination of hydraulic and pneumatic fluid power.

HYDROSTATICS: Engineering science pertaining to the energy of liquids at rest.

PNEUMATICS: Engineering science pertaining to gaseous pressure and flow.

SECTION 3

FLUID POWER LAWS AND RELATED TERMS

(Letter symbols at end of Section)

BERNOULLI'S LAW: If no work is done on or by a flowing frictionless liquid its energy due to pressure and velocity remains constant at all points along the streamline.

BOYLE'S LAW: The absolute pressure of a fixed mass of gas varies inversely as the volume, provided the temperature remains constant.

CHARLES' LAW: The volume of a fixed mass of gas varies directly with absolute temperature, provided the pressure remains constant.

CONTINUITY EQUATION: The mass rate of fluid flow into any fixed space is equal to the mass flow rate out. Hence, the mass flow rate of fluid past all cross sections of a conduit is equal.

DARCY'S FORMULA: A formula used to determine the pressure drop due to flow friction through a conduit.

$$h_f = \frac{fLv^2}{2Dg}$$

HAGEN POISEUILLE LAW: The friction factor of Darcy's Formula is a ratio of 64 to the Reynolds Numbers when flow is laminar.

$$f = \frac{64}{N_r}$$

PASCAL'S LAW: A pressure applied to a confined fluid at rest is transmitted with equal intensity throughout the fluid.

REYNOLDS NUMBERS: A numerical ratio of the dynamic forces of mass flow to the shear stress due to viscosity. Flow usually changes from laminar to turbulent between Reynolds Numbers 2,000 and 4,000.

$$N_r = \frac{\rho vD}{\mu} \qquad N_r = \frac{vD}{v}$$

TORICELLI'S THEOREM: The liquid velocity at an outlet discharging into the free atmosphere is proportional to the square root of the head.

$$v = \sqrt{2gh}$$

LETTER SYMBOLS

D   = Internal diameter of conduit — feet

f   = Friction factor

g   = Acceleration due to gravity — nominal 32.2 feet per second squared

h   = Head — feet

h_f = Head loss — feet

L   = Length of conduit — feet

N_r = Reynolds Number

v   = Mean velocity of flow — feet per second

(Greek Letters)

μ (Mu) Absolute viscosity — pounds mass per foot second

v (Nu) Kinematic viscosity — square feet per second

ρ (Rho) Mass density of fluid — pounds mass per cubic foot

SECTION 5

FLOW TERMS

CAVITATION: A localized gaseous condition within a liquid stream which occurs where the pressure is reduced to the vapor pressure.

FLOW, Laminar: A flow situation in which fluid moves in parallel lamina or layers.

Flow, Metered: Flow at a controlled rate.

Flow, Steady State: A flow situation wherein conditions such as pressure, temperature, and velocity at any point in the fluid do not change.

Flow, Streamline: (See FLOW, Laminar)

Flow, Turbulent: A flow situation in which the fluid particles move in a random manner.

Flow, Unsteady: A flow situation wherein conditions such as pressure, temperature and velocity at points in the liquid change.

SHOCK WAVE: A pressure wave front which moves at a supersonic velocity.

SURGE: A momentary rise of pressure in a circuit.

SECTION 6

MENSURATION TERMS

AIR, STANDARD: Air at a temperature of 68 °F, a pressure of 14.70 pounds per square inch absolute, and a relative humidity of 36% (0.0750 pounds per cubic foot). In gas industries the temperature of "standard air" is usually given as 60 °F.

ANILINE POINT: The lowest temperature at which a liquid is completely miscible with an equal volume of freshly distilled aniline (ASTM Designation D611-55T).

BULK MODULUS: The measure of resistance to compressibility of a fluid. It is the reciprocal of the compressibility.

COMPRESSIBILITY: The change in volume of a unit volume of a fluid when subjected to a unit change of pressure.

FLASH POINT: The temperature to which a liquid must be heated under specified conditions of the test method to give off sufficient vapor to form a mixture with air that can be ignited momentarily by a specified flame.

FLOW RATE: The volume, mass, or weight of a fluid passing through any conductor per unit of time.

FLUID FRICTION: Friction due to the viscosity of fluids.

HEAD: The height of a column or body of fluid above a given point expressed in linear units. Head is often used to indicate gage pressure. Pressure is equal to the height times the density of the fluid.

Head, Friction: The head required to overcome the friction at the interior surface of a conductor and between fluid particles in motion. It varies with flow, size, type and condition of conductors and fittings, and the fluid characteristics.

Head, Static: The height of a column or body of fluid above a given point.

Head, Static Discharge: The static head from the centerline of the pump to the free discharge surface.

Head, Static Suction: The head from the surface of the supply source to the centerline of the pump.

Head, Total Static: The static head from the surface of the supply source to the free discharge surface.

Head, Velocity: The equivalent head through which the liquid would have to fall to attain a given velocity. Mathematically it is equal to the square of the velocity (in feet) divided by 64.4 feet per second square.

$$h = \frac{v^2}{2g}$$

HYDRAULIC HORSEPOWER: Horsepower computed from flow rate and pressure differential.

Hydraulic Horsepower = .000583Qp

Q = Flow rate — gallons per minute

p = Pressure — pounds per square inch

LIFT: The height of a column or body of fluid below a given point expressed in linear units. Lift is often used to indicate vacuum or pressure below atmospheric.

Lift, Static Suction: The lift from the centerline of the pump to the surface of the supply source. (See Head, Static Sunction)

MICRON: A millionth of a meter or about 0.00004 inch.

NEUTRALIZATION NUMBER: A measure of the total acidity or bascity of an oil; this includes organic or inorganic acids or bases or a combination thereof (ASTM Designation D974-58T).

NEWT: The standard unit of kinematic viscosity in the English system. It is expressed in square inches per second.

POISE: The standard unit of absolute viscosity in the c.g.s. (centimeter-gram-second) system. It is the ratio of the shearing stress to the shear rate of a fluid and is expressed in dyne seconds per square centimeter; 1 centipoise equals .01 poise.

POUR POINT: The lowest temperature at which a liquid will flow under specified conditions (ASTM Designation D97-57).

PRECIPITATION NUMBER: The number of milliliters of precipitate formed when 10 ml. of lubricating oil are mixed with 90 ml. of ASTM precipitation naptha and centrifuged under prescribed conditions (ASTM Method D91-61).

PRESSURE: Force per unit area, usually expressed in pounds per square inch.

Pressure, Absolute: The sum of atmospheric and gage pressures.

Pressure, Atmospheric: Pressure exerted by the atmosphere at any specific location. (Sea level pressure is approximately 14.7 pounds per square inch absolute.)

Pressure, Back: The pressure encountered on the return side of a system.

Pressure, Breakloose: The minimum pressure which initiates movement.

Pressure, Breakout: (See Pressure, Breakloose)

Pressure, Burst: The pressure which causes rupture.

Pressure, Charge: The pressure at which replenishing fluid is forced into a fluid power system.

Pressure, Cracking: The pressure at which a pressure operated valve begins to pass fluid.

Pressure, Differential: The difference in pressure between any two points of a system or a component.

Pressure, Gage: Pressure differential above or below atmospheric pressure.

Pressure Head: The pressure due to the height of a column or body of fluid. It is usually expressed in feet.

Pressure, Operating: The pressure at which a system is operated.

Pressure Override: The difference between the cracking pressure of a valve and the pressure reached when the valve is passing full flow.

Pressure, Pilot: The pressure in the pilot circuit.

Pressure, Precharge: The pressure of compressed gas in an accumulator prior to the admission of a liquid.

Pressure, Proof: The non-destructive test pressure in excess of the maximum rated operating pressure.

Pressure, Rated: The qualified operating pressure which is recommended for a component or a system by the manufacturer.

Pressure, Shock: The pressure existing in a wave moving at supersonic velocity.

Pressure, Static: The pressure in a fluid at rest.

Pressure, Suction: The absolute pressure of the fluid at the inlet of a pump.

Pressure, Surge: The pressure existing from surge conditions.

Pressure, System: The pressure which overcomes the total resistances in a system. It includes all losses as well as useful work.

Pressure, Vapor: The pressure, at a given fluid temperature, in which the liquid and gaseous phases are in equilibrium.

Pressure, Working: The pressure which overcomes the resistance of the working device.

PRESSURE DROP: (See Pressure, Differential)

REYN: The standard unit of absolute viscosity in the English system. It is expressed in pound-seconds per square inch.

SPECIFIC GRAVITY (LIQUID): The ratio of the weight of a given volume of liquid to the weight of an equal volume of water.

STOKE: The standard unit of kinematic viscosity in the c.g.s. (centimeter-gram-second) system. It is expressed in square centimeters per second; 1 centistoke equals .01 stoke.

SURFACE TENSION: The contractile surface force of a liquid in contact with a fluid by which it tends to assume a spherical form and to present the least possible surface. It is expressed in pounds per foot or dynes per centimeter.

VACUUM: Pressure less than atmospheric pressure. It can be expressed in absolute or gage pressure.

VISCOSITY: A measure of the internal friction or the resistance of a fluid to flow.

Viscosity, Absolute: The ratio of the shearing stress to the shear rate of a fluid. It is usually expressed in centipoise.

Viscosity, Kinematic: The absolute viscosity divided by the density of the fluid. It is usually expressed in centistokes.

Viscosity, SAE Number: The Society of Automotive Engineers arbitrary numbers for classifying fluids according to their viscosities. The numbers in no way indicate the viscosity index of fluids.

Viscosity, SUS: Saybolt Universal Seconds (SUS), which is the time in seconds for 60 milliliters of oil to flow through a standard orifice at a given temperature (ASTM Designation D88-56).

VISCOSITY INDEX: A measure of the viscosity-temperature characteristics of a fluid as referred to that of two arbitrary reference fluids (ASTM Designation D567-53).

SECTION 7

CIRCUITRY AND RELATED TERMS

CIRCUIT: An arrangement of interconnected component parts.

Circuit, Pilot: A circuit used to control a main circuit or component.

Circuit, Pressure Control: Any circuit whose main purpose is to adjust or regulate fluid pressure in the system or any branch of the system.

Circuit, Regenerative: A circuit in which pressurized fluid discharged from a component is returned to the system to reduce power input requirements. On single rod end cylinders the discharge from the rod end is often directed to the opposite end to increase rod extension speed.

Circuit, Safety: A circuit which prevents accidental operation, protects against overloads, or otherwise assures safe operation.

Circuit, Sequence: A circuit which establishes the order in which two or more phases of a circuit occur.

Circuit, Servo: A circuit which is controlled by automatic feed back; i.e., the output of the system is sensed or measured and is compared with the input signal. The difference (error) between the actual output and the input controls the circuit. The controls attempt to minimize the error. The system output may be position, velocity, force, pressure, level, flow rate, or temperature.

Circuit, Speed Control: Any circuit where components are arranged to regulate speed of operation.

Circuit, Synchronizing: A circuit in which multiple operations are controlled to occur at the same time.

Circuit, Unloading: A circuit in which pump volume is returned to reservoir at near zero gage pressure whenever delivery to the system is not required.

CYCLE: A single complete operation consisting of progressive phases starting and ending at the neutral position.

Cycle, Automatic: A cycle of operation

which once started is repeated indefinitely until stopped.

Cycle, Manual: A cycle which is manually started and controlled through all phases.

Cycle, Semi-automatic: A cycle which is started upon a given signal, proceeds through a predetermined sequence, and stops with all elements in their initial position.

DIAGRAM, Combination: A drawing utilizing a combination of graphical, cutaway and pictorial symbols showing interconnected lines.

Diagram, Cutaway: A drawing showing principle internal parts of all components, controls and actuating mechanisms, all interconnecting lines and functions of individual components.

Diagram, Graphical: A drawing or drawings showing each piece of apparatus including all interconnecting lines by means of approved (ASA, JIC) standard symbols.

Diagram, Pictorial: A drawing showing each component in its actual shape according to the manufacturer's installation.

Diagram, Pressure-Time: A graphical presentation of pressure plotted against time for a complete cycle of the equipment.

Diagram. Schematic: (See Diagram, Graphical)

PHASE: A distinct functional operation during a cycle. Some typical sequential phases are: neutral, rapid advance, feed or pressure stroke, dwell and rapid return.

Phase, Dwell: The phase of a cycle where a specified motion is stopped for a pre-determined length of time.

Phase, Feed: The phase of a cycle where work is performed on the workpiece.

Phase, Neutral: The phase of a cycle from which the work sequence begins.

Phase, Rapid Advance: The phase of a cycle where tools or workpiece approach at high speed to the feed position.

Phase, Rapid Return: The phase of a cycle where tools or workpiece returns at high speed to the cycle starting position.

SECTION 9

GENERAL TERMS

BACK CONNECTED: Where connections are made to normally unexposed surfaces of components.

FRONT CONNECTED: Where connections are made to normally exposed surfaces of components.

POSITION REVERSAL: A reversal of direction of movement initiated by a signal given at some predetermined point of the movement.

POSITIVE POSITION STOP: A structural member which accurately stops motion.

POSITIVE SAFETY STOP: A structural member which confines maximum travel to safe limits.

PRESSURE REVERSAL: A reversal of direction of movement initiated by a signal responsive to rise in pressure.

SECTION 11

FLUIDS AND RELATED TERMS

ADDITIVE: A chemical compound or compounds added to a fluid to change its properties.

AIR, Compressed: Air at any pressure greater than atmospheric pressure.

Air. Free: Air under the pressure due to atmospheric conditions at any specific location.

Air, Pressure: (See Air, Compressed)

Air, Standard: (See MENSURATION SECTION)

FLUID: A liquid or a gas.

Fluid, Emulsion, Water-Oil: A stabilized mixture of two immiscible components, water and oil. It may contain additives. There are two types: OIL IN WATER and WATER IN OIL.

Fluid, Emulsion, Oil in Water: A dispersion of oil in a continuous phase of water.

Fluid, Emulsion, Water in Oil: A dispersion of water in a continuous phase of oil.

Fluid, Fatty Oil: A fluid composed of fats derived from animal, marine or vegetable origin. It may contain additives.

Fluid, Fire Resistant: A fluid difficult to ignite which shows little tendency to propagate flame.

Fluid, Halogenated: A fluid composed of halogenated organic materials. It may contain additives.

Fluid, Hydraulic: A fluid suitable for use in a hydraulic system.

Fluid, Non-Flammable: (Deprecated)

Fluid, Organic Ester: A fluid composed of esters which are compounds of carbon, hydrogen, and oxygen only. It may contain additives.

Fluid, Petroleum: A fluid composed of petroleum oil. It may contain additives.

Fluid, Phosphate Ester: A fluid composed of phosphate esters. It may contain additives.

Fluid, Phosphate Ester Base: A fluid which contains a phosphate ester as one of the major components.

Fluid, Polyglycol: A non-aqueous fluid composed of polyglycols or polyglycol derivatives. It may contain additives.

Fluid. Silicate Ester: A fluid composed of organic silicates. It may contain additives.

Fluid, Silicone: A fluid composed of silicones. It may contain additives.

Fluid, Water-Glycol: A fluid whose major constituents are water and one or more glycols or polyglycols.

FLUID STABILITY: Resistance of a fluid to permanent change in properties.

Fluid Stability, Chemical: Resistance of a fluid to chemical change.

Fluid Stability, Hydrolytic: Resistance of a fluid to permanent changes in properties caused by chemical reaction with water.

Fluid Stability, Oxidation: Resistance of a fluid to permanent changes caused by chemical reaction with oxygen.

Fluid Stability, Thermal: Resistance of a fluid to permanent changes caused solely by heat.

INHIBITOR: Any substance which slows or prevents such chemical reactions as corrosion or oxidation.

SECTION 21

FLUID CONDITIONERS AND RELATED TERMS

AIR BREATHER: A device permitting air movement between atmosphere and the component in which it is installed.

BLEEDER: A device for removal of pressurized fluid.

Bleeder, Air: A bleeder for the removal of air.

CONTAMINANT: Detrimental matter in a fluid.

COOLER: A heat exchanger which removes heat from a fluid.

Cooler, Aftercooler: A device which cools a gas after it has been compressed.

Cooler, Intercooler: A device which cools a gas between the compressive steps of a multiple stage compressor.

Cooler, Precooler: A device which cools a gas before it is compressed.

FILTER: A device whose primary function is the retention by a porous media of insoluable contaminants from a fluid.

FILTER ELEMENT: The porous device which performs the actual process of filtration.

FILTER MEDIA: The porous materials which perform the actual process of filtration.

Filter Media, Depth: Porous materials which primarily retain contaminant within a tortuous path.

Filter Media, Effective Area: The total funtional area of the porous media.

Filter Media, Surface: Porous materials which primarily retain contaminants on the influent face.

FLUID CONDITIONER: A device which controls the physical characteristics of a fluid.

HEAT EXCHANGER: A device which transfers heat through a conducting wall from one fluid to another.

LUBRICATOR: A device which adds controlled or metered amounts of lubricant into a fluid power system.

MUFFLER: A device for reducing gas flow noise. Noise is decreased by back pressure control of gas expansion.

RESERVOIR: A container for storage of liquid in a fluid power system.

Reservoir, Atmospheric: A reservoir for storage of fluid media at atmospheric pressure.

Reservoir, Sealed: A reservoir for storage of fluids isolated from atmospheric conditions.

Reservoir, Sealed, Pressure: A sealed reservoir for storage of fluids under pressure.

SEPARATOR: A device whose primary function is to isolate undesirable fluids and.or contaminants by physical properties other than size.

SILENCER: A device for reducing gas flow noise. Noise is decreased by tuned resonant control of gas expansion.

STRAINER: A coarse filter.

TANK: A container for the storage of fluid in a fluid power system.

Tank, Vacuum: A container for gas at less than atmospheric pressure.

SECTION 31

COMPRESSORS — INTENSIFIERS — PUMPS

COMPRESSOR: A device which converts mechanical force and motion into pneumatic fluid power.

Compressor, Multiple Stage: A compressor having two or more compressive steps in which the discharge from each supplies the next in series.

Compressor, Single Stage: A compressor having only one compressive step between inlet and outlet.

INTENSIFIER: A device which converts low pressure fluid power into higher pressure fluid power.

POWER UNIT: A combination of pump, pump drive, reservoir, controls and conditioning components which may be required for its application.

PUMP: A device which converts mechanical force and motion into hydraulic fluid power.

Pump, Axial Piston: A pump having multiple pistons disposed with their axes parallel.

Pump, Centrifugal: A pump which produces fluid velocity and converts it to pressure head.

Pump, Centrifugal, Concentric: (See Diffuser Centrifugal Pump).

Pump, Centrifugal, Diffuser: A centrifugal pump in which fluid enters at the center of the impeller, is accelerated radially, and leaves through vanes arranged to provide a gradually enlarging flow passage.

Pump, Centrifugal, Peripheral: A centrifugal pump in which fluid enters, follows, and leaves the periphery of the impeller.

Pump, Centrifugal, Spiral: (See Pump, Centrifugal, Volute)

Pump, Centrifugal, Volute: A centrifugal pump in which fluid enters at the center of the impeller, is accelerated radially, and leaves through a gradually enlarging flow passage.

Pump, Fixed Displacement: A pump in which the displacement per cycle cannot be varied.

Pump, Gear: A pump having two or more intermeshed rotating members enclosed in a housing.

Pump, Hand: A hand operated pump.

Pump, Multiple Stage: Two or more pumps in series.

Pump, Radial Piston: A pump having multiple pistons disposed radially actuated by an eccentric element.

Pump, Reciprocating Duplex: A pump having two reciprocating pistons.

Pump, Reciprocating Single Piston: A pump having a single reciprocating piston.

Pump, Screw: A pump having one or more screws rotating in a housing.

Pump, Vane: A pump having multiple radial vanes within a supporting rotor.

Pump, Variable Displacement: A pump in which the displacement per cycle can be varied.

SECTION 36

ACCUMULATORS
AND AIR RECEIVERS

ACCUMULATOR: A container in which fluid is stored under pressure as a source of fluid power.

Accumulator, Hydropneumatic: An accumulator in which compressed gas applies force to the stored liquid.

Accumulator, Hydropneumatic, Bladder: A hydropneumatic accumulator in which the liquid and gas are separated by an elastic bag or bladder.

Accumulator, Hydropneumatic, Diaphragm: A hydropneumatic accumulator in which the liquid and gas are separated by a flexible diaphragm.

Accumulator, Hydropneumatic, Non-Separator: A hydropneumatic accumulator in which the compressed gas operates dirctly on liquid within the pressure chamber.

Accumulator, Hydropneumatic, Piston: A hydropneumatic accumulator in which the liquid and gas are separated by a floating piston.

Accumulator, Mechanical: An accumulator incorporating a mechanical device which applies force to the stored fluid.

Accumulator, Mechanical, Spring: A mechanical accumulator in which springs apply force to the stored fluid.

Accumulator, Mechanical, Weighted: A mechanical accumulator in which the gravitational force acting upon weights applies force to the stored fluid.

AIR RECEIVER: A container in which gas is stored under pressure as a source of pneumatic fluid power.

PRESSURE VESSEL: A container which holds fluid under pressure.

SECTION 41

CONDUCTORS

CHANNEL: A fluid passage, the length of which is large with respect to its cross-sectional area.

CONDUCTOR: A component whose primary function is to contain and direct fluid.

CONDUIT: Any confining element employed to transfer a fluid.

HOSE: A flexible line.

Hose, Wire Braided: Hose consisting of a flexible material reinforced with woven wire braid.

LINE: A tube, pipe, or hose for conducting fluid.

Line, Drain: A line returning leakage fluid independently to the reservoir or vented manifold.

Line, Exhaust: A line returning power or control fluid back to the reservoir or atmosphere.

Line, Pilot: A line which conducts control fluid.

Line, Suction: A supply line at sub-atmospheric pressure to a pump, compressor, or other component.

Line, Working: A line which conducts fluid power.

LINES: Two or more fluid power lines.

Lines, Joining: Lines which connect in a circuit.

Lines, Passing: Lines which cross but do not connect in a circuit.

MANIFOLD: A conductor which provides multiple connection ports.

Manifold, Vented: A manifold which is open to the atmosphere and returns fluid to the reservoir.

NIPPLE: A short length of pipe or tube.

PASSAGE: A machined or cored fluid-conducting path which lies within or passes through a component.

PIPE: A line whose outside diameter is standardized for threading. Pipe is available in Standard, Extra Strong, Double Extra Strong or Schedule wall thicknesses.

SUBPLATE (BACK PLATE): An auxiliary ported plate for mounting components.

TUBE: A line whose size is its outside diameter. Tube is available in varied wall thicknesses.

SECTION 42

CONNECTORS AND CLOSURES

BUSHING: A short externally threaded connector with a smaller size internal thread.

CAP: A cover for fluid passages.

CLOSURE: A cap or a plug.

CONNECTOR: A device for joining a conductor to a component port or to one or more other conductors.

COUPLING: A straight connector for fluid lines.

Coupling, Quick Disconnect: A coupling which can quickly join or separate lines.

CROSS: A connector with four ports arranged in pairs, each pair on one axis, and the axes at right angles.

ELBOW: A connector that makes an angle between mating lines. The angle is always 90 degrees unless another angle is specified.

FITTING: A connector or closure for fluid power lines and passages.

Fitting, Compression: A fitting which seals and grips by manual adjustable deformation.

Fitting, Flange: A fitting which utilizes a radially extending collar for sealing and connection.

Fitting, Flared: A fitting which seals and grips by a preformed flare at the end of the tube.

Fitting, Flared, AN: A United States Air Force—Navy 37° flared tube fitting Design Standard.

Fitting, Flareless: A fitting which seals and grips by means other than a flare.

Fitting, Reusable Hose: A hose fitting that can be removed from a hose and reused.

Fitting, Welded: A fitting attached by welding.

JOINT: A line positioning connector.

Joint, Rotary: A joint connecting lines which have relative operational rotation.

Joint, Swivel: A joint which permits variable operational positioning of lines.

PLUG: A closure which fits into a fluid passage.

Plug, Dryseal Pipe: A plug made with a thread which conforms to Dryseal Pipe Thread Standards.

Plug, Short Pipe Thread: A plug which conforms in all respects to standard pipe threads except that the full thread has been shortened one full thread from the small end.

Plug, Standard Pipe Thread: A plug with American (National) tapered pipe threads.

Plug, Straight Thread: A plug with straight thread conforming to unified thread standards.

REDUCER: A connector having a smaller line size at one end than the other.

TEE: A connector with three ports, a pair on one axis with one side outlet at right angles to this axis.

UNION: A connector which permits lines to be joined or separated without requiring the lines to be rotated.

WYE (Y): A connector with three ports, a pair on one axis with one side outlet at any angle other than right angles to this axis. The side outlet angle is usually 45°, unless another angle is specified.

SECTION 43

PORTS AND THREADS

PORT: An internal or external terminus of a passage in a component.

Port, AND10050: A United States Air Force —Navy Aeronautical Design Standard in which a straight thread port is used to attach tube fittings to various components. It employs an "O" ring seal compressed in a special cavity.

Port, Bleed: A port which provides a passage for the purging of gas from a system or component.

Port, Control: A port which provides a passage for control fluid.

Port, Cylinder: A port which provides a passage to or from an actuator.

Port, Discharge: A port which provides a passage for fluid power to the system.

Port, Drain: A port which provides a free passage for the removal of slippage fluid from a hydraulic system or moisture from a pneumatic system.

Port, Exhaust: A port which provides a passage to the atmosphere.

Port, Fill: A port which provides a passage for filling purposes.

Port, Inlet: A port which provides a passage from the source of fluid.

Port, Pipe: A port which conforms to pipe thread standards.

Port, Plain "O" Ring: A port which uses an "O" ring in a groove located on the port face.

Port, Pressure: A port which provides a passage from the source of fluid.

Port, Reservoir: (See Port, Tank)

Port, Return: (See Port, Tank)

Port, SAE: A straight thread port used to attach tube and hose fittings. It employs an "O" ring compressed in a wedge-shaped cavity. A standard of the Society of Automotive Engineers.

Port, Suction: A port which provides a passage for atmospheric charging of a pump or compressor.

Port, Tank: A port which provides a passage to the fluid source.

Port, Vent: A port which provides a passage to the atmosphere.

PIPE THREADS: Screw threads for joining pipe.

Pipe Threads, Dryseal: Pipe threads in which sealing is a function of root and crest interference.

SECTION 46

PANELS — COMPARTMENTS — ENCLOSURES

COMPARTMENT: A space within the base, frame or column of the equipment.

ENCLOSURE: A housing for components.

PANEL: A plate or a surface for mounting components.

Panel, Control: A grouping of components mounted on a panel or integrally built into an assembled unit having a single mounting surface.

Panel, Mounting: A panel on which a number of components may be mounted.

SECTION 51

ACTUATORS — CYLINDERS — MOTORS

ACTUATOR: A device which converts fluid power into mechanical force and motion.

CYLINDER: A device which converts fluid power into linear mechanical force and motion. It usually consists of a movable element such as a piston and piston rod, plunger or ram, operating within a cylindrical bore.

Cylinder, Adjustable Stroke: A cylinder equipped with adjustable stops at one or both ends to limit piston travel.

Cylinder, Cushioned: A cylinder with a piston-assembly deceleration device at one or both ends of the stroke.

Cylinder, Double Acting: A cylinder in which fluid force can be applied to the movable element in either direction.

Cylinder, Double Rod: A cylinder with a single piston and a piston rod extending from each end.

Cylinder, Dual Stroke: A cylinder combination which provides two working strokes.

Cylinder, Piston: A cylinder in which the movable element has a greater cross-sectional area than the piston rod.

Cylinder, Plunger: A cylinder in which the movable element has the same cross-sectional area as the piston rod.

Cylinder, Ram: (See Cylinder, Plunger)

Cylinder, Retractable Stroke: An adjustable-stroke cylinder in which the stop can be temporarily changed to permit full retraction of the piston assembly.

Cylinder, Single Acting: A cylinder in which the fluid force can be applied to the movable element in only one direction.

Cylinder, Single Rod: A cylinder with a piston rod extending from one end.

Cylinder, Spring Return: A cylinder in which a spring returns the piston assembly.

Cylinder, Tandem: Two or more cylinders with interconnected piston assemblies.

Cylinder, Telescoping: A cylinder with nested multiple tubular rod segments which provide a long working stroke in a short retracted envelope.

MOTOR: A device which converts fluid power into mechanical force and motion. It usually provides rotary mechanical motion.

Motor, Fixed Displacement: A motor in which the displacement per unit of output motion cannot be varied.

Motor, Linear: (See CYLINDER)

Motor, Rotary: A motor capable of continuous rotary motion.

Motor, Rotary, Limited: A rotary motor having limited motion.

Motor, Variable Displacement: A motor in which the displacement per unit of output motion can be varied.

SECTION 56

CYLINDER PARTS

(Wherever the term "piston rod" appears it is intended to also mean "plunger" or "ram").

ANGLE: A mounting device which is angular in cross section. It is usually made from a 90° structural angle.

BACK END: (See CAP)

BASE: (See SIDE)

BLIND END: (See CAP)

BLIND HEAD: (See CAP)

CAP: A cylinder end closure which completely covers the bore area.

CLEVIS: A "U" shaped mounting device which contains a common pin hole at right angle or normal to the axis of symmetry through each extension. A clevis usually connects with an eye.

END: Either of two envelope surfaces at right angle or normal to the piston rod centerline.

EYE: A mounting device consisting of a single extension which contains a mounting pin hole at right angle or normal to the axis of symmetry. An eye usually connects with a clevis.

FLANGE: A mounting device consisting of a plate or collar extending past the basic cylinder profile to provide clearance area for mounting bolts. A flange is usually at right angle or normal to the piston rod centerline.

FOOT: (See LUG)

FRONT END: (See HEAD)

FRONT FACE: (See HEAD)

FRONT HEAD: (See HEAD)

HEAD: The cylinder end closure which covers the differential area between the bore area and the piston rod area.

HINGE: (See CLEVIS and EYE)

LUG: A mounting device consisting of a block extending past the basic cylinder profile. The block usually has a tapped or through mounting hole at right angles to the cylinder axis.

PENDULUM: (See CLEVIS and EYE)

RABBET: A mounting device which utilizes matching male and female forms (usually coaxial circular) between the cylinder and its mating element.

REAR END: (See CAP)

REAR HEAD: (See CAP)

ROD END: (See HEAD)

ROD, HEAD: (See HEAD)

SIDE: An envelope surface which is parallel to the piston rod centerline.

TIE ROD: An axial external cylinder element which traverses the length of the cylinder. It is pre-stressed at assembly to hold the ends of the cylinder against the tubing. Tie rod extensions can be a mounting device.

TRUNNION: A mounting device consisting of a pair of opposite projecting cylindrical pivots. The cylindrical pivot pins are at right angle or normal to the piston rod centerline to permit the cylinder to swing in a plane.

SECTION 57

CYLINDER MOUNTINGS

(See Section 56 for detail definitions—wherever the term "piston rod" appears it is intended to also mean "plunger" or "ram").

MOUNTING: A device by which a cylinder is fastened to its mating element.

Mounting, Centerline: A mounting which permits connection on a plane in line with the piston rod centerline.

Mounting, Centerline, Lug: A centerline mounting consisting of two opposite lugs at each end of the cylinder.

Mounting, End: A mounting which permits connection at either or both ends of a cylinder.

Mounting, End, Both: A mounting at both ends of the cylinder.

Mounting, End, Both, Tie Rods Extended: A cylinder mounted at both ends by means of extended tie rods.

Mounting, End, Cap (Back End, Blind End, Blind End Head, Rear End, Rear Head): A mounting which permits connection at the cap end.

Mounting, End, Cap, Circular: A direct circular cap mounting.

Mounting, End, Cap, Circular Flange: A cap mounting consisting of a supplementary circular flange plate.

Mounting, End, Cap, Detachable Clevis: A cap mounting consisting of a clevis which can be removed or rotated.

Mounting, End, Cap, Detachable Eye: A cap mounting consisting of an eye which can be removed or rotated.

Mounting, End, Cap, Fixed Clevis: A cap mounting consisting of a clevis integral with the cap to maintain a fixed clevis-port relationship.

Mounting, End Cap, Fixed Eye: A cap mounting consisting of an eye integral with the cap to maintain a fixed eye-port relationship.

Mounting, End, Cap, Rectangular Flange: A cap mounting consisting of a supplementary rectangular flange plate.

Mounting, End, Cap, Square: A direct square cap mounting.

Mounting, End, Cap, Square Flange: A cap mounting consisting of a supplementary square flange plate.

Mounting, End, Cap, Tie Rods Extended: A cap mounting consisting of extended tie rods.

Mounting, End, Cap, Trunnion: A cap mounting consisting of trunnion pins near or at the cap end of the cylinder.

Mounting, End, Head (Front End, Front Face, Front Head, Rod End, Rod End Head:) A mounting which provides cylinder connection at the head end.

Mounting, End, Head, Circular: A direct circular head mounting.

Mounting, End, Head, Circular Flange: A head mounting consisting of a supplementary circular flange plate.

Mounting, End, Head, Female Rabbet: A head mounting consisting of a female pilot recess.

Mounting, End, Head, Male Rabbet: A head mounting consisting of a male pilot extension.

Mounting, End, Head, Rectangular Flange: A head mounting consisting of a supplementary rectangular flange plate.

Mounting, End, Head, Square: A direct square head mounting.

Mounting, End, Head, Square Flange: A head mounting consisting of a supplementary square flange plate.

Mounting, End, Head, Tie Rods Extended: A head mounting consisting of extended tie rods.

Mounting, End, Head, Trunnion: A head mounting consisting of trunnion pins near or at the head end of a cylinder.

Mounting, Fixed: A mounting which provides rigid connection between the cylinder and the mating element wherein the piston rod reciprocates in a fixed line.

Mounting, Intermediate: A mounting which provides cylinder connection at an intermediate external position along the piston rod centerline between ends.

Mounting, Intermediate, Fixed Trunnion: An intermediate mounting consisting of trunnion pins which cannot be repositioned.

Mounting, Intermediate, Movable, Trunnion: An intermediate mounting consisting of trunnion pins which can be repositioned.

Mounting, Pivot: A mounting which permits a cylinder to change its alignment in a plane.

Mounting, Side (Base): A mounting which provides cylinder connection at one of its sides.

Mounting, Side (Base), End Angles: A side mounting consisting of an angle at each end of the cylinder with the free legs facing a common side.

Mounting, Side (Base), End Lugs: A side mounting consisting of one or more lugs at each end of the cylinder and facing a common side.

Mounting, Side (Base), End Plates: A side mounting consisting of an extending plate at each cylinder end and facing a common side.

Mounting, Side (Base), Lugs (Foot): A side mounting consisting of two opposite lugs at each cylinder end facing a common side.

Mounting, Side (Base), Tapped: A side mounting consisting of one or more tapped holes at each cylinder end facing a common side.

Mounting, Side (Base), Through Holes: A side mounting consisting of holes drilled across both ends to a common side.

Mounting, Universal: A mounting which permits a cylinder to change its alignment in all directions.

SECTION 58

CUSHIONS AND SNUBBERS

CUSHION: A device which provides controlled resistance to motion.

Cushion, Cylinder: A cushion built into a cylinder to restrict flow at the outlet port thereby arresting the motion of the piston rod.

Cushion, Die: A cushion installed with a die on a press to provide controlled resistance against the work. The return motion of the cushion is sometimes used to eject the work.

Cushion, Hydraulic: A cushion in which resistance is developed hydraulically.

Cushion, Hydropneumatic: A cushion in which resistance is developed hydraulically and pneumatically.

Cushion, Pneumatic: A cushion in which resistance is developed pneumatically.

SECTION 61

SEALING DEVICES
AND RELATED TERMS

DUROMETER HARDNESS: A comparative indication of elastomer hardness determined by a durometer.

PACK: To install a packing.

PACKING: A sealing device consisting of bulk deformable material or one or more mating deformable elements, reshaped by manually adjustable compression to obtain and maintain effectiveness. It usually uses axial compression to obtain radial sealing.

Packing, Coil: Packing in coil form.

Packing, "U": A packing in which the deformable element has a "U" shaped cross-section.

Packing, "V": A packing in which the deformable element has a "V" shaped cross-section.

Packing, "W": A packing in which the deformable element has a "W" shaped cross-section.

RING, "O": A ring which has a round cross-section.

Ring, Piston: A piston sealing ring. It is usually one of a series and is often split to facilitate expansion or contraction.

Ring, Scraper: A ring which removes material by a scraping action.

Ring, "U": A ring which has a "U" shaped cross-section.

Ring, "V": A ring which has a "V" shaped cross-section.

Ring, "W": A ring which has a "W" shaped cross-section.

Ring, Wiper: A ring which removes material by a wiping action.

SEALING DEVICE: A device which prevents or controls the escape of a fluid or entry of a foreign material.

SEAL: (See SEALING DEVICE)

Seal, Axial: A sealing device which seals by axial contact pressure.

Seal, Cup: A sealing device with a radial base integral with an axial cylindrical projection at its outer diameter.

Seal, Diaphragm: A relatively thin, flat or molded sealing device fastened and sealed at its periphery with its inner portion free to move.

Seal, Dished Diaphragm: A diaphragm in which the central area is depressed in a free state. It permits longer travel than a flat comparable diaphragm.

Seal, Flat Diaphragm: (See Seal, Diaphragm)

Seal, Dynamic: A sealing device used between parts that have relative motion.

Seal, End: (See Seal, Axial)

Seal, Face: (See Seal, Axial)

Seal, Flange: A sealing device with a radial base integral with an axial projection at its inner diameter.

Seal, Gasket: (See Seal, Static)

Seal, Hat: (See Seal, Flange)

Seal, Lip: A sealing device which has a flexible sealing projection.

Seal, Lubricant: A sealing device which uses a lubricant as a sealing barrier.

Seal, Mechanical: A sealing device in which sealing action is aided by mechanical force.

Seal, Oil: A sealing device which retains oil.

Seal, "O" Ring: A sealing ring which has a round cross-section.

Seal, Piston: A sealing device installed on a piston to maintain a sealing fit with a cylinder bore.

Seal, Plunger: (See Seal, Ram)

Seal, Pressure: A sealing device in which sealing action is aided by fluid pressure.

Seal, Radial: A sealing device which seals by radial contact pressure.

Seal, Ram: A sealing device which seals the periphery of a ram.

Seal, Rod: A sealing device which seals the periphery of a piston rod.

Seal, Rotary: A sealing device used between parts that have relative rotary motion.

Seal Shaft: (See Seal, Rod and Seal, Rotary)

Seal, Shoulder: (See Seal, Axial)

Seal, Sliding: A sealing device used between

parts that have relative reciprocating motion.

Seal, Static: A sealing device used between parts that have no relative motion.

Seal, Stem: (See Seal, Rod and Seal, Shaft)

Seal, Water: A sealing device which uses water as a sealing barrier.

Seal, Wiper: A sealing device which operates by a wiping action.

SECTION 63

AUXILIARY DEVICES
FOR SEALS

ADAPTER: A seal support shaped to conform with the contour of the seal and the mating element.

Adapter, Female: An adapter with a concave seal support.

Adapter, Male: An adapter with a convex seal support.

Adapter, Pedestal: A male adapter of "T" shaped cross section usually used to support a "U" seal.

ANTI-EXTRUSION RING: A ring which bridges a clearance to minimize seal extrusion.

FILLER RING: A ring which fills the recess of a "V" or "U" seal.

GLAND: The cavity of a stuffing box.

GLAND FOLLOWER: The closure for a stuffing box.

JUNKET RING: (See ANTI-EXTRUSION RING).

LANTERN RING: A ring which provides support and venting for adjacent sealing devices.

SHELL: A structural form to which a sealing element is assembled or bonded.

SPRING, Expander: A spring which produces outward radial sealing force.

Spring, Finger: A spring with flexible fingers which produce sealing force.

Spring, Garter: A compression or tension ring formed from a long helical wire spring with connected ends to produce sealing force.

Spring, Lug: (See Spring, Finger)

Spring, Marcel: (See Spring, Wave)

Spring, Spreader: A spring which produces sealing force against both lips of "U" or "V" seals.

Spring, Wave: A compression spring of waved configuration which produces sealing force.

STUFFING BOX: A cavity and closure with manual adjustment for a sealing device.

SUPPORT RING: (See ADAPTER).

SECTION 71

CONTROLS AND RELATED TERMS

CONTROL: A device used to regulate the function of a component or system.

Control, Automatic: A control which actuates equipment in a predetermined manner.

Control, Combination: A combination of more than one basic control.

Control, Cylinder: A control in which a fluid cylinder is the actuating device.

Control, Electric: A control actuated electrically.

Control, Hydraulic: A control actuated by a liquid.

Control, Liquid-Level: A device which controls the liquid level by a float switch or other means.

Control, Manual: A control actuated by the operator.

Control, Mechanical: A control actuated by linkages, gears, screws, cams or other mechanical elements.

Control, Pneumatic: A control actuated by air or other gas pressure.

Control, Pressure Compensated: A control in which a pressure signal operates a compensating device.

Control, Servo: A control actuated by a feed back system which compares the output with the reference signal and makes corrections to reduce the difference.

CONTROLS, PUMP: Controls applied to positive displacement variable delivery pumps to adjust their volumetric output or direction of flow.

RESTRICTOR: A device which reduces the cross-sectional flow area.

Restrictor, Choke: A restrictor, the length of which is relatively large with respect to its cross-sectional area.

Restrictor, Orifice: A restrictor, the length of which is relatively small with respect to its cross-sectional area. The orifice may be fixed or variable. Variable types are non-compensated, pressure compensated, or pressure and temperature compensated.

SECTION 72

VALVES — FUNCTIONAL TYPES

VALVE: A device which controls fluid flow direction, pressure, or flow rate.

Valve, Air: A valve for controlling air.

Valve, Directional Control: A valve whose primary function is to direct or prevent flow through selected passages.

Valve, Directional Control, Check: A directional control valve which permits flow of fluid in only one direction.

Valve, Directional Control, Diversion: (See Valve, Directional Control, Selector)

Valve, Directional Control, Four Way: A directional control valve whose primary function is to alternately pressurize and exhaust two working ports.

Valve, Directional Control, Selector: A directional control valve whose primary function is to selectively interconnect two or more ports.

Valve, Directional Control, Servo: A directional control valve which modulates flow or pressure as a function of its input signal.

Valve, Directional Control, Straightway: A two port directional control valve.

Valve, Directional Control, Three Way: A directional control valve whose primary function is to alternately pressurize and exhaust a working port.

Valve, Flow Control: A valve whose primary function is to control flow rate.

Valve, Flow Control, Deceleration: A flow control valve which gradually reduces flow rate to provide deceleration.

Valve, Flow Control, Pressure Compensated: A flow control valve which controls the rate of flow independent of system pressure.

Valve, Flow Control, Pressure-Temperature Compensated: A pressure compensated flow control valve which controls the rate of flow independent of fluid temperature.

Valve, Flow Dividing: A valve which divides the flow from a single source into two or more branches.

Valve, Flow Dividing, Pressure Compensated: A flow dividing valve which divides the flow at constant ratio regardless of the difference in the resistances of the branches.

Valve, Flow Metering: (See Valve, Flow Control)

Valve, Hydraulic: A valve for controlling liquid.

Valve, Pilot: A valve applied to operate another valve or control.

Valve, Pneumatic: A valve for controlling gas.

Valve, Prefill: A valve which permits full flow from a tank to a "working" cylinder during the advance portion of a cycle, permits the operating pressure to be applied to the cylinder during the working portion of the cycle, and permits free flow from the cylinder to the tank during the return portion of the cycle.

Valve, Pressure Control: A valve whose primary function is to control pressure.

Valve, Pressure Control, Counterbalance: A pressure control valve which maintains back pressure to prevent a load from falling.

Valve, Pressure Control, Decompression: A pressure control valve that controls the rate at which the contained energy of the compressed fluid is released.

Valve, Pressure Control, Load Dividing: A pressure control valve used to proportion pressure between two pumps in series.

Valve, Pressure Control, Pressure Reducing: A pressure control valve whose primary function is to limit outlet pressure.

Valve, Pressure Control, Relief: A pressure control valve whose primary function is to limit system pressure.

Valve, Pressure Control, Relief, Safety: A relief valve whose primary function is to provide pressure limitation after malfunction.

Valve, Pressure Control, Unloading: A pressure control valve whose primary function is to permit a pump or compressor to operate at minimum load.

Valve, Priority: A valve which directs flow to one operating circuit at a fixed rate and directs excess flow to another operating circuit.

Valve, Sequence: A valve whose primary function is to direct flow in a pre-determined sequence.

Valve, Shutoff: A valve which operates fully open or fully closed.

Valve, Shuttle: A connective valve which selects one of two or more circuits because flow or pressure changes between the circuits.

Valve, Surge Damping: A valve which reduces shock by limiting the rate of acceleration of fluid flow.

Valve, Time Delay: A valve in which the change of flow occurs only after a desired time interval has elapsed.

SECTION 73

VALVES — BASIC DESIGNS

FLAPPER ACTION: A valve design in which output control pressure is regulated by a pivoted flapper in relation to one or two orifices.

JET ACTION: A valve design in which flow effect is controlled by the relative position of a nozzle and a receiver.

SEATING ACTION: A valve design in which flow is stopped by a seated obstruction in the flow path.

Seating Action, Ball: A seating action valve design which utilizes a solid ball to obstruct the flow path.

Seating Action, Diaphragm: A seating action valve design which utilizes a diaphragm to obstruct the flow path.

Seating Action, Disc: A seating action valve design which utilizes a disc to obstruct the flow path.

Seating Action, Disc, Swing: A seating action valve design which utilizes a hinged disc to obstruct the flow path.

Seating Action, Gate: A seating action valve design which utilizes a solid gate to obstruct the flow path.

Seating Action, Gate, Spreader: A gate valve which utilizes two companion discs which are positively seated by common spreaders to obstruct the flow path.

Seating Action, Gate, Wedge: A gate valve which utilizes a solid wedge shaped gate to obstruct the flow path.

Seating Action, Globe: (See Seating Action, Disc)

Seating Action, Needle: A seating action valve design which utilizes an externally adjustable tapered closure to obstruct the flow path.

Seating Action, Plug: A seating action valve design which utilizes a plug to obstruct the flow path.

Seating Action, Poppet: A seating action valve design in which the seating element pops open to obtain free flow in one direction and immediately reseats when flow reverses.

SHEAR ACTION: A valve design in which flow is modulated by an element which slides across the flow path.

Shear Action, Ball: A shear action valve

design which utilizes a ported ball that rotates on an axis normal to the flow path.

Shear Action, Plug: A shear action valve design which utilizes a ported plug that rotates on an axis normal to the flow path.

Shear Action, Plunger: (See Shear Action, Spool)

Shear Action, Spool: A shear action valve design which utilizes a spool that slides through the flow path.

Shear Action, Sliding Plate: A shear action valve design which utilizes a plate that slides across the flow path.

Shear Action, Sliding Plate, Linear: A sliding plate shear action valve design in which the motion of the plate is linear.

Shear Action, Sliding Plate, Rotary: A sliding plate shear action valve design in which the motion of the plate is rotary.

SECTION 74

VALVE POSITIONS

VALVE POSITION: The point at which flow directing elements provide a specific flow condition in a valve.

Valve Position, Center: The selective mid-position in a directional control valve.

Valve Position, Detent: A predetermined position maintained by a holding device acting on the flow-directing elements of a directional control valve.

Valve Position, Normal: The valve position when signal or actuating force is not being applied.

Valve Position, Offset: An off-center position in a directional control valve.

Valve Position, Return: The initial valve position.

Valve, Four Position: A directional control valve having four positions to give four selections of flow conditions.

Valve, Three Position: A directional control valve having three positions to give three selections of flow conditions.

Valve, Two Position: A directional control valve having two positions to give two selections of flow conditions.

SECTION 75

VALVE FLOW CONDITIONS

VALVE FLOW CONDITION: A flow pattern in a directional control valve.

Valve Flow Condition, Closed: All ports are closed.

Valve Flow Condition, Float: Working ports are connected to exhaust or reservoir.

Valve Flow Condition, Hold: Working ports are blocked to hold a powered device in a fixed position.

Valve Flow Condition, Open: All ports are open.

Valve Flow Condition, Regenerative: Working ports are connected to supply.

Valve Flow Condition, Tandem: Working ports are blocked and supply is connected to the reservoir port.

SECTION 76

VALVE ACTUATORS

VALVE ACTUATOR: The valve part(s) through which force is applied to move or position flow-directing elements.

Valve Actuator, Manual: A valve actuator consisting of a hand lever, palm button, foot treadle, or other manual energizing devices.

Valve Actuator, Mechanical: A valve actuator consisting of a cam, lever, roller, screw, spring, stem, or other mechanical energizing devices.

Valve Actuator, Pilot: A valve actuator which utilizes pilot fluid.

Valve Actuator, Pilot, Barrier: A pilot valve actuator wherein the working fluid is isolated from the actuator.

Valve Actuator, Pilot, Differential Area:

A pilot valve actuator wherein pilot fluid acts on unequal areas.

Valve Actuator, Pilot, Differential Pressure: A pilot valve actuator wherein pilot fluid acts at unequal pressure.

Valve Actuator, Pilot, External: A pilot valve actuator wherein fluid is received from an external source.

Valve Actuator, Pilot, Internal: A pilot valve actuator wherein pilot fluid is received from within the valve.

Valve Actuator, Pilot, Solenoid Controlled: A pilot valve actuator wherein pilot fluid is controlled by the action of one or more solenoids.

Valve Actuator, Solenoid: A valve actuaator which utilizes one or more solenoids.

SECTION 77

VALVE MOUNTINGS
AND RELATED TERMS

Valve Mounting, Base: The valve is mounted to a plate which has top and side ports.

Valve Mounting, Line: The valve is mounted directly to system lines.

Valve Mounting, Manifold: The valve is mounted to a plate which provides multiple connection ports for two or more valves.

Valve Mounting, Sub-Plate: The valve is mounted to a plate which provides straight-through top and bottom ports.

SECTION 91

INSTRUMENTS

FLOWMETER: A device which indicates either flow rate, total flow, or a combination of both.

GAGE: An instrument or device for measuring, indicating, or comparing a physical characteristic.

Gage, Bellows: A gage in which the sensing element is a convoluted closed cylinder. A pressure differential between outside and inside causes the cylinder to expand or contract axially.

Gage, Bourdon Tube: A pressure gage in which the sensing element is a curved tube that tends to straighten out when subjected to internal fluid pressure.

Gage, Diaphragm: A gage in which the sensing element is relatively thin and its inner portion is free to deflect with respect to its periphery.

Gage, Fluid Level: A gage which indicates the fluid level at all times.

Gage, Piston: A pressure gage in which the sensing element is a piston operating against a spring.

Gage, Pressure: A gage which indicates the pressure in the system to which it is connected.

Gage, Vacuum: A pressure gage for pressures less than atmospheric.

MANOMETER: A differential pressure gage in which pressure is indicated by the height of a liquid column of known density. Pressure is equal to the difference in vertical height between two connected columns multiplied by the density of the manometer liquid. Some forms of manometers are "U" tube, inclined tube, well, and bell types.

PITOT TUBE: A velocity-sensing tubular probe with one end facing fluid flow and the other end connected to a gage.

PRESSURE SWITCH: An electric switch operated by fluid pressure.

SECTION **II**

PNEUMATICS

CHAPTER 11

BASIC PRINCIPLES OF PNEUMATICS

The term *pneumatics* as treated in this chapter pertains to *power or control signal transmitted and controlled through the use of a pressurized gas.*

Pneumatics, according to Webster's dictionary, is the branch of physics that deals with the physical properties of air and other gases.

The art of applying gaseous fluid power is expanding in both power and control systems (see Fig. 11–1).

Most manufacturing and processing plants maintain compressed air for a variety of operations. Because air is convenient and universally available, new uses for it are constantly being discovered.

For the application of pneumatic power and control systems, there are a number of important basic principles which are discussed in this chapter.

Properties of Air

Air is a mechanical mixture of gases containing, by volume, approximately 78 per cent nitrogen, 21 per cent oxygen, and about 1 per cent of other gases, including argon and carbon dioxide.

Water vapor, called humidity, is present and varies in amounts almost constantly from hour to hour.

Life on earth depends upon air for survival, and man harnesses its forces to do useful work.

Air is compressible and has weight. The weight of air is easily shown in the laboratory by first weighing an airtight container from which all the air has been removed, and then weighing it again after allowing it to fill with air. Because air is heavier than some other gases, balloons inflated with the lighter gases can float in air.

The earth is surrounded by atmosphere and at *any point* exerts a pressure, because of the column of air above *that point*. The reference point is sea level, where air exerts a pressure of 14.7 psi. At 1000 ft above sea level the atmospheric pressure is about 14.2 psi. This indicates that the air pressure varies about one-half pound for each 1000 ft of altitude. In addition, the air pressure varies constantly, depending upon atmospheric conditions. Weather forecasters use these changes to predict weather conditions.

Atmospheric pressure of 14.7 at sea level is used as a standard for making air circuit computations. The accepted weight for air is 0.08071 lb per cu ft at 14.7 psi and at 32°F. (Dry air weighs about 0.0746 lb/cu ft at 72°F.) Atmospheric pressure is equal in every direction at any given point of contact. This can be easily demonstrated in the laboratory by ex-

Pneumatic controls have contributed much to safety, efficiency and economy of diesel-powered craft. Remote panel-mounted devices provide fingertip control assuring split-second response for any desired maneuver or speed. Interlocking logic control prevents damage to engines and assures proper sequence of operations. Marine console unit. *(Courtesy of WABCO.)*

Fig. 11-1. Shows how vacuum is created by the rotary motion of the off-center rotor whose four vanes each "bite" off a slice of air at the inlet side, carry it down and around the bottom of the pump, and then compress it so it goes out of the exhaust. Centrifugal force keeps the vanes in contact with the inside wall of the rotary pump. Notice how the vanes slide out to the greater length at the bottom and are pushed in for shorter length when they get to the top. Each "bite" past the inlet removes more air from the reserve tank, creating the desired vacuum for the milking machine. *(Courtesy of Jamesway Manufacturing Co., Inc.)*

Fig. 11-2 (a), (b), (c). Vac-U-Lift 180° turn over unit. *(Courtesy of Vac-U-Lift Company, Division of Lear Siegler, Inc.)*

Fig. 11-3. Can crushed by atmospheric pressure.

hausting all the air from the inside of an airtight container with thin walls and watching it collapse. When the air is removed from the inside, reducing atmospheric pressure, the outer pressure due to the atmosphere creates a force against the walls and the container is crushed. If the sides of this container were 144 sq in., the force of the atmosphere would be

Fig. 11-4. One pound of air occupies 12.39 cu ft.

more than a ton on each side. This shows the tremendous forces available by utilizing atmospheric pressure.

$$\text{Force} = \text{pressure} \times \text{area}$$

$$\text{Force} = 14.7 \times 144$$

$$\text{Force} = 2116.8 \text{ lb (one side)}$$

(One pound of air occupies 12.39 cu ft)

$$= \frac{1 \text{ ft}^3}{0.08071 \text{ lb}} = 12.39 \frac{\text{cu ft}}{\text{lb}}$$

1 cu ft of air at 32°F

weighs 0.08071 lb

(Dry air at 72°F weighs 0.0746 lb.)

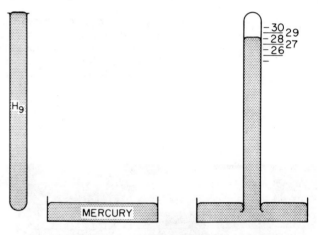

Fig. 11-5. Mercury barometer.

Atmospheric pressure is measured with a barometer, as shown in Fig. 11–5. The barometer consists of a straight glass tube over 30 in. long, having the upper end sealed. It is completely filled with mercury, and the open end is submerged into an open vessel of mercury. The level of the mercury in the tube will be about 30 in. above the level of the mercury in the open vessel. This is because the weight of the mercury column in the tube is in balance with the 14.7 psi of the atmosphere. The atmospheric pressure acting on the mercury in the open vessel supports the column of mercury in the tube because there is no air pressure in the sealed portion of the tube above the mercury (perfect vacuum).

The height of the mercury column remains at approximately 29.92 in. at 14.7 psi. Any changes in atmospheric conditions will cause the column of mercury to vary in height.

For accurate readings with mercury gauges, tubes should be clean and dry, using pure distilled mercury.

WATER MANOMETER   MERCURY U-TUBE   McLEOD GAGE

Fig. 11-6. Devices for measuring air pressure.

After the tubes have been filled, the mercury should be boiled in the tube to expel all moisture.

Other devices for measuring air pressure include the water monometer, the mercury "U" tube monometer, and the McLeod gauge, as shown in Fig. 11–6.

Pressure readings above zero (perfect vacuum) are called absolute pressures. It could also be stated that absolute pressure is atmospheric pressure plus gauge pressure. Gauge pressure is pressure above atmospheric pressure as measured by the standard bourdon or spring-type gauge.

Absolute pressure = gauge pressure + atmospheric pressure

Gauge pressure = absolute pressure − atmospheric pressure

Absolute pressures may be determined by adding atmospheric pressure at any given elevation to gauge pressure. For example, if the gauge of an air tank reads 80 psi, then at sea level the absolute pressure would be 80 + 14.7 = 94.7 psi.

Another example: Suppose a vacuum gauge shows 5 in. of mercury and absolute pressure reading is required. Barometric readings in inches of mercury vacuum can be converted to psi by multiplying by the factor of 0.491 psi/in. of Hg. To find absolute pressure:

$$5 \times 0.491 = 2.455 \text{ psi}$$

$$14.7 - 2.455 = 12.245 \text{ psi absolute pressure}$$

The term *vacuum* is usually applied to any degree of pressure less than atmospheric pressure. If the pressure in an airtight container is 10 psi absolute pressure, it is a partial vacuum. If the pressure is reduced further until it is zero, it would be full vacuum or perfect vacuum. A high vacuum means a very low absolute pressure. Vacuum is measured with a mercury monometer gauge and is usually expressed in inches of mercury.

Bourdon tube gauges are also used for measuring vacuum. A compound bourdon tube gauge reads both vacuum and gauge pressure.

Fluid power in vacuum form (see Fig. 11–7) is being used for many jobs in industry, such as chucking, bottle filling, instrument testing, forming, lifting, feeding, printing machines, milking machines, and farming operations, plus others.

Figure 11–8 shows the effect of vacuum. With the vacuum pump in operation a degree of vacuum will build up in the container, but it will not disturb the dirt. A static condition of vacuum exists, and there is no motion in the air.

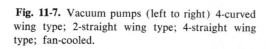

Fig. 11-7. Vacuum pumps (left to right) 4-curved wing type; 2-straight wing type; 4-straight wing type; fan-cooled.

Fig. 11-7. (*cont'd*).   Vacuum pumps (left to right)
radiator-cooled; rotary vane; vacuum and pressure,

TABLE 11-1
ATMOSPHERIC PRESSURE AT
ALTITUDES ABOVE SEA LEVEL

Altitude in feet	Absolute pressure of atmosphere in lb per sq in.	Mercury gauge reading in inches
Sea level	14.7	0
1,000	14.2	1.0
2,000	13.7	2.1
3,000	13.2	3.1
4,000	12.7	4.1
5,000	12.2	5.0
6,000	11.7	6.0
7,000	11.3	6.9
8,000	10.9	7.7
9,000	10.5	8.6
10,000	10.1	9.4
15,000	8.29	13.0
20,000	6.75	16.3
30,000	4.36	21.1
40,000	2.72	24.5
50,000	1.70	26.5
60,000	1.05	27.79
70,000	0.65	28.60
80,000	0.40	29.10
90,000	0.25	29.41
100,000	0.15	29.60
110,000	0.10	29.72
120,000	0.065	29.792
130,000	0.045	29.833
140,000	0.030	29.865
150,000	0.020	29.885
160,000	0.014	29.897
170,000	0.010	29.906
180,000	0.007	29.912
190,000	0.005	29.917
200,000	0.003	29.920

Fig. 11-9. Effect of vacuum on dirt in tank.

If an orifice is put into the end of the container, the vacuum will cause air to enter through the orifice.

The velocity of the air through the container increases as the vacuum increases. The high velocity disturbs the dirt, causing it to flow toward the pump. Figure 11–9 shows that vacuum alone cannot move material, but it will cause air velocity that can move material.

When a sheet is being picked up with vacuum, the pickup force is proportional to the hole area.

The illustration shown in Fig. 11–10 is a practical application using vacuum power.

Example. If a vacuum of 20.4 in. Hg is used with a sucker having an area of 2 sq in., what is the lifting force?

$$\text{Lifting pressure} = \frac{\text{vacuum in inches of mercury}}{2.04}$$

$$\text{Lifting pressure psi} = \frac{20.4}{2.04}$$

$$(2.04 = \text{reciprocal of } 0.491)$$

$$\text{Lifting pressure} = 10 \text{ psi}$$

$$\text{Lifting force} = \text{pressure} \times \text{area}$$

$$\text{Lifting force} = 10 \times 2$$

Fig. 11-8.

Fig. 11-10. Effect of air entering vacuum tank.

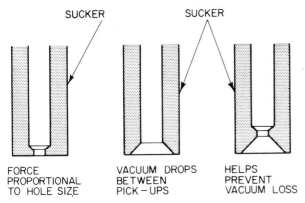

FORCE PROPORTIONAL TO HOLE SIZE

VACUUM DROPS BETWEEN PICK-UPS

HELPS PREVENT VACUUM LOSS

Fig. 11-11. Vacuum sucker design.

Lifting force = 20 lb

Example. How much pull will a piston exert if the area is 10 sq in. and the vacuum is 20.4 in. Hg?

$$\text{Lifting pressure psi} = \frac{\text{vacuum in inches of mercury}}{2.04}$$

$$= \frac{20.4}{2.04}$$

$$= 10 \text{ psi}$$

Pulling force = pressure × area

Pulling force = 10 psi × 10 sq in.

Pulling force = 100 lb

Thus, it can be seen that absolute pressures below zero gauge pressures can be used to perform useful work.

Fig. 11-12. Example problem—vacuum cylinder.

20.4 INCHES OF MERCURY VACUUM

LOAD

10 sq. in.

TABLE 11-2
HOLDING FORCE CREATED BY VACUUM ON SUCKER

Vacuum inches of mercury	Diameter of sucker					
	1 in.	1½ in.	2 in.	3 in.	4 in.	5 in.
5	1.9	4.4	7.8	17.7	31.4	49
10	3.8	8.7	15.4	34.6	62.0	96
15	5.8	13.1	23.3	52.2	93.0	145
20	7.6	17.3	30.8	69.0	123.0	192
25	9.6	21.8	38.6	87.0	154.0	241

Measuring Air Flow

Impact Tube

The impact tube is an instrument for measuring low pressure air flow through pipes or ducts. The air impinging on the tube opening at point *A* causes the water in the U tube to react to the impact pressure, and the static pressure is applied at point *B* (see Fig. 11–13). The instrument indicates the difference between the two pressures, which is the velocity pres-

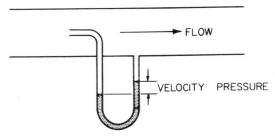

FLOW

VELOCITY PRESSURE

Fig. 11-13. Impact tube.

sure. The static pressure reading is subtracted from the impact pressure reading to get the velocity pressure.

The velocity in lineal feet per minute can be obtained by the following equation:

$$\text{LFM} = 4005\sqrt{\text{velocity pressure}}$$

To convert inches of water to pounds per square inch:

$$\text{Inches of water} = 27.686 \times \text{psig}$$

$$\text{Psig} = \frac{\text{inches of water}}{27.686}$$

Pitot Tube

The Pitot tube (see Fig. 11–14) is based on the same principle as the impact tube. The tube is placed inside of the static pressure tube, which eliminates the need of two holes in the pipe or duct. The dif-

Fig. 11-14. Pitot tube.

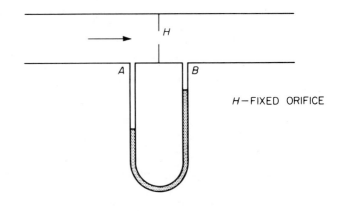

H—FIXED ORIFICE

Fig. 11-15. Flow meter.

ference would again indicate the velocity pressure for calculating LFM.

Flow Meter

The principle of the flow meter is shown in Fig. 11–15. The instrument measures pressure drop across the fixed orifice, which has a known coefficient. The pressure upstream is picked up at *A* and the pressure downstream registers at *B*. The U tube gives the difference in pressures across the orifice.

The number of cubic feet per minute is then found by the following formula:

$$\text{CFM} = \frac{C}{60}\sqrt{I \times P}$$

where   *C* = coefficient of orifice.

I = inches of water (1 in. of water = 0.0361 psi).

P = absolute pressure on the upstream side.

CFM = cubic feet of free air per minute.

Figure 11–16 shows typical orifices and their coefficients.

Gasometer

A gasometer provides accurate measure of free air leaving a pipe or compressor. It consists of an inverted tank which is counterbalanced in a tank of water so that it floats freely to any position (see Fig. 11–17).

The air is directed by a pipe to the inside, where it causes the inverted tank to rise. A calibrated scale,

Fig. 11-16. Various orifices and their coefficients.

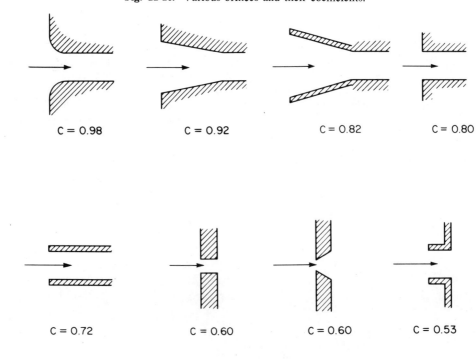

C = 0.98   C = 0.92   C = 0.82   C = 0.80

C = 0.72   C = 0.60   C = 0.60   C = 0.53

based on the volume per inch of rise, is attached to the apparatus. The number of cubic feet per minute is obtained by timing the rising tank with a stop watch.

Free Air and Standard Air

/Free air is air at normal atmospheric conditions./ Because the atmosphere is subject to change of pressure and temperature, free air characteristics may change from day to day. The term *"standard air"* is derived from an average atmospheric condition with the air at a temperature of 68°F, atmospheric pressure at 14.7 psi, and a relative humidity of 36 per cent. (Relative humidity is the percentage of saturation of air with water vapor.)

Compressed Air and the Gas Laws

When air is being compressed for use in pneumatic systems, pressure, temperature, and volume have an effect on the operation. Thus, it is important to know the effect of these three variables in applying pneu-

Fig. 11-17. Gasometer.

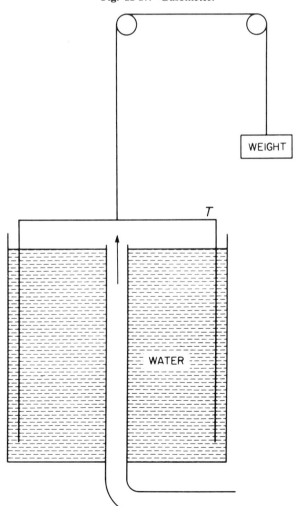

matic power and control. Air resembles a perfect gas, because it follows closely the laws of a perfect gas in expansion, contraction, and absorbing and releasing heat.

Boyle's law states: "When the temperature of a confined gas remains constant, the volume varies inversely as its absolute pressure." This can be stated mathematically as a simple formula:

$$\frac{\text{Initial absolute pressure}}{\text{Final absolute pressure}} = \frac{\text{final volume}}{\text{initial volume}}$$

$$\left(\frac{P_1}{P_2} = \frac{V_2}{V_1}\right)$$

or as

Initial absolute pressure × initial volume = final absolute pressure × final volume

$$(P_1 \times V_1 = P_2 \times V_2)$$

(psia = pounds per square inch absolute)

(psig = pounds per square inch gauge)

Example 1. An air tank contains four cubic feet of air at 20 psi. The tank is charged with a liquid until the air occupies only one-half its original volume. What is the final pressure in the tank?

$$P_1 V_1 = P_2 V_2$$

Transpose the formula to proper form.

$$P_2 = \frac{P_1 V_1}{V_2}$$

$$P_1 = 20 \text{ psi} + 14.7 = 34.7$$

$$V_1 = 4$$

$$V_2 = 2$$

$$P_2 = \text{unknown}$$

$$P_2 = \frac{34.7 \times 4}{2}$$

$$P_2 = 69.4 \text{ psia}$$

(Remember, if gauge pressure is desired from absolute pressure, subtract 14.7 from the absolute pressure.)

Example 2. How many cubic feet of free air per minute are required to operate an air cylinder (single-acting) 100 cycles per minute if the volume of the cylinder is 1.5 cu ft and the operating gauge pressure is 60 psi?

$$\frac{\text{Free air}}{\text{CFM}} = \frac{\text{vol. of cylinder}}{\text{cu ft}} \times \frac{\text{cycles}}{\text{minute}}$$

$$\times \frac{(\text{gauge pressure} + 14.7)}{14.7}$$

Fig. 11-18. Example problem 2a.

Thus, it is easy to see that the basic transposed formula $V_2 = (V_1 \times P_1)/P_2$ is flexible in use, and units of time may be added.

$$\frac{\text{Free air/cu ft}}{\text{Time}} = \text{vol. of cylinder/cu ft}$$

$$\times \text{ no. of cycles/min.}$$

$$\times \frac{(\text{gauge psi} + 14.7)}{14.7}$$

$V_1 = 1.5$ cu ft

$P_1 = 60$ psi $+ 14.7 = 74.7$ psia

$P_2 = $ zero gauge or 14.7 atmospheric pressure

$V_2 = $ unknown

$$V_2 = \frac{1.5 \times 74.7}{14.7}$$

$V_2 = 6.2$ cu ft/min of free air

Many of the basic laws pertaining to liquids also apply to gases. For instance, the cylinder in Example 2 must be large enough to overcome the force of the work. Pascal's law would apply as follows: Suppose the force on the cylinder due to the work load is 900 lb. What piston area would be required?

Given:

$$\text{Force} = 900 \text{ lb}$$

$$\text{Pressure} = 60 \text{ psi}$$

$$\text{Area} = \text{unknown}$$

Thus:

$$\text{Area} = \frac{\text{force in pounds}}{\text{pressure in.}^2}$$

$$\text{Area in.}^2 = \frac{900 \text{ lb force}}{60 \text{ psi}}$$

$$\text{Area in.}^2 = 15 \text{ sq in.}$$

Example 3. How many cubic feet of free air are

Fig. 11-19. Example problem 2b.

required to raise the receiver from zero gauge to final pressure? Suppose the receiver is 12 in. in diameter and 24 in. long and that 80 psi is the final pressure.

$$\text{Volume of receiver} = \text{area} \times \text{length}$$

First, find the area:

$$\text{Area} = 0.7854 \ D^2$$

$$\text{Area} = 0.7854 \times 12 \times 12$$

$$\text{Area} = 113 \text{ sq in.}$$

The volume in cubic inches $= 113$ sq in. $\times 24$ in.

$$V = 2712 \text{ cu in.}$$

To change cubic inches to cubic feet we must multiply by 1 ft³/1728 in.³

$$\text{Vol. in cu ft} = \tfrac{2712}{1728} = 1.57 \text{ ft}^3$$

$$\text{Vol. in cu ft} = 1.57 \text{ ft}^3$$

Fig. 11-20. Example problem 3.

Boyle's law

$P_1 = 14.7$ $P_1 V_1 = P_2 V_2$

$P_2 = 80 + 14.7$ $V_2 = \dfrac{P_1 V_1}{P_2}$

$V_1 = 1.57 \times X$

$X = $ free air unknown

$V_2 = 1.57$ $V_2 = \dfrac{1.57 \text{ ft}^3 \times 80\text{lb/in}^2}{14.7\text{lb/in}^2}$

Substituting gives

$$(14.7)(1.57 + X) = (80 + 14.7)(1.57)$$

$$1.57 + X = \frac{(80 + 14.7)(1.57)}{14.7}$$

$$X = \frac{(80 + 14.7)(1.57)}{14.7} - 1.57$$

$$X = \frac{(94.7)(1.57) - 1.57}{14.7} = 8.54 \text{ ft}^3$$

Example 4. Suppose the air in the receiver which is discussed in Example 3 were used until the pressure dropped from 80 psig to 30 psig. How much free air would be required to bring the pressure back to 80 psig? We now have the following relationship:

Cu ft of free air required to
raise receiver from 30 psig
to 80 psig

$$= \frac{\text{Vol. of receiver}}{\text{cu ft}} \times \frac{(\text{final psig} - \text{initial psig})}{14.7}$$

$$X \text{ cu ft} = 1.57 \text{ cu ft} \times \frac{(80 \text{ psig} - 30 \text{ psig})}{14.7}$$

$$X \text{ cu ft} = 1.57 \times \frac{50}{14.7}$$

$$X = 5.34 \text{ cu ft}$$

Charles' law. When the pressure of confined gas remains constant, the volume of the gas is directly proportional to its absolute temperature. This can be stated mathematically with a simple formula:

$$\frac{\text{Initial volume}}{\text{Final volume}} = \frac{\text{initial absolute temperature}}{\text{final absolute temperature}}$$

or

$$\frac{\text{Initial absolute temperature}}{\text{Initial volume}}$$
$$= \frac{\text{final absolute temperature}}{\text{final volume}}$$

or

$$\frac{V_1}{V_2} = \frac{T_1}{T_2} \text{ or } V_2 T_1 = T_2 V_1$$

Example 1. A tank holding 300 cu ft of air at atmospheric pressure and 60°F is heated to 120°F. What is the volume of air that escapes from the tank?

$$V_2 = \frac{V_1 T_2}{T_1}$$

$T_1 = 60 + 460 = 520°$ F initial absolute temperature

$T_2 = 120 + 460 = 580°F$ final absolute temperature

(Absolute temperature in degrees Rankine
= F° + 460)

$$V_2/\text{cu ft} = \frac{300 \times 580}{520}$$

$$V_2 = 334.6$$

Vol. escaping = 334.6 − 300

Vol. escaping = 34.6 cu ft

Guy-Lussac found that if the volume of a gas remains constant, the pressure exerted by the confined gas is directly proportional to the absolute temperature. This law is expressed as follows:

$$\frac{\text{Initial pressure}}{\text{Final pressure}} = \frac{\text{initial absolute temperature}}{\text{final pressure}}$$

or

$$\frac{\text{Initial absolute temperature}}{\text{Initial pressure}}$$
$$= \frac{\text{final absolute temperature}}{\text{final pressure}}$$

or

$$\frac{P_1}{P_2} = \frac{T_1}{T_2} \text{ or } \frac{T_1}{P_1} = \frac{T_2}{P_2}$$

or

$$P_2 T_1 = T_2 P_1$$

Example 1. A *closed* tank holding 300 cu ft of air at 14.7 psi and 60°F is heated to 120°F. What will be the final pressure? (Remember to use absolute temperatures)

$$T_1 = 460 + 60 = 520°F$$

$$T_2 = 460 + 120 = 580°F$$

$$P_2 = \frac{P_1 \times T_2}{T_1}$$

$$P_2 = \frac{14.7 \times 580}{520}$$

$$P_2 = 16.2 \text{ psi}$$

The gas laws may be combined with the following results:

a. Final pressure:

$$P_2 = \frac{P_1 V_1 T_2}{V_2 T_1}$$

b. Final volume:

$$V_2 = \frac{P_1 V_1 T_2}{P_2 T_1}$$

c. Final absolute temperature:

$$T_2 = \frac{P_2 V_2 T_1}{P_1 V_1}$$

Example 1. 50 cubic feet of air at atmospheric pressure and 40°F is compressed to 10 cu ft at a temperature of 160°F. What is the final pressure?

$$P_2 = \frac{P_1 V_1 T_2}{V_2 T_1}$$

$$P_1 = 14.7$$

$$V_1 = 50 \text{ cu ft}$$

$T_2 = 160 + 460 = 620°F$ absolute temperature

$T_1 = 40 + 460 = 500°F$ absolute temperature

$V_2 = 10$ cu ft

$$P_2 = \frac{14.7 \times 50 \times 620}{10 \times 500}$$

$P_2 = 91.14$ psi

Example 2. 60 cu ft of air at 30 psig and 50°F is compressed to 80 psig and 110°F. What is the final volume?

$$V_2 = \frac{P_1 V_1 T_2}{P_2 T_1}$$

$P_1 = 30 + 14.7 = 44.7$ psia

$V_1 = 60$ cu ft

$T_2 = 110 + 460 = 570°F$ absolute temperature

$P_2 = 80 + 14.7 = 94.7$ psia

$T_1 = 50 + 460 = 510°F$ absolute temperature

$$V_2 = \frac{44.7 \times 60 \times 570}{94.7 \times 510}$$

$V_2 = 31.6$ cu ft

Example 3. The air space above the oil in a closed hydraulic reservoir contains 160 cu in. of air at 14.7 and 60°F. Return oil coming back from the system compresses the air into a space of 60 cu in. What is the final temperature of the air if the final pressure is 30 psi?

$$T_2 = \frac{P_2 V_2 T_1}{P_1 V_1}$$

$P_2 = 30 + 14.7 = 44.7$

$V_2 = 60$ cu in.

$T_1 = 60°F + 460 = 520°F$ absolute temperature

$P_1 = 14.7$

$V_1 = 160$ cu in.

$$T_2 = \frac{44.7 \times 60 \times 520}{14.7 \times 160}$$

$T_2 = 593 - 460$

$T_2 = 133°F$

The basic laws and principles discussed in Section I are also applicable to pneumatics since both gases and liquids are classified as fluids.

Air and water set the standards and all other fluids are compared with them for such things as specific gravity, density, and other physical properties.

Questions

1. What is the most common instrument for measuring the pressure of the atmosphere?
2. Write a comparison of the physical characteristics of gases with liquids.
3. What is meant by absolute pressure?
4. What is meant by gauge pressure?
5. Explain how vacuum can be used to perform work.
6. If the gauge on a receiver tank reads 96 psi, what would the absolute pressure be at sea level?
7. Express 8 inches of mercury vacuum in psia.
8. How much pull in pounds would a vacuum cylinder lift provide if the piston area is 12 sq in. with a vacuum pump operating at 26 in. of mercury?
9. If one psi supports a column of water 27.686 in. high, what length of tube would be necessary to measure atmospheric pressure at sea level?
10. Draw the diagram of a "gasometer."
11. What is meant by free air?
12. Explain the importance of Boyle's law.
13. A closed air tank contains 16 cu ft of air at 14.7 psi; oil is forced into the tank until the air occupies only one-fourth its original volume. What is the final pressure?
14. How many cu ft of free air are required to raise the receiver from zero gauge to 80 psi if the receiver is 20 in. in diameter and 36 in. long?
15. Explain the importance of Charles' law.
16. An accumulator that holds 864 cu in. of N2 at 60°F is heated to 140°F. What is the final pressure?
17. Sixty-four cubic feet of air at atmospheric pressure and 50°F is compressed to 12 cu ft at 154°F. What is the final pressure?

CHAPTER **12**

HOW AIR IS COMPRESSED

A compressor is a machine for compressing air or other gases from an initial intake pressure, usually atmospheric pressure, to a higher pressure. It may also be stated that a compressor increases pressure exerted by a gas by reducing the volume.

The two general classifications of compressors are positive displacement and nonpositive displacement.

Compressors used for pneumatic power are generally reciprocating piston or rotary designs which are positive displacement units.

Reciprocating compressors have one or more pistons which reciprocate within a cylindrical bore. In a single piston unit air is compressed from its initial inlet pressure to the desired final pressure in one compressive step. This is known as a single-stage air compressor (see Fig. 12–1).

A compressor having two or more compressive steps in which the discharge from each supplies the next in series is a multistage compressor (see Fig. 12–1).

In a single-stage unit the size, stroke, and speed of the reciprocating piston determines the volume displacement, generally expressed in cubic feet per minute (cfm).

For multistage compressors, the piston displacement of the first stage is commonly used as that of the entire machine.

Rotary compressors are available in both single-from inlet to outlet. The vane design contains a number of rectangular vanes fitted in a slotted rotor. The rotor is off center to the housing, and as the shaft turns, centrifugal force causes the vanes to move out against the contour of the retaining ring. Air is trapped between the vanes during one-half of one revolution as the space between the rotor and ring increases. Air is pushed out during the other one-half of one revolution as the space decreases (see Fig. 12–2).

Rotary compressors are available in both single- and multistage units.

The vanes in a rotary compressor are generally made of an asbestos cloth impregnated with a phenolic resin. The vanes take on a very high polish during operation and provide excellent wear characteristics. When air is compressed, it gains heat as the molecules of the gas become closer together and bounce off each

273

SINGLE-STAGE

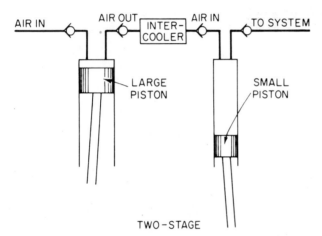

Fig. 12-1. Compressor cylinders–Single- and two-stage.

other faster and faster. Because of the difficulty in removing heat with a single-stage unit, there is an economical limit to the pressure level that can be attained in one compressive step. Generally this limit

Fig. 12-2. Sliding vane compressor.

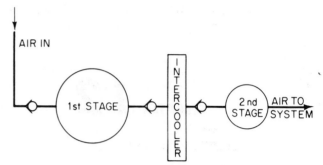

Fig. 12-3. Rotary compressor.

is between 100 and 150 psig. Multistage compressors are used when high pressures are required, because better cooling between stages can effectively increase the efficiency and reduce the input power requirements.

Type	Capacity, psig
Single-stage compressor	100–150
Two-stage compressor	150–500
Three-stage compressor	500–2500
Four-stage compressor	2500up

Adiabatic and Isothermal Compression

Theoretically, air may be compressed with all the heat of compression retained (adiabatically) or by removing all the heat of compression (isothermally). However, in practical application air compressors give performance between these two theoretical concepts.

Students should keep in mind that the relationship of these two concepts are limits within which the actual compression always occurs. This relationship may be expressed as the ratio of specific heat at constant pressure to the specific heat at constant volume.

Specific heat is expressed by the number of BTU (British Thermal Units) necessary to raise the temperature of one pound of air (or any substance) one degree Fahrenheit.

Through experiments, the specific heat of air has been determined as 0.2375 BTU per lb. (Specific heat for air=1 or 0.2375 BTU.)

Isothermal compression occurs when air undergoes a change in pressure or volume at constant temperature, which means that the specific heat equals 1 or 0.2375 BTU.

Adiabatic compression occurs when air undergoes a change in pressure or volume while retaining heat of compression, which means a change in specific heat. This change or difference has been determined by Regnault's experiments to be 0.0686 BTU per lb, which gives a difference $(0.2375-0.0686)=0.1689$ BTU per lb. The 0.0686 BTU represents energy lost when air is compressed adiabatically.

The ratio of difference between the two concepts may now be expressed as

$$\frac{0.2375}{0.1689} = 1.41$$

(Boyle's law modified)　　　　(Boyle's law)
Adiabatic compression　　　Isothermal compression
$$P_1 V_1^{1.4} = P_2 V_2^{1.4} \qquad P_1 V_1 = P_2 V_2$$

This concept could also be illustrated with an indicator diagram, which is a graphic representation of the relationship of adiabatic to isothermal compression (see Fig. 12–4).

ADE represents the additional power required for adiabatic compression that is greater and above the power required for isothermal compression. In fact, about one-third more power is required to compress air to 100 psia adiabatically than isothermally, which indicates the importance of cooling during compression.

Figure 12–5 shows a typical idealized indicator card for a single-stage piston-type compressor. At point 1 the piston is at the end of its stroke and a certain percentage of air at discharge pressure is trapped in the clearance volume. The clearance volume is the volume of air in the cylinder over and above the volume swept by the piston. The clearance volume includes air left between the piston and the head and in and under the check valves used for displacement.

When the piston begins its return stroke, the

Fig. 12-5.　Indicator card for a single-stage piston-type compressor.

clearance air expands until the pressure in the cylinder is slightly lower than the pressure in the intake line. This occurs at point 2, where the inlet valves open, and the cylinder takes in air between 2 and 3 on the diagram.

At point 3 the piston is at the opposite end of its stroke and it now returns, compressing the air from point 3 to 4. When it reaches point 4, the air is compressed to a point slightly higher than the pressure in the receiver, and the outlet check valve opens, allowing discharge to the system. When the air is discharged from the cylinder, the piston is again back to point 1 and the cycle repeats.

Intercoolers normally cool air by the use of the same cooling medium as the compressor itself (air or water). Air-cooled intercoolers may consist of finned tubes through which air is forced, or it may be of the radiator type. Cooling air is blown over the outer surface of the tubes to effect the transfer of heat caused by compression.

A water-cooled intercooler consists of a nest of tubes through which water is passed to effect the cooling. The tubes are enclosed within a shell, and

Fig. 12-4.　Relationship of adiabatic to isothermal compression.

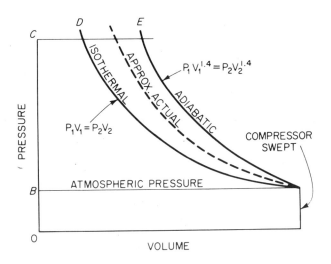

the air being cooled passes over the outside of the tubes. Baffles may be used to control the paths of both the air being cooled and the coolant for maximum efficiency.

The power savings effected by the staging technique of positive displacement compressors depends upon factors such as the ratio between the suction and discharge pressures, the cooling medium, and the effectiveness of the intercooler.

The capacity of positive displacement compressors is determined by the actual amount of fluid that is delivered through the discharge valves in one minute. This is referred to as the "actual delivered capacity." The piston displacement volume for one minute divided into the actual delivered capacity of the compressor is called the *volumetric efficiency* and is usually expressed as a percentage.

If the fluid were incompressible, it might be supposed that a compressor's capacity would be nearly the same as the piston displacement. However, since air expands and contracts in volume with changes in pressure and temperature, the actual capacity is less than piston displacement. This is mainly because of the clearance between piston and cylinder head at the end of each stroke and because of the valve opening in the cylinder.

At the end of each compression stroke as shown in Fig. 12–6, the clearance volume is filled with compressed air that has not been delivered through the discharge valves. On the intake stroke this air will re-expand in the cylinder until the pressure is below the suction pressure to allow new air into the cylinder through the inlet valve. The greater the differential pressure between inlet and outlet, the lower the volumetric efficiency will be. However, this does not represent a direct power loss, because the expanding fluid does work on the intake stroke by exerting pres-

Fig. 12-6. Clearances.

Fig. 12-7.   Hand air pump.

sure on the piston, which is, in turn, transferred to the drive arrangement.

$$\text{Volumetric efficiency} = \frac{\text{free air}}{\text{piston displacement}}$$

The volumetric efficiency involves all things that cause a compressor to deliver less air than piston displacement. This includes reduction in capacity due to preheating of the fluid, slippage of air past piston seals, and reversal of air flow during valve operation. Actual delivered capacity is determined by measuring the volume of air discharged from the compressor and referring it to suction conditions. Volumetric efficiency is not a measure of compression efficiency. Compression efficiency, when heat is contained, is the ratio of the horsepower theoretically required to the horsepower actually expended in the cylinder (known as indicated horsepower). This indicated horsepower is the brake horsepower at the compressor shaft minus mechanical horsepower lost.

Compressor efficiency (adiabatic) is the ratio of the theoretical horsepower to the shaft horsepower input. It is equal to the product of the compression efficiency times the mechanical efficiency.

Types of Compressors

One of the simplest designs of the positive displacement type is the hand pump (see Fig. 12–7).

The basic principle of operation is the same in all reciprocating-type compressors, and the understanding of this simple design can be transferred to other more complex designs of the reciprocating type.

The hand pump consists of a piston secured by a rod to a handle, a cylinder, a cup seal, and a one-way valve or check. With the handle pulled all the way to the top, air enters a small hole in the pump cylinder. As the handle is pushed downward, the air now trapped within the cylinder is compressed. As the piston approaches the bottom, the pressure is sufficient to force open the spring-held check valve

Fig. 12-8. Single-stage compressor. (*Courtesy of Compressed Air & Gas Institute.*)

Fig. 12-9. Two-stage compressor. (*Courtesy of Quincy Compressor.*)

and enter the system where it is being used. When the piston reaches the bottom of the cylinder and the pressure on each side of the check valve is equalized, the spring closes the check valve.

The upward stroke of the piston creates a partial vacuum inside the cylinder until the inlet port is reached near the top of the stroke.

Popular Industrial Types

A single-stage compressor (see Fig. 12–8) has one or more cylinders and operates on the following principle: On the downward stroke of each piston air is taken in at atmospheric pressure. On the upward stroke the compressed air is discharged into the air receiver. Single-stage compressors usually operate at pressures under 100 psi and for intermittent-type duty.

A two-stage compressor (see Fig. 12–9) has a minimum of two cylinders, generally a low pressure cylinder (largest) and one high pressure cylinder (smallest). When the LP (large) cylinder piston goes downward, air is taken in at atmospheric pressure through a one-way valve. On the up stroke, the LP piston discharges the air through an intercooler and into the HP (small) cylinder which is on the down stroke. On the upward stroke of the HP cylinder piston air is discharged into the air receiver. Two-stage compressors are generally recommended for use where operating pressures are above 100 psi. The two-stage compressors are more efficient, since they provide some cooling during compression and deliver

air to the receiver at lower discharge temperature for the high pressures. Generally, the normal operating pressure for this design is 175 psi maximum.

Reciprocating compressors of the type shown in Fig. 12–10 operate on the same principle as the hand pump but with refinements toward greater efficiency, durability, and practicability. This shows a horizontal single-stage double-acting compressor which is water-cooled. The rotary motion of a prime mover, such as an electric motor or internal combustion engine, is converted into reciprocating motion by use of a crank and connecting rod. One end of the connecting rod is fastened to the crank pin, the other to a crosshead which reciprocates as the shaft rotates. A piston rod connects the crosshead to the piston, which reciprocates within the cylinder as in the hand pump.

Notice that the compressor in Fig. 12–11 is fitted with one inlet check valve and one discharge check valve on each end of the cylinder. As the piston moves to the left, the inlet valve to the right allows atmospheric pressure to enter that end of the cylinder. In the same direction, air in the opposite end of the cylinder is being discharged out through the discharge valve on the left side and into the system. On the return stroke to the right, air enters through the left side inlet check, and the delivery of

Fig. 12-10. Water-cooled single-stage compressor. *(Courtesy of Compressed Air & Gas Institute.)*

Fig. 12-11. Double-acting compressor. *(Courtesy of Compressed Air & Gas Institute.)*

Fig. 12-12. 3-cylinder, single-acting air-cooled compound compressor. *(Courtesy of Compressed Air & Gas Institute.)*

air to the system is now through the discharge check valve on the right end of the cylinder.

In other words, a double-acting compressor delivers air to the system on each stroke. In most cases, the discharge is to a receiver or flask from which it can be used as needed in the system. The use of storage devices helps to give smoother operation, and to provide some cooling of the hot discharged air.

Figure 12–12 shows a three-cylinder single-acting air-cooled compound compressor. The compounding is effected by having two cylinders which discharge into a single high pressure cylinder through an air-cooled intercooler. The cylinders are arranged in "W" formation with three connecting rods on a single crank. The locations of the air inlet and discharge valves are in the heads of each cylinder. The lubrication of the cylinder walls results from the splash of oil from the crankcase. Compressors of this type are available in single-stage or two-stage designs. The usual pressure range runs as high as 125 psig.

There are also several other cylinder arrangements for "angle-type" reciprocating compressors such as multicylinder vertical two-cylinder "Y" or "V" type, four-cylinder radial type, and the two-cylinder angle compressor with one cylinder vertical and one horizontal.

The portable air compressor which is illustrated in Fig. 12–13 has a self-contained compressed air power plant. The unit consists of an air compressor, a prime mover and an air receiver, and is complete with cooling, lubricating, regulating, and starting systems, all enclosed on a running gear for ready movement.

The prime mover may be an internal combustion engine or an electric motor. Standard discharge pressures for portable compressors essentially match the

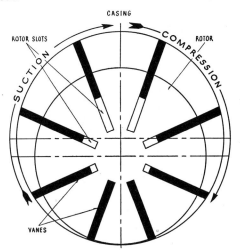

Fig. 12-14. Sliding vane type of rotary compressor. *(Courtesy of Compressed Air & Gas Institute.)*

requirements of portable air tools. Sizes are available as high as 100 hp.

Figure 12–14 illustrates the operating principle of the sliding vane type of rotary compressor. A cylindrical slotted rotor turns within a larger diameter casing. In the rotor slots are rectangular vanes, shown by heavy black lines, which slide in and out of the rotor slots due to centrifugal force. Thus, turning clockwise, air is trapped and compressed between the vanes and discharged through a port on the com-

Fig. 12-15. 2-impeller positive type compressor.

Fig. 12-13. Portable air compressor. *(Courtesy of Compressed Air & Gas Institute.)*

Fig. 12-16. Liquid piston type compressor. *(Courtesy of Compressed Air & Gas Institute.)*

pression or left-hand side. Rotary sliding vane compressors are used for compressing up to approximately 50 psig in a single-stage, which may be increased to 150 psi in a two-stage, etc. Sizes cover as high as 700 hp.

Another design of rotary compressor is the two-impeller positive type shown by Fig. 12–15. The two identical impellers are kept in proper rotative position by a pair of external gears. There is no internal contact between the impellers, so no lubrication is required. These compressors are used for lower pressure ranges and for capacities of about 50,000 cu ft per minute.

Figure 12–16 illustrates a third type of rotary compressor known as the liquid piston type, which employs a liquid, usually water, as the compressing medium. The blades of the turning rotor form a series of buckets carrying the liquid around the inside of the elliptical casing. Since the liquid following the contour of the inside of the casing surges back into the buckets at the narrow point of the ellipse, the air in the buckets is compressed and discharged through properly located ports. Compressors of this type develop about 75 psi pressure in one stage. The capacity range is up to 5000 cu ft per minute.

Non-positive Displacement Types

Figure 12–17 shows the construction of a four-stage centrifugal compressor. Air enters through the large inlet flange at the left end, and enters the eye of the first stage impeller. The rotative speed of the impeller imparts a high velocity to the air as it moves radially outward into the diffuser, which converts part

of the velocity head to static pressure head. The air then moves radially inward to the eye of the next impeller and the process is repeated, increasing the pressure of the air with each stage of compression. Multistage centrifugal air compressors are generally used for handling large volumes of air at lower pressures, but are occasionally designed for pressures of 150 psi or higher. Capacities range as high as 150,-000 cu ft per minute.

Figure 12–18 shows a multistage axial flow, uncooled blower-type compressor with six rows of rotation vanes. Both centrifugal and axial machines inherently operate at high speed. Axial machines are characterized by substantially constant capacity delivery at variable pressures, whereas centrifugal machines deliver practically constant pressure over a considerable range of capacities. These characteristics must be considered when equipment is selected for a given application. Compressors of this type have ratings over 1,000,000 cu ft per minute with pressure ranges similar to the centrifugal types. (This material is used through the courtesy of the Committee on Engineering Education of Compressed Air and Gas Institute.)

Fig. 12-17. Multi-stage centrifugal compressor. *(Courtesy of Compressed Air & Gas Institute.)*

Fig. 12-18. Multi-stage axial compressor. *(Courtesy of Compressed Air & Gas Institute.)*

Questions

1. What are the two general classes of compressors?
2. List several design types under each class.
3. What is generally considered the maximum pressure and capacity of a single-stage compressor?
4. Draw the compression curve that illustrates the difference between adiabatic and isothermal compression. Explain.
5. Show the modified Boyle's law for adiabatic compression.
6. What effect does an intercooler have in compressing air?
7. What effect does an aftercooler have on compressed air?
8. Why is the actual delivery from a positive displacement compressor less than piston displacement?
9. What is clearance volume?
10. Explain how the volumetric efficiency of a compressor is determined.

CHAPTER 13

THE COMPRESSED AIR SYSTEM

Compressed air is a major source of industrial power. It is safe, flexible, and easy to use for countless jobs throughout modern industry. A well planned and developed system can provide excellent trouble-free service, with flexibility to expand for the future.

Air-operated equipment is classified as continuous or intermittent duty in operation, much the same as hydraulic equipment.

The type of duty generally indicates the type of compressor controls to use. Constant speed controls are recommended for continuous operations where the compressor operates about 50 per cent of the time. This prevents excessive wear and prevents power losses due to frequent starts and stops.

Dual controls offering automatic stops and starts with a high and low range setting of a pressure switch or continuous speed may be needed in plants working only one shift in 24 hours. If automatic start and stop controls are used, the compressor should have enough capacity to operate only one-third of the time.

The number of cubic feet of air per minute depends upon the various tools and circuits that require air.

Single-stage compressors are generally used for

TABLE 13-1
AVERAGE FREE AIR CONSUMPTION OF TYPICAL TOOLS AND EQUIPMENT
80–150 psi

Air hammer	approx.	16	cfm
Air hoist (cylinder)	″	1–4	″
Air hoist (motor)	″	2–4	″
Rotary drills $\frac{1}{16}''-\frac{5}{8}''$	″	4–7	″
Rotary drills $\frac{1}{4}''$	″	18–20	″
Rotary drills $\frac{3}{8}''$	″	20–40	″
Rotary drills $\frac{1}{2}''-\frac{3}{4}''$	″	70	″
Rotary drills $\frac{7}{8}''-1''$	″	80	″
Piston drills $\frac{1}{2}''-1\frac{1}{4}''$	″	45	″
Piston drills $\frac{7}{8}''-1\frac{1}{4}''$	″	75–80	″
Piston drills $1\frac{1}{2}''-2''$	″	89–90	″
Piston drills $2''-3''$	″	100–110	″
Grinders vertical and horizontal	″	10–20	″
1 hp air motor	″	10	″
2 hp air motor	″	15	″
3 hp air motor	″	20	″

continuous operation when the maximum pressure is below 80 psi; also for intermittent operation up to 125 psi if less than 50 per cent of the compressor's capacity is being used.

Two-stage compressors are generally used for

continuous operation when the maximum pressure is greater than 80 psi; also for intermittent operations above a maximum pressure of 125 psi.

Air compressors are generally rated in terms of cubic feet per minute of free air. Free air is defined as the normal air at atmospheric conditions at any given place.

Since compressor capacity is given in terms of free air, it is convenient for making calculations as follows:

$$\frac{P_1 V_1}{T_1} = \frac{P_2 V_2}{T_2}$$

Cu ft per min. (cfm) $= V_1 = \dfrac{P_2 V_2 T_1}{T_2 P_1}$

Example. Air is used at the rate of 20 cfm from a receiver at 80°F and 100 psi. If the barometer reads 14.5 psia and the air inlet temperature is 60°F, how many cfm must the compressor provide?

$$V_1 \, \text{cfm} = \frac{114.5 \times 20 \times 520}{540 \times 14.5}$$

$V_1 = ?$

$V_2 = 20$ cu ft

$P_1 = 14.5$

$P_2 = 100 + 14.5$

$T_1 = 60 + 460$

$T_2 = 80 + 460$

cfm $= 152$ cu ft of free air/min.

If the atmospheric pressure dropped to 14.0 psia, how much would the compressor have to deliver?

$$\text{cfm} = \frac{114.5 \times 20 \times 520}{540 \times 14}$$

cfm $= 157.5$ cu ft of free air/min

This indicates the difference when normal atmospheric conditions change.

The total capacity required for an air system is determined in a large way by the previous experience with usage of air. The consumption of air by motors or cylinders can be computed as follows:

Cylinders:

$$\text{cfm} = \frac{\text{area} \times \text{length}}{1728} \times \text{no. of strokes per minute}$$

How much air would a 2 in. bore single-acting cylinder use that has a stroke length of 12 in. and makes four strokes a minute?

$$\text{cfm} = \frac{3.14 \times 12}{1728} \times 4$$

$$A = 0.7854 D^2$$

$$A = 3.14 \qquad \text{cfm} = 0.0872$$

(1728 = cu in. in a cu ft)

The 0.0872 cfm is at working pressure and would have to be converted to free air for compressor capacity calculations.

The following equation is used:

$$\text{cfm} = V_1 = \frac{P_2 V_2 T_1}{T_2 P_1}$$

The following Force Chart may be useful in selecting cylinders for a given application.

When calculating air motors, the displacement per revolution would be needed.

Example. If an air motor is operating at 2000 rpm and has a displacement of 0.02 cu in. of air per revolution, how much air is being consumed?

Cylinder diameter—inches	$\frac{3}{8}$	$\frac{1}{2}$	$\frac{9}{16}$	$\frac{5}{8}$	$\frac{3}{4}$	$\frac{7}{8}$	1	$1\frac{1}{8}$	$1\frac{1}{4}$	$1\frac{3}{8}$	$1\frac{1}{2}$	$1\frac{3}{4}$	2	$2\frac{1}{4}$	$2\frac{1}{2}$	3	4	5	6	8	10
Cylinder area—square inches	.11	.20	.25	.31	.44	.60	.78	1.0	1.2	1.5	1.8	2.4	3.1	4.0	4.9	7.1	13	20	28	50	79
Supply air psig	\multicolumn{21}{c}{Force delivered by cylinder—pounds}
50	5	10	12	15	22	30	39	50	60	75	90	120	160	200	250	360	650	1000	1400	2500	4000
75	8	15	17	23	33	45	60	75	90	110	120	180	240	300	370	530	1000	1500	2100	3700	6000
100	11	20	25	31	44	60	78	100	120	150	180	240	310	400	490	710	1300	2000	2800	5000	8000
150	16	30	37	45	66	90	120	150	180	220	270	360	460	600	740	1050	2000	3000	4200	7500	12000
	\multicolumn{21}{c}{Double-acting cylinder air usage—cubic feet (free basis) per inch of stroke}
50				.002	.003	.004	.005	.006	.008	.009	.013	.016	.021	.026	.037	.068	.10	.15	.26	.41	
75				.003	.004	.006	.007	.009	.011	.013	.017	.022	.029	.035	.051	.094	.14	.20	.36	.57	
100	.001	.002	.002	.003	.004	.005	.007	.009	.011	.014	.017	.022	.029	.037	.045	.065	.12	.18	.26	.46	.73
150				.006	.008	.010	.013	.016	.020	.024	.032	.041	.053	.065	.094	.17	.26	.37	.66	1.04	

Note 1. Divide the above figures by 2 for single-acting cylinders.

Note 2. To allow for loss of force, or reduction in air usage, because of space taken by the rod of double-acting cylinders, subtract the respective values for a cylinder of the same diameter as the rod.

$$\text{cfm} = \frac{\text{disp in in}^3 \text{ per rev.}}{1728}$$

$$\text{cfm} = \frac{0.02 \times 2000}{1728}$$

$$\text{cfm} = 0.023 \text{ cu ft/min}$$

The general gas law equation would again be used to convert this figure into free air for purposes of compressor capacity.

TABLE 13-2

Selector Chart—Single-stage System (125 psi or less)

System requirements for continuous operation		System requirements for intermittent operations	
Free air needed by system (cfm)	Hp for compressor	Free air needed by system (cfm)	Hp for compressor
Up to 1.9	$\frac{1}{2}$	Up to 6.6	$\frac{1}{2}$
2.0 to 3.0	$\frac{3}{4}$	6.7 to 10.5	$\frac{3}{4}$
3.1 to 3.9	1	10.6 to 13.6	1
4.0 to 5.8	$1\frac{1}{2}$	13.7 to 20.3	$1\frac{1}{2}$
5.9 to 7.6	2	20.4 to 26.6	2
7.7 to 10.2	3	26.7 to 32.5	3
10.3 to 18.0	5	32.6 to 38.0	5

Two-stage System (over 125 psi)			
Up to 4.2	1	Up to 14.7	1
4.3 to 6.4	$1\frac{1}{2}$	14.8 to 22.4	$1\frac{1}{2}$
6.5 to 8.7	2	22.5 to 30.4	2
8.8 to 13.2	3	30.5 to 46.2	3
13.3 to 20.0	5	46.3 to 60.0	5
20.1 to 29.2	$7\frac{1}{2}$	60.1 to 73.0	$7\frac{1}{2}$
29.3 to 40.0	10	73.1 to 100.0	10
40.1 to 60.0	15	101.0 to 150.0	15
60.1 to 80.0	20	151.0 to 200.0	20

Courtesy of Champion Pneumatic Machinery Co. Inc.

Air lines for the distribution of air power must be properly installed and must be sized to provide an adequate supply of air for each work station. The two principal factors affecting pressure drop in air systems are leaks and friction.

Air leaks can be prevented through proper maintenance procedures. Because compressed air is not cheap and leaks cut down on production, air distribution systems should be thoroughly inspected at regular intervals. Soap solution and candle flame are two common methods of pinpointing air leaks. Some plants inject essence of peppermint into the air system to find leaks.

The effects of friction have been accurately computed through the years of using compressed air, and charts and tables are available for determining this important loss in air systems. The coefficient of friction depends mainly on the diameter or size of the air line and the velocity of flow. The density or viscosity of the air also affects friction, but changes very little.

TABLE 13-3
EFFECT OF AIR LEAKS

Size of opening (inches)	Cu ft wasted per month based on 100 psi and nozzle coefficient of 0.65	Cost of air wasted per month based on 10¢ per 1000 cu ft
$\frac{1}{32}$	45,508	4.56
$\frac{1}{16}$	182,272	18.21
$\frac{1}{8}$	740,210	74.01
$\frac{1}{4}$	2,920,840	292.09
$\frac{3}{8}$	6,671,890	667.19

TABLE 13-4
FRICTION LOSS IN PIPES
N VALUES
Nominal Diameter in Inches

Cu ft free air per minute	$\frac{1}{2}$ in. dia.	$\frac{3}{4}$ in. dia.	1 in. dia.	$1\frac{1}{4}$ in. dia.	$1\frac{1}{2}$ in. dia.	2 in. dia.	$2\frac{1}{2}$ in. dia.
5	0.127	0.12					
10	0.507	0.78	0.22				
15	1.14	1.76	6.49				
20		3.04	0.87	0.20			
25		5.0	1.36	0.32			
30		7.04	1.96	0.45			
35		9.59	2.66	0.62	0.27		
40		12.53	3.48	0.81	0.36		
45			4.40	1.02	0.45		
50			5.44	1.20	0.56		
60			7.83	1.82	0.80	0.22	
70			10.66	2.47	1.09	0.29	
80			13.92	3.25	1.43	0.38	
90				4.09	1.81	0.48	
100				5.05	2.23	0.60	
110				6.11	2.70	0.72	0.28
120				7.27	3.22	0.86	0.33
130				8.53	3.78	1.01	0.39
140				9.89	4.38	1.17	0.46
150				11.36	5.03	1.34	0.52

(*N* factors derived through experiment.)

The pressure loss due to friction in 100 ft of standard pipe is found by substituting the *N* value into the following formula:

$$P_1 = \text{initial psig}$$

$$\text{pressure drop (psi)} = \frac{N \times 14.7}{P_1 + 14.7}$$

Example. What is the pressure drop for 90 cfm flowing through 100 feet of $1\frac{1}{4}$ in. pipe with the receiver at 40 psig?

$$\text{Pressure drop} = \frac{N \times 14.7}{P_1 + 14.7}$$

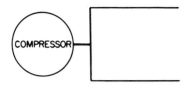

CENTRALIZED   SYSTEM – 1 COMPRESSOR

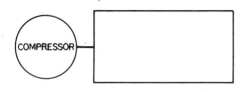

CENTRALIZED LOOP SYSTEM – 1 COMPRESSOR

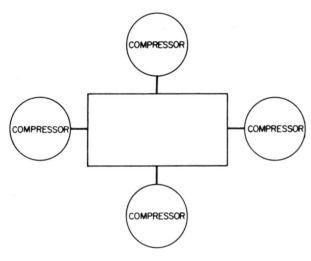

DECENTRALIZED LOOP SYSTEM – 2 OR MORE COMPRESSORS

Fig. 13-1. Air system pipings.

$$= \frac{4.09 \times 14.7}{40 + 14.7}$$

$$= \frac{60.12}{54.7}$$

$$= 1.09$$

Pressure at point of work:

Initial   40.00 psig

Drop −1.09 psig
        38.91 psig

In long runs of pipe, as pressure drop increases, the friction loss increases. In other words, there is greater friction loss in the last few feet than at the beginning of the line. Since the pressure decreases toward the pipe outlet, the velocity must increase. Pressure drop should be controlled by selecting pipe sizes to prevent a loss greater than 10 per cent between the receiver and point of use. In addition to the proper sizing of pipes, fittings, and valves, a number of techniques are used to ensure proper air supply.

The loop system provides two-way distribution of air to increase efficiency of use (see Fig. 13–1). Long distribution lines may be equipped with receivers or storage tanks near the point of usage to dampen compressor pulsations or to decrease line losses for intermittent operations.

The Header system is also effective for an adequate distribution of air to several tools simultaneously (see Fig. 13–2).

When air flows through pipes, the flow is retarded by friction due to the walls of the pipe. Air flow next to the walls of the pipe slows down, while the velocity in the center of the pipe is high. This loss is proportionately greater when the air pressure is at lower values. The loss is much less if the proper pipe diameters are used.

TO STATIONS

FROM
RECEIVER

Fig. 13-2. Manifold system.

Table 13–5 is useful as a guide in determining size of equipment in compressed air work.

For convenience in making calculations, loss through fittings is given in terms of equivalent feet of straight pipe. (This has been developed through actual testing.)

Example. A circuit requiring 65 cu ft per minute free air and operating at initial pressure of 100 lb psi could feasibly use a $\frac{1}{2}$ in. nominal standard pipe size. However, it must be remembered that for every 100 ft of run a pressure drop of 19 psi would occur. This means that the next largest nominal standard pipe size would be selected where long lines are necessary between supply and the pneumatic circuit. In the

TABLE 13-5

MAXIMUM RECOMMENDED FLOW THROUGH STANDARD WEIGHT PIPE

	Nominal Standard Pipe Size							
	$\frac{1}{8}''$	$\frac{1}{4}''$	$\frac{3}{8}''$	$\frac{1}{2}''$	$\frac{3}{4}''$	$1''$	$1\frac{1}{4}''$	$1\frac{1}{2}''$
Pressure drop in 100 ft.	54.0	31.7	25.4	19.0	12.4	9.2	5.94	4.68
Initial psi	Maximum Recommended flow (SCFM—free air)							
10	5.5	9.2	18.6	30.2	51.1	84.	139.	186.
20	6.5	11.	22.	35.7	60.6	100.	165.	220.
30	7.4	12.5	25.	40.7	68.4	112.	187.	250.
40	8.2	13.3	27.6	44.9	76.	125.	207.	276.
50	8.9	15.0	30.0	48.8	82.2	135.	226.	301.
60	9.6	16.1	32.2	52.3	89.0	146	242.	323.
70	10.4	17.7	34.3	56.0	94.	154.	258.	345.
80	10.8	18.2	36.3	59.0	100.	165.	273.	364.
90	11.3	19.1	38.2	62.2	105.	172.	286.	382.
100	12.0	20.0	40.	65.	110	180.	300.	400.
110	12.5	20.8	41.8	67.7	114.	188.	312.	417.
120	12.8	21.6	43.4	70.4	119.	195.	325.	434.
130	13.3	22.4	45.	73.0	123.	202.	337.	450.
140	13.8	23.2	46.5	75.5	127.	209.	348.	465.
150	14.2	23.9	48.	77.7	132.	218.	360.	479.

case mentioned, if the run were only a short distance, say ten feet, less than 2 psi pressure drop would occur, which could be tolerated. The designer must remember that circumstances sometimes alter a choice to provide adequate supply of air to the circuit under consideration.

Table 13–6 shows the equivalent feet of straight pipe loss for the various pipe fittings that are used in connecting the air circuit. For example, if a right-angle gate valve were used to shut off the branch circuit, the pressure drop through this valve would

TABLE 13-6

FRICTION LOSS IN PIPE FITTINGS (EQUIVALENT FEET OF STRAIGHT PIPE)

	$\frac{1}{4}$	$\frac{3}{8}$	$\frac{1}{2}$	$\frac{3}{4}$	1	$1\frac{1}{4}$	$1\frac{1}{2}$	2
Gate Valve (full open)	.30	.30	.35	.44	.56	.74	.86	1.10
Tee (straight through)	.50	.50	.70	1.10	1.50	1.80	2.20	3.00
Tee (Side outlet)	2.50	2.50	3.30	4.20	5.30	7.00	8.10	10.4
90° Ell	1.40	1.40	1.70	2.10	2.60	3.50	4.10	5.20
45° Ell	.50	.50	.78	.97	1.23	1.60	1.90	2.40
Angle Valve (full open)	8.00	8.00	9.30	11.5	14.7	19.3	22.6	29.0
Globe Valve (full open)	14.00	14.00	18.6	23.1	29.4	38.6	45.2	58.0

(The number of fittings should be kept to a minimum to keep down losses.)

be equivalent to 9.3 ft of loss in the $\frac{1}{2}$ in. standard pipe size. It is necessary for the circuit designer to calculate all frictional losses in straight run pipe, fittings, and valves in order to determine how much initial pressure is needed to overcome these resistances and still end up with sufficient pressure to accomplish the work.

The flow of air through pipes is generally expressed in cfm, cubic feet per minute. Flow of air can be measured in lineal feet per minute and expressed as the velocity of flow past a certain point. The relationship is given in the following equation:

Cubic feet per minute = lineal ft/min × area

Example. What is the velocity in ft/min of 50 cfm flowing through a hole having an area of $\frac{1}{10}$ sq ft?

$$\text{cfm} = \text{lfm} \times \text{area ft}^2$$

$$(\text{Velocity}) \ \text{lfm} = \frac{\text{cfm}}{\text{area}}$$

$$= \frac{50}{1/10}$$

$$(\text{Velocity}) \ \text{lfm} = 500 \text{ ft per min}$$

C_v Capacity Coefficent

1. Originally used as a means of comparing capacity of liquid passing or carrying valves. A valve with a C_v rating of 1 would pass 1 U.S. gallon of water at 60°F, in one minute, with a 1 lb per sq in. loss in pressure across the valve.

   The formula for liquid flow is

   $$C_v = \frac{\text{gpm}}{\sqrt{\frac{\Delta P}{G}}}$$

   where

   gpm = gal/min liquid flow.

   ΔP = pressure drop across valve.

   g = specific gravity of liquid.

2. Research has proven that same C_v rating applicable to liquid or noncompressible fluids is applicable to gases or compressible fluids by revising the formula to take into consideration the various factors applicable to flow of gases. There is some discussion relative to which factors should be used in the formula. For working air, we feel the following is a good formula:

TABLE 13-7
**CAPACITY CONSTANTS FOR VARIOUS
INLET PRESSURES AND PRESSURE DROPS**

Inlet pressure P (psig)	Capacity constant (A) for various pressure drops (ΔP)		
	$\Delta P = 2$ psi	$\Delta P = 5$ psi	$\Delta P = 10$ psi
40	0.0986	0.0642	0.0479
50	0.0905	0.0586	0.0433
60	0.0840	0.0543	0.0398
70	0.0788	0.0508	0.0371
80	0.0745	0.0478	0.0348
90	0.0708	0.0454	0.0329
100	0.0676	0.0433	0.0313
110	0.0648	0.0414	0.0299
120	0.0623	0.0398	0.0287
130	0.0600	0.0384	0.0276
140	0.0581	0.0371	0.0267
150	0.0563	0.0359	0.0258

$C_v = A$ cfm

$C_v =$ capacity coefficient.
$A =$ constant obtained from chart.
cfm $=$ cu ft/min (free air).
$P =$ valve inlet pressure (gauge).
$\Delta P =$ allowable pressure drop across the valve.
Temp. $= 70°$F.

$$C_v = \frac{\text{cfm}}{22.67} \sqrt{\frac{GT}{(P_1 - P_2)P_2}}$$

where

cfm $=$ air flow in cubic feet/min *of free air*.

$G =$ specific gravity—1 for air

$T = $ *absolute* temperature $= °$F $+ 460$.

$P_1 = $ *inlet* or *upsream* pressure.

$P_2 = $ *outlet* or *down-stream* pressure.

} *Both* lb/in.^{2} absolute or psig $+ 14.7$ or gauge pressure $+ 14.7$

Remember, cfm *free air* or S. C. F. M. is displacement or swept volume in cubic feet per minute multiplied by the ratio of

$$\frac{\text{gauge pressure} + 14.7}{14.7}.$$

3. Very important. After a given ΔP (or $P_1 - P_2$) is reached, no more flow (cfm) can be pushed through a valve having a given C_v rating. This "terminal capacity" is reached when $P_2 = 0.53 \times P_1$, or when P_2 is 53 per cent of P_1. There-

TABLE 13-8

Size of orifice	Flow of Air Through Orifice in Cu Ft of Free Air per Minute																		
	Receiver gauge pressure																		
	2	5	10	15	20	25	30	35	40	45	50	60	70	80	90	100	125	150	200
$\frac{1}{64}''$	0.038	0.0597	0.0842	0.103	0.119	0.133	0.156	0.173	0.19	0.208	0.225	0.26	0.295	0.33	0.364	0.40	0.486	0.57	0.76
$\frac{1}{32}''$	0.153	0.242	0.342	0.418	0.485	0.54	0.632	0.71	0.77	0.843	0.914	1.05	1.19	1.33	1.47	1.61	1.97	2.33	3.07
$\frac{3}{64}''$	0.342	0.545	0.77	0.94	1.07	1.21	1.4	1.56	1.71	1.91	2.05	2.35	2.68	2.97	3.28	3.66	4.12	5.20	6.9
$\frac{1}{16}''$	0.647	0.965	1.36	1.67	1.93	2.16	2.52	2.80	3.07	3.36	3.64	4.2	4.76	5.32	5.87	6.45	7.85	9.20	12.2
$\frac{3}{32}''$	1.37	2.18	3.06	3.75	4.28	4.84	5.6	6.24	6.84	7.6	8.2	9.4	10.7	11.88	13.1	14.5	17.5	20.8	27.5
$\frac{1}{8}''$	2.435	3.86	5.45	6.65	7.7	8.6	10.0	11.2	12.27	13.4	14.5	16.8	19.0	21.2	23.5	25.8	31.4	36.7	48.7
$\frac{3}{16}''$	5.45	8.72	12.3	15.0	17.1	19.4	22.4	25.0	27.4	30.4	32.8	37.5	42.9	47.5	52.4	58.0	70.0	83.2	110.0
$\frac{1}{4}''$	9.74	15.4	21.8	26.7	30.8	34.5	40.0	44.7	49.0	53.8	58.2	67.0	76.0	85.0	94.0	103.2	125.5	147.0	195.8
$\frac{3}{8}''$	21.95	34.6	49.0	60.0	69.0	77.0	90.0	100.0	110.5	121.0	130.0	151.0	171.0	191.0	211.0	231.0	282.0	330.0	440.0
$\frac{1}{2}''$	39.0	61.6	87.0	107.0	123.0	138.0	161.0	179.0	196.4	215.0	232.0	268.0	304.0	340.0	376.0	412.0	502.0	588.0	782.0
$\frac{5}{8}''$	61.0	96.5	136.0	167.0	193.0	216.0	252.0	280.0	306.8	336.0	364.0	420.0	476.0	532.0	587.0	645.0	785.0	920.0	1220.0
$\frac{3}{4}''$	87.6	133.0	196.0	240.0	277.0	310.0	362.0	400.0	442.0	482.0	521.0	604.0	685.0	765.0	843.0	925.0	1127.0	1322.0	1760.0
$\frac{7}{8}''$	119.5	189.0	267.0	326.0	378.0	422.0	493.0	550.0	601.0	658.0	710.0	722.0	930.0	1004.0	1145.0	1260.0	1531.0	1804.0	
$1''$	156.0	247.0	350.0	427.0	494.0	550.0	645.0	715.0	786.0	860.0	930.0	1070.0	1215.0	1360.0	1500.0	1648.0	2000.0	2350.0	
$1\frac{1}{8}''$	196.0	310.0	440.0	538.0	620.0	692.0	812.0	900.0	987.0	1082.0	1170.0	1350.0	1530.0	1710.0	1890.0				
$1\frac{1}{4}''$	242.0	384.0	543.0	665.0	770.0	860.0	1000.0	1120.0	1230.0	1345.0	1455.0	1680.0	1900.0	2130.0	2350.0				
$1\frac{3}{8}''$	291.0	464.0	660.0	803.0	929.0	1040.0	1200.0	1350.0	1480.0	1630.0	1760.0								
$1\frac{1}{2}''$	350.0	550.0	780.0	960.0	1102.0	1240.0	1435.0	1600.0	1770.0	1930.0	2085.0								
$1\frac{3}{4}''$	473.0	752.0	1068.0	1304.0	1505.0	1680.0	1960.0												
$2''$	625.0	985.0	1395.0	1700.0	1967.0	2200.0	2586.0												

Courtesy of Quincy Compressor Co.

Fig. 13-3. Typical compressed air system.

fore, never assume you can get more speed out of a system just because you would be willing to allow more pressure drop or ΔP.

For example, with 100 psi (114.7 psia) line pressure, when the pressure in the cylinder drops to 114.7×0.53 or 60.7 psia (46 psi gauge) cfm flow is at its maximum. No higher flow rate can be attained.

This does not mean that the pressure will not drop lower or that piston speed will not increase. But, if these things happen, they are due to *external* causes, *not* to air flow rate increasing.

The piping installation should be designed to provide compressed air to all parts of the plant in proper condition for the power and control that may be required. The diagram shows the proper installation of a compressor system with receiver tanks, moisture traps, drains, distribution lines, and branch connections for pneumatic power.

In larger capacity systems it may be necessary to use additional equipment, such as aftercoolers and air driers. Figure 13-3 illustrates a typical air system and some of the equipment that is used.

Free air from the atmosphere enters the compressor through a filter, which is essential to keep dust and contamination from entering the system. The air passes from the first compressive step through an intercooler, where it loses some of its heat of compression before entering the second stage. High pressure air discharged from the second stage then passes through the aftercooler to the air drier (a moisture separator is also used). From the drier air passes into the receiver, where it is stored as pressure energy until it is needed by the pneumatic circuit.

When the four-way valve is energized, the air passes from the main line through a filter, regulator, and lubricator combination unit before entering the cylinder to do work. A silencer or muffler is generally used to exhaust the spent air back to atmosphere from the four-way valve as the cylinder reciprocates.

Moisture in Air

Free air from the atmosphere contains varying amounts of moisture in vapor form. The water vapor content is measured in terms of relative humidity. Relative humidity is the ratio of the amount of water vapor actually present in the atmosphere to that which would be present if the air were saturated.

Vapor in air condenses when the temperature of saturated air is reduced. It is the purpose of an aftercooler, to reduce the temperature of the compressed air and separate the moisture out of the air before it enters the system. Moisture in an air system can result in harmful effects, such as washing away lubricants, excessive wear, and corrosion.

When air lines extend outside, pockets of water in the piping may freeze in cold weather. Water also freezes at the exhaust of pneumatic tools, causing sluggish operation.

If the air system provides air for both power systems and control systems, a drier may be needed. Control air must be dry and clean when used in precision instruments. Automatic air control systems are used to regulate power plant equipment, furnaces, pneumatic power systems, hydraulic power systems, data processing machines, computers, and many others.

Air driers use a chemical substance, called a *desiccant,* which is very dry and attracts the moisture. Some driers use a porous filtering material that much be recharged periodically by heating.

Table 13–8 is based on the moisture content of free air at atmospheric conditions and is useful for determining separators, driers, and moisture traps for an installation.

Piping the System

Standard schedule 40 pipe, tools, and fittings are adequate for piping up the system. Pipe and fittings should be selected of proper size to maintain less than a 3 psi pressure drop in main distribution and branch lines. Oversized pipe may be a wise investment for growing companies.

Piping should be properly supported and installed to prevent sag and vibration. Lines should be installed at a slight angle slope to facilitate drain for any condensate. At low points and at the ends of the system, "water legs" with drains may be provided to collect and dispose of condensate.

Table 13–5, p. 284, previously mentioned, is useful in determining the maximum recommended flow through standard pipe and fittings to prevent excessive friction and pressure drop.

Installation should avoid excess fittings and sharp bends that increase pressure drop and loss of power.

The compressed air system should be provided with protection against overloads by installing a relief in the system. This is usually located in the receiver.

The loop system of distribution is generally recommended where possible to help provide a more even distribution of pneumatic power to all circuits.

A Summary of System Design Factors

A careful study of immediate and future needs should be made before the compressed air system components are selected. Volume and pressure requirements are essential, as well as a detailed analysis of each pneumatic device to be operated.

If 100 cfm at 100 psig are required and a 50 cfm compressor is installed, the pressure will soon drop lower than the effective amount needed, since the volume produced is only half the amount required for effective tool operation. In such a case, the compressor will run continuously, trying to keep up, and will not reach the unload or cutoff point until the air demand is below the normal work load.

A safety factor of approximately 25 per cent is desirable for efficient operation to reduce maintenance costs. This applies to constant speed control units particularly. With automatic start and stop service, the percentage of intermittent use should be approximately one-third running time and two-thirds down time. This provides for fewer starts per hour for the electric motors, and prevents overheating with subsequent kickout of the starter thermal overloads. It is far more practical to oversize than to undersize a compressed air system, and higher benefits will result over a period of time.

If altitude location is not considered, the unit may fall far short of actual requirements. Losses run approximately 3 per cent for each 1000 feet of eleva-

TABLE 13-9
MOISTURE IN AIR
Air Pressure—psig

Temp. °F.	30	40	60	80	100	150
	Pints of water per 1000 cubic feet saturated air (free basis)					
40	0.094	0.075	0.054	0.041	0.036	0.026
60	0.20	0.16	0.12	0.089	0.069	0.050
80	0.36	0.30	0.22	0.18	0.14	0.10
100	0.67	0.55	0.40	0.32	0.27	0.20
120	1.2	0.93	0.70	0.55	0.46	0.32

tion above sea level because of rarified air at lower barometric pressure. This condition also affects cooling of electric motors and has serious implications on the sizing of engine-driven units.

An orifice $\frac{1}{4}$ in. in diameter passing air at 80 psig requires approximately 20 hp. This indicates the importance of accurate calculations in designing air systems. A listing of all application requirements including the suggested safety factor of 25 per cent can eliminate costly errors.

The following questions may be helpful in determining specific needs of the compressed air system:

1. What elevation is to be considered?
2. What is the ambient operating temperature?
3. Is the compressor to run one, two, or three shifts per day?
4. What are the actual uses of compressed air?
5. Should the compressor be constant speed, automatic start and stop, or dual control?
6. If motor driven, what is the electrical service available?
7. Are special electrical controls required?
8. In the case of dual units, is an electrical automatic alternator advisable?
9. What is the maximum pressure requirement?
10. Will the unit operate in a hazardous, dusty, or gaseous atmosphere where totally enclosed fan cooling and electrical equipment are necessary?
11. Will oil and moisture in the air present a problem?
12. What accessories are needed, such as coolers, silencers, filters, or driers?

Questions

1. Explain the use of constant speed controls.
2. Explain the use of dual controls.
3. What are some of the important considerations in selecting a single-stage compressor? When should a multistage compressor be used over a single-stage compressor?
4. Why are compressors rated in terms of cubic feet of free air per minute?
5. Air is used at the rate of 24 cfm from a receiver at 80°F and 100 psi. If the barometer reads 14.6 psia and the inlet temperature is 56°F, how many cfm must the compressor provide?
6. How many cfm of air would a 4 in. bore single-acting cylinder use that has a stroke length of 12 in. and makes 800 strokes a minute?
7. How much air would the compressor take in if the above cylinder operates at 110 psi?
8. If air at 100 psi escapes from an opening of $\frac{1}{16}$ in., what would the approximate cost be if it continued to leak for a period of six months?
9. How is pressure drop controlled in a pneumatic installation?
10. Explain the advantages of the loop system of air distribution over the header system.
11. What size of nominal standard pipe would be selected for a pressure of 100 psi at a maximum flow of 300 cu ft per minute?
12. An air motor uses 10 cfm at 3000 rpm. What is its displacement?
13. How are leaks detected in an air system?
14. How are friction losses determined in an air system?
15. Explain why friction loss increases as the pressure drop increases.
16. How are friction losses controlled in an air system?
17. Diagram a loop system of air distribution and show how a connection is made for a branch line.
18. Diagram or explain the "header" system of air distribution.
19. What is the velocity in feet per minute of 60 cfm flowing through a hole $\frac{1}{12}$ sq fit in area?
20. What effect does moisture have on an air distribution system?
21. What is the principle of a chemical air drier?
22. List ten factors that are important in the selection of a compressed air system.

CHAPTER **14**

CONTROLLING PNEUMATIC POWER

Pneumatic circuits are easily applied for both linear and rotary motion.

Highlighting Pneumatics Versus Hydraulics

The main difference between applying pneumatic and hydraulic fluid power is the compressibility factor of the two media.

The weight of the hydraulic fluid is of concern when sudden acceleration or deceleration is experienced. Since force is equal to mass times acceleration when a hydraulic motor reverses under load, the resultant pressure created will be many times that created if the fluid were air.

Hydraulic fluid averages about 58 lb per cu ft. Dry air weighs 0.0746 lb per cu ft at atmospheric pressure and 72°F. It increases to about 5 lb/cu ft at 1000 psi and about 15 lb/cu ft at 3000 psi. It is apparent that pneumatic circuits are less concerned with shock loads in the system.

The kinetic energy of moving fluid is shown by

$$EK = \tfrac{1}{2}mv^2$$

where EK = kinetic energy, ft lb.

m = mass of fluid, lb/32.2 ft/sec^2

v = velocity of fluid, ft per sec (fps).

Pressure drop and temperature effects are also quite different when the two media are compared.

Hydraulic fluids exchange pressure energy for velocity and heat due to friction when fluid by-passes over a relief valve or flow control. The increase in temperature changes the viscosity of the fluid, which may have an effect on the operation of a machine during warmups, etc.

In pneumatic circuits when pressure energy is exchanged for velocity energy, generally a decrease in temperature results (frost or snow at the exhaust side of an air motor). Pneumatic power is immediately available for quick response at low temperatures. Air compressed to psia, if suddenly released to the atmosphere, travels approximately 17,000 ft/min and about 35,000 ft/min if compressed to 50 psi. In an actual circuit velocity is a function of conductor size in area.

Pneumatics will usually require the use of lubricating devices for the prevention of wear on integral moving parts.

Operating pressures for industrial pneumatic ap-

Fig. 14-1. Combo unit. *(Courtesy of Bastian-Blessing Company.)*

plications are generally between 28 in. Hg (mercury) and 150 psi, which limits the force capabilities unless very large cylinders are used. Aircraft pneumatic systems have found that high pressure pneumatic systems have certain weight advantages, as previously mentioned. Aircraft systems use pressures up to 5000 psi and higher for some military uses. Missile launching devices may reach 30,000 psi.

Industrial uses of high pressure pneumatic systems have been limited because of the cost of producing air pressure above 150 psi in large quantities.

Pneumatic Controls

Branch lines leading to each pneumatic circuit are generally equipped with a filter regulator and lubricator combination unit (see Fig. 14-1).

Filter

The purpose of the filter is to remove contamination from the air before it reaches the directional valve and actuating cylinder or air motor. Air line filters are generally fitted with a filter element that removes contaminants in the 25 to 50 micron range (see Fig. 14-2).

Air flow entering the filter is directed downward with a swirling motion that forces moisture and heavier particles to the wall of the bowl. When removed from the main air stream, these contaminants are pulled out of suspension and fall to the bottom of the bowl and become trapped. When the drain cock is opened, they are forced out of the filter.

Pressure drop through the filter should be less than 2 psi in most designs.

There are a number of different designs from which to select a filter to keep circuits operating without trouble. The filter design in Fig. 14-3 drains contaminants during circuit operation and automatically closes when the air flow stops. This helps to prevent sediment in the bowl from getting back into the system.

Regulators

The maximum system pressure is controlled by the compressor control system, which generally maintains a pressure range. In other words, the compressor may start automatically when the system pressure drops to 120 psi, and automatically stop again when the pressure in the receiver reaches 140 psi. The compressor is generally controlled with pressure switches and protected by a relief valve.

In order to maintain a continuous pressure at a work station and eliminate compressor fluctuations, or to utilize pressures lower than compressor output, a pressure control valve is used.

The pressure regulator is fitted with an adjustable spring that allows the valve to hold a given pressure on the downstream side (see Fig. 14-4). The force of the spring on top of the valve spool is set for the required pressure downstream. This force holds the valve open until the pressure downstream starts to exceed the spring force. The pressure downstream acting against the area on the bottom of the spool causes it to move up, throttling the supply air to control the proper pressure downstream.

A number of excellent designs are available for accurate pressure control of pneumatic circuits. It should be understood that a regulator at each work station helps to keep the entire system stable and provides the correct air supply for each job without wasting power.

Figure 14-5 shows a regulator with a diaphragm operating against the adjustable spring. When the diaphragm overcomes the adjusted spring force, it allows the push rod to move up, and the spring-loaded valve at the bottom throttles the supply air to the controlled pressure side.

Fig. 14-2. Air line filter. (*Courtesy of Bastian-Blessing Company.*)

Fig. 14-3. Filter design. (*Courtesy of WABCO.*)

Fig. 14-4. Pressure regulator.

Fig. 14-5. Air line regulator. *(Courtesy of Bastian-Blessing Company.)*

Lubricators

Lubricators of various designs are available to assure proper lubrication of integral moving parts of pneumatic components. Figure 14–6 explains the principle of operation.

Mufflers

Mufflers (or silencers) are used to control the noise caused by rapidly exhausting (expanding) air. Dust filters may be used at air intakes to minimize the entry of contaminants into the system.

Functional Types of Pneumatic Directional Control Valves

Pneumatic directional control valves are functionally described by the number of their flow ports, the number of valve positions, and the internal connections of their flow ports in each position. The most common designations are listed below.

Two-way Valve

This is an on-off type of device. This valve is usually provided with two external flow ports—a supply port and an exhaust port. A normally open two-way valve permits flow in its normal or "at rest" position and blocks flow when shifted. The normally closed two-way valve blocks flow in its normal position and permits flow when actuated (see Figs. 14–8 and 14–9).

Three-way Valve

This device is a flow path selector. One flow port is internally connected to either of two other flow ports. It may be used alternately to pressurize or exhaust one port, or to direct pressure alternately from one source to either of two other ports. Three-way valves may be constructed either normally open or normally closed. These valves can be used as pilot relays to operate other valves. Three-way valves may be used singly to control single-acting cylinders, or in pairs to control double-acting cylinders (see Figs. 14–10 and 14–11).

Four-way Valve

This is a flow-reversing device. Usually, four external flow ports are provided—a supply port, two working ports, and an exhaust port. In one position, the valve allows air to flow from the supply source to

LOW AIR FLOW   HIGH AIR FLOW

OIL ADJUSTING SCREW

OIL CHANNEL

AUTOMATIC SELF-ADJUSTING FLOW GUIDE

FELT FILTER PAD

SIGHT GAUGE

LOW AIR FLOW

HIGH AIR FLOW

BOWL PRESSURIZING INLET VALVE

Fig. 14-6. Lubricators. *(Courtesy of WABCO.)*

END VIEW OF LUBRICATOR INLET

AIR

FLOW GUIDE COLLAPSED

END VIEW OF LUBRICATOR INLET

AIR

FLOW GUIDE EXTENDED

Fig. 14-7. Application of filter regulator and lubricator. *(Courtesy of Bastian-Blessing Company.)*

NORMALLY CLOSED

NORMALLY OPEN

Fig. 14-8. Symbols—two-position and two-way valves.

Fig. 14-9. Pilot-operated, 2-position; 2-way air valve; normally closed.

Fig. 14-10. 2-position, 3-way air valves.

NORMALLY OPEN   NORMALLY CLOSED

Fig. 14-11. 2-position, 3-way, manually operated air valve.

Fig. 14-12. (a) 2-position, 4-way air valve; (b) 2-position 5-way air valve with adjustable exhaust ports.

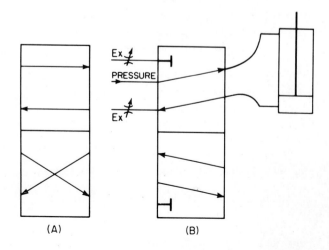

(A)   (B)

one of the working ports. Simultaneously, air is allowed to flow from the other working port to exhaust. When the valve is shifted, the flow paths are reversed. That is, supply is directed to the second working port, while the first working port is open to exhaust. The most common application of four-way valves is to control the motion of double-acting cylinders (see Fig. 14–13).

Four-way valves are also available with five external ports. This permits use of either dual supply or dual exhaust. These five-ported versions provide the same basic control of flow paths of four-ported versions do. In addition, dual supply ports permit the use of different pressures for cylinder retraction.

Fig. 14-13. 4-way valve solenoid showing variable orifice control of exhaust ports. *(Courtesy of Numatics Inc.)*

Dual exhaust ports conveniently permit individual speed control of the exhaust from each of the working ports (see Figs. 14–14 and 14–15).

The foregoing discussions of pneumatic directional control valves have been confined to two position devices—a normal and one actuated position. Four-way valves are also available in three-position varieties (either four- or five-ported). These three-position types have a normal and two actuated positions. The *normal position* may permit the following flow arrangements:

1. Both working ports connected to supply pressure.
2. All ports blocked (closed center)—this is a "hold" position, and permits "inching" of a cylinder.
3. Both working ports open to exhaust—this provides a "float" condition, and enables positioning of the cylinder by external forces.

Other Valve Types

In addition to the basic directional control valves described, other devices provide optimum flexibility of pneumatic circuitry.

Speed (flow) control valves are self-contained devices which provide "free flow" in one direction,

and an adjustable restriction to flow in the reverse direction. Free flow is accomplished by means of a built-in check valve. Restricted, or metered, flow in the reverse direction is accomplished by means of a variable orifice that by-passes the check valve. Speed control of pneumatic cylinders and equipment may require special design considerations, because of the compressibility factor of air. A pneumatic cylinder performing work, for instance, could "jump" forward if the work resistance suddenly diminished. This rapid

Fig. 14-14. Speed control with 2 exhaust ports.

Fig. 14-15. Mufflers or dust filters. *(Courtesy of Air-Mite Inc.)*

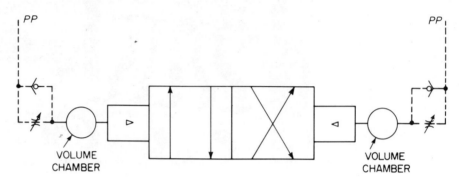

Fig. 14-16. Time-delay valves—volume chambers.

partial stroking of the cylinder is caused by the stored energy of the compressed air on the driving side of the cylinder. A similar occurrence of "jump" could be encountered if a directional valve is shifted before the cylinder completes its stroke, or if the valve is shifted too soon after the cylinder stroke is completed. Speed controls are most often installed to allow free flow of working air into the cylinder and to "meter out" or restrict the air flowing from the cylinder. Some applications may require use of the speed controls to "meter in," or perhaps a combination of both "meter out" and "meter in" control.

A special purpose valve peculiar to pneumatic circuitry is the quick-exhaust valve. The quick-exhaust valve is used close-connected to a cylinder. Air from a directional control valve (which may be remotely positioned) passes through the connecting lines and the quick-exhaust valve into a cylinder. When the directional control valve is shifted, only the air in the connecting lines between the directional valve and the quick-exhaust valve is exhausted through the directional valve. A built-in valving mechanism allows the air from the cylinder to exhaust directly through the quick-exhaust valve to atmosphere. This eliminates the need for exhaust air from the cylinder to flow through long or restricted lines to the main control valve. The quick-exhaust valve permits in-

creased cylinder velocities and/or smaller directional control valves.

Pneumatic time delay valves may be used to adjust the timing of various phases of a machine cycle. This is accomplished with a pilot-operated directional control valve that receives pilot pressure from an external source and either meters air flow into or out of the pilot chamber to create the delay.

Other special-purpose devices used in pneumatic circuitry include shuttle valves, sequence valves, and deceleration valves.

Basic Designs of Pneumatic Directional Control Valves

Carefully consider the basic types of valve design when selecting directional control valves. Each design provides advantages for certain applications. The basic designs are shear action (D-slide or rotary disc), metal-to-metal spool, packed spool, and poppet type.

Shear Action (D-slide or Rotary Disc)

An element slides across the flow path. Frequently, pressure unbalance is used to force the element against a mating surface to effect a metal-to-metal seal. This valve readily provides two-, three-, or four-way action, but does not lend itself to the large number of flow path variations which are often required in modern pneumatic circuitry. Some important advantages of the shear action type of design are as follows:

1. Valve mechanisms "wear in" with use, enabling these valves to provide millions of trouble-free cycles, even under adverse conditions.
2. Valve materials may be selected which are not affected by contaminants in compressed air lines.
3. The shear action of this valve tends to exclude foreign material from lodging between the mating surfaces.
4. The valve is not seriously affected by temperature extremes.
5. The valve can be used as a throttling device to control flow rates.

Consider the following factors of shear-action-type valves to avoid circuit problems:

1. Large forces may be required to shift the valve slide, especially at high pressures when designs are used which are not pressure-balanced.
2. Units generally require long stroke lengths to obtain full flow capabilities.
3. Valve must be continuously lubricated for maximum life.
4. The sealing member (D-slide or rotary disc) can be forced from its mating surface if the pressure under the plate (working pressure) exceeds supply pressure. Valves may not be suitable for cylinder applications where large external forces could create such "reverse" pressures.
5. This design usually does not provide "bubble-tight" sealing ability.

Metal-to-metal Spool

A variation of the shear action design is the metal-to-metal spool-type valve. This valve has cylindrical flow and sealing paths rather than flat (single-plane) paths. This design depends upon close fits between a metal spool and its matching bore. Useful features of metal-to-metal spool-type valves include

1. Design is capable of providing almost any flow path configuration desired, in a variety of porting and actuating configurations.
2. External forces required to shift a balanced spool are low.
3. Spool lands can be made wide enough to over-

lap bore openings and thus prevent undesirable interconnection of flow ports while the valve is being shifted from one flow path arrangement to another. This ability to eliminate "crossover" flow during shift reduces the chance of shift failure when a pilot-operated valve at low supply pressures is being used.

To avoid circuit problems with metal-to-metal spool valves, consider these factors:

1. Foreign matter may get between mating parts, causing wear and leakage, and even possibly sticking of the spool.
2. This type of unit requires good filtration and lubrication.
3. Oxidized airborne lubricant from a compressor or other foreign air line material may accumulate in the small valve clearances, and cause the spool to "varnish" in place. This occurs particularly when valves are idle, as over weekends.

Packed Spool

Packed spool valve designs have seals to provide leakproof sealing ability. Usually, the seals are an integral part of the moving spool and cross over the flow passages. Selection of the seal materials is critical. The elastomer must be resilient to provide tight sealing, yet tough enough to resist abrasion for millions of trouble-free cycles. It must be dimensionally stable—and remain resilient—under operating temperature extremes. It must also maintain its dimensional stability in the presence of various contaminant fluid particles in air lines. Unwanted shrinking of the seal could cause loss of sealing interference, and excessive swelling could cause sticking of the spool. Advantages of the packed spool type of valve design are

1. Design is available for most porting and actuating configurations.
2. Forces on the spool are balanced. Shifting forces are usually only slightly higher than for metal-to-metal designs.
3. Resilient seals provide "leak-tight" sealing ability.
4. Resilient seals make the valve less vulnerable to abrasion by foreign material.
5. Valve spools have limited contact areas and less tendency to "varnish" in place than metal-to-metal designs.
6. Lubrication requirements are not as important as for metal-to-metal types.

Fig. 14-17. Valvair valve. *(Courtesy of Bellows-Valvair.)*

Lightweight moving members require only a short stroke to provide maximum flow opening.

2. Short stroke of the valve gives minimum wear and maximum life capabilities.
3. Self-cleaning resilient seats provide "bubble-tight" sealing capabilities.
4. Lubrication requirements are not so important as with metal-to-metal valves.
5. Dimensional stability of the sealing member is not as critical as with packed spool valves.

Fig. 14-18. Cutaway view—symbol. *(Courtesy of Bellows-Valvair.)*

7. A spool may be replaced without changing mating parts.

Consider the following factors to avoid circuit problems with packed spool valve designs:

1. Improper selection of seal material may mean that the elastomers can adhere to the surfaces against which they seal.
2. Environmental temperatures and air line fluids must not be allowed to affect dimensional stability of the seals excessively.

Poppet Type

Poppet-type valves are desirable for use in circuits where high flow capacity and rapid response are required. These valves usually have resilient seals to provide tight sealing and help absorb the kinetic energy of the moving member.

The poppet type is easily adopted for three-way valve designs. Four-way valve designs are achieved with a dual poppet arrangement—that is, really two three-way valves in a common body. Some important advantages of the poppet-type valve design are

1. Poppet valves offer rapid cycling capabilities.

To avoid circuit problems with poppet-type valve designs, consider the following:

1. Poppet designs with direct operated actuators are bulky in large port sizes.
2. The poppet valve design allows flow from the pressure port to escape to exhaust as the valve is shifting. This "crossover" flow may drop the supply pressure below the minimum operating pressure, causing the valve to fail to complete its shift.
3. Some flow path configurations are not available in poppet valves.

Solenoid Valves

Figure 14–17 shows the actual valve, which is a poppet-type, two-position, three-way valve that operates at very high speeds.

Figure 14–18 shows the internal construction of the value and the symbol which gives the operation information. Solenoid differential pilot valves can provide about 3000 cycles/min.

Questions

1. Highlight the advantages and disadvantages between using air or oil fluid power.
2. Does an air system have a relief valve? If yes, where is it located?
3. Describe the design and operation of a filter-regulator-lubricator combination unit.
4. What are the main differences between hydraulic controls and air controls?
5. What is the pressure drop through a four-way valve that is passing 30 cfm per hour (60°F = 14.7 psia); specific gravity of air = 1; temperature = 72°F, and the pressure downstream is 80 psi. (The C_v factor for the valve is 5.)
6. What is the main purpose of a pressure regulator?
7. Why are lubricators used with pneumatic circuits?
8. How is noise controlled with air circuitry?
9. Explain the function of a two-position four-way valve.
10. Why do some four-way valves have five external ports?
11. What is the normal position of any control valve?
12. What kind of valve is used to control the speed of cylinders or air motors?
13. What is the purpose of a quick-exhaust valve?
14. Name the construction types of directional control valves available. Name several considerations when the valve types available are being selected.
15. What is a shuttle valve used for?

CHAPTER **15**

USING
PNEUMATIC
POWER

The application of pneumatics is expanding in two main categories: pneumatic power and pneumatic control.

Pneumatic power uses include punching, stamping, shearing, trimming, pressing, lifting, hoisting, clamping, and many others.

The application of pneumatics for jobs of this nature involves the same calculations as previously covered throughout Section I and in earlier chapters of Section II.

Both actuating cylinders and air motors can be applied to provide linear and rotary motion.

The same general design of cylinders described in Section I is used for pneumatic circuits.

Examples of Applications

Air presses are popular for light press assembly work, etc. The press shown in Fig. 15–1 can develop about $7\frac{1}{2}$ tons of force.

Air hoists as shown in Fig. 15–2 are convenient to apply for lifting parts and machine members in manufacturing processes.

The linear motor shown in Fig. 15–3 is air-pow-ered and hydraulically controlled for feed speed. The large cylinder on the left provides the power, and the small cylinder directly behind it contains hydraulic oil.

When the forward solenoid is energized and the cylinder extends, fluid from the rod side of the small hydraulic cylinder is metered back to the large area side through a needle valve. When the air cylinder is reversed, oil in the large area side of the hydraulic cylinder is returned to the rod side. A small spring-loaded accumulator (shown with the short stem) provides for the difference in volume of the two sides of the hydraulic cylinder because of the space taken by the rod. Linear motors without the hydraulic control can be applied as air cylinders would be, offering the simplicity of mounting obtained by having the control valve integrally mounted.

Precise feed speeds can be obtained with linear motors, which makes them valuable for uses such as feeder gates, fan controls, drilling and tapping machines, and the positioning of valves to control water systems, chemicals, etc.

Precise feed control is given to the cut-off saw in Fig. 15–4 with a linear motor.

302

Fig. 15-1. Air press. *(Courtesy of Flick-Reedy Corporation.)*

Fig. 15-3. Linear motor. *(Courtesy of Bellows-Valvair.)*

1. Overhead plant rail
2. Air hoist eye
3. 3-way air control valve
4. Case-hardened (50-54 Rockwell C), hard chrome plated piston rod
5. Hook swivels 360° on ball joint that also permits lateral movement in any direction
6. Check valve
7. Flexible air hose is all you need to connect hoist to your shop air supply

Fig. 15-2. Air hoist. *(Courtesy of Flick-Reedy Corporation.)*

1. Lifting the load. Shifting 3-way valve control lever to lift position opens cap end port to exhaust. Air pressure at head end lifts load.

2. Stopping the load. Shifting 3-way valve control lever to neutral position traps pressurized air in cap end of hoist. This results in a balance between the forces at cap end and head end, thus stopping and holding the load.

Fig. 15-4. Standard cut-off saw.

In Fig. 15–5, both hands must be used to punch out the port, which protects the operator from injury and possible loss of fingers.

Air Power Emergency Power Unit

The emergency power unit shown in Fig. 15–6 uses air over oil boosters to provide a crushing force to crimp the casing and drilling rod closed in case of an oil gusher. The operator trips an emergency blowout preventer, as he retires from the scene, to actuate the boosters and crush the casing around the drill rod. The system develops 5000 psi. Air is supplied by the tanks to operate the boosters. The four valves are also air-operated with cylinders.

Fig. 15-5. Safety 2-hand machine operation.

Fig. 15-6. Emergency power unit.

Rotary Air Motors

Rotary air motors provide a smooth source of power and are not subject to damage due to overloads. Air motors can stall out for long periods of time, and there are no heat problems. Air motors stop and start with quick response. Pressure regulating and metering of flow provide infinitely variable torque and speeds.

The two most popular types are the vane and piston design. The radial piston designs are low speed, high torque motors and are generally fitted with four, five, and six reciprocating pistons (see Fig. 15–7).

$$\text{hp} = \frac{T \times \text{rpm}}{5252}$$

Power developed is dependent on inlet pressure, number of pistons, the area of the pistons, stroke, and speed. Horsepower ranges are available up to about 25 hp. The timing valve is generally linked to the crank shaft and automatically supplies air to the cylinder or exhausts the spent air from the cylinder.

The large reciprocating piston air motor is limited in speed by the inertia of the mechanical parts.

Axial piston air motors are smaller and develop greater speeds but are only available in smaller units up to about 3 hp. They are equipped with built-in lubrication.

Vane air motors are high speed units and deliver more horsepower per pound than piston motors.

Fig. 15-7. Reciprocating piston air motor.

Some vane motors operate up to 25,000 rpm, depending on the diameter of the rotor.

Generally, vane motors are equipped with a governor to prevent overspeed when loss of load occurs.

Air motor speeds are easy to adjust with an ordinary globe or needle valve.

Vane-type Air Motor

Figure 15–8 shows a rotary vane air motor and performance curve for its operating range.

Pneumatic Control Systems

In this age of space ships and computers, pneumatic control is taking on real significance for future developments.

There are many divisions of control which range from direct mechanical connection between the operator and the parts operated to "logic control," which is a relatively new field holding much promise as it is further developed and applied.

Logic control involves pneumatic devices such as actuators and pilot valves that provide means for obtaining the control functions of AND, OR, NOT, MEMORY, and TIME.

For practical purposes, logic might be compared to an actuating cylinder that advances but is blocked because the previous phase of the cycle failed to eject the machined part. If the cylinder stops and the machine jams, no logic is involved. However, if the control system senses the obstruction and automatically retracts, causing the previous failure to recycle, then logic is involved.

Fig. 15-8. Rotary vane air motor. *(Courtesy of Gast Manufacturing Co.)*

Logic, then, is the control system's ability to determine an alternative course of action when the course directed is proved undesirable.

Logic control can become extremely complex when the number of functions and components used increases.

The art of employing electricity for control has gained much experience throughout the years. People working in electronics have developed symbols and diagrams that can easily be read to trace down any small malfunctions within a highly complex control system.

Pneumatic control has many advantages and some disadvantages compared to electrical control. One of the major factors in favor of air is that it does not have the self-destructive characteristics of other systems that are subject to resistance and heat.

Specifically, air-piloted valves can operate indefinitely without malfunction compared to solenoids, which have a certain life expectancy and can burn out if overheated.

Air can provide a positive holding power for long periods of time. It will hold while the control system performs other functions or is static.

Air performs the function of work as well as control and does not require additional equipment to convert its energy to motive power.

It is reasonable to believe that if some of the diagraming techniques used for electrical control were employed for logic circuits, an improved method of control is possible that can be relatively uncomplicated.

Through the use of thermoplastic tubing and recently designed small components, air controls can be easily installed in compact metal boxes just as electrical switches, relays, etc. can.

The following example of logic control has been made available through the courtesy of Mr. Edward Holbrook, Clippard Instrument Laboratory, Inc. He is one of the outstanding authorities in pneumatic logic control systems.

Pneumatic Logic Control

The purpose of this section is not to produce all the answers or to offer pat solutions, but to indicate methods whereby control answers can be attained. Too, much is yet being learned about pneumatic systems that, in time, will lead to easier methods. Also, it is not intended to be a deep technical treatise, since it would serve no useful purpose and much of the mathematics is still more approximate than accurate.

It will be obvious that many variations and/or combinations can be made to suit the conditions that confront the engineer.

The diagrams illustrate means for obtaining the control functions of AND, OR, NOT, MEMORY, and TIME so that circuits can be assembled with a minimum of time. For the most part, these functions are produced with standard types of devices, such as two-way, three-way, four-way, shuttle, regulator valves, etc. However, the author would caution that varying spring values, friction differences, lubrication or lack of it, etc. can create problems that can be irritating, giving reactions that make a system inoperable or unduly sensitive. Most, if not all, of these problems can be solved, but without experience and prior knowledge the answer is not always readily apparent. A device can sometimes be improperly located relative to upstream and downstream volumes, for example, thus causing false signals that cause damage if sequencing is necessary and is used as the means for protection.

Much has been made of the comparison between pneumatic and electrical circuitry to a decided advantage, but the engineer must not lose sight of the fact that air as a medium operates differently, so that both its good and bad qualities must be taken into account. Actually, very few valves on the market today are good signal valves, because little has been done except to miniaturize known power valve structures except where special applications were handled with a device designed for that specific purpose.

Figure 1 illustrates the type of symbols used in the circuitry diagrams which are based on the block and hinge principle, with the control block diagram including the hinge to indicate flow paths.

A valuable point is to remember that all the control functions we get by pressuring lines can be obtained by venting lines. This can often be a means of simplifying an otherwise complicated system. It can be noted that the AND functions illustrated by Figure 2, when the lines are pressurized, become OR functions when these lines are vented. That is, in Figure 2a, all four incoming lines have to be filled prior to getting a pressure signal out at line 5, yet venting 1, 2, 3, or 4 will shut off the pressure at line 5.

The OR circuit, Figure 3b, becomes an AND circuit when the incoming signals are exhausted. That is, lines 1, 2, 3, or 4 will create pressure signal at 5, but *all* four incoming signals, if pressurized, must be vented before line 5 can be vented.

Figure 3b illustrates one of the many uses

DIAGRAM SYMBOLS

Figure 1

"AND" FUNCTION

1,2,3 AND 4 MUST BE
PRESSURIZED TO GET A
SIGNAL AT 5

(a)

SAME AS (a) EXCEPT TO ILLUSTRATE
THAT ANY ONE OF THE FOUR
INCOMING SIGNALS CAN EMERGE AS
THE OUTPUT

(b)

BOTH VALVES MUST BE ACTUATED
TO RELEASE 4 AND FILL 5

(c)

ALL 4 VALVES MUST BE ACTUATED TO PRODUCE
A WORKING SIGNAL AT 1

(d)

Figure 2

for NOT circuitry. If supply were introduced through valve A, air would fill lines 4, 5, and 6. Air in 6 would trip B to block line 1 and connect line 2 to the atmosphere. If, then, valve A is shifted rapidly, it would connect 4 to the atmosphere and fill line 1. Air would *not* flow through valve B until the pressure dropped below the spring value which, if very low (3 to 5 psi), would insure the full release of line 5 before 2 could be pressurized.

The shuttle valve OR function of Fig. 3b is obvious. Fig. 3d illustrates the use of two four-way valves to produce the effect of a three-way electric switch combination. As shown, supply flows through B into 1 and through A into 2. If A is moved, 2 will vent into 5 and 4 through B to the atmosphere, while the supply in 4 is directed to 3. If, from the position shown, B is shifted, line 1, therefore 2, is vented

while the supply in 1 is directed to 3. No matter what position A or B is in, the other valve will alter the air condition in lines 2 and 3.

In Fig. 3e, supply through lines 5 and 6 is maintained on both sides of valve G. If any one of bleeder valves A, B, or C is tripped, pressure in line 4 will drop (replacement retarded by choke H), causing a pressure differential across valve G which will shift to the left. Valves D, E, or F will cause G to shift to the right.

To indicate the "false signal" problem, refer to Fig. 4a and assume this to be a one-shot cycle. The signal (50 psi or more) coming into line 1 is from a cam valve that, when depressed, will be held for a period of time. The outgoing signal at 2 is to shift a double-piloted four-way valve, and we want that valve to be free to return *before* the signal is vented from line 1—hence the limited memory function. We will also assume that the three-way valve (B) available to us has a 40 psi return spring and that the pressure to shift the double-piloted four-way valve, due to detent or friction, is 10 psi.

Air enters line 1, 1-b and through B to line 2, to shift the four-way valve. Pressure, past flow control B, builds up above 40 psi to shift B and vent 2. This gives us a satisfactory circuit function. Later, the cam valve supplying line 1 is relieved and 1 begins to vent. Just below 40 psi, valve B shifts, and any air in 1 or 1-b expands into line 2. If we have placed valve B such that line 2 is 2 ft long and line 1 is that or greater, the pressure resulting from this expansion may exceed 10 psi so that the four-way valve, attached to line 2, will shift and, if the sequenced circuit is critical, damage can result. If the values are known, the easiest means to control this false signal is to be assured that line 2 is about four times the length of line 1 so that the balanced pressure, after expansion, is less than 10 psi, or below that which it takes to shift the four-way valve. The ideal signal valve for B would be one that resets at a much lower psi than it takes to trip, so that timing is easier and the critical positioning requirement is removed. Control functions like this can often occur in pneumatic circuitry and, although troublesome, can usually be removed by properly associating its position with that of other valves with which it must work—or by giving full consideration to the proper design of signal valves.

Figure 3

If the signal is momentary (line 3 of Fig. 4b), the small volume will be filled then as indicated in line 3a, valve B will trip, causing flow between 1 and 2. Flow control A allows free flow in but choked flow out, so this connection between 1 and 2 can be timed to meet the requirements.

Figure 4c illustrates a use for the limited memory in creating two complete cycles of a cylinder from the actuation of one valve. A is depressed to put air in line 1, through C and past shuttle (OR) H to extend the cylinder. B times the action at the end of which C is tripped to vent 5 and fill 2, through E and past shuttle I to return the cylinder. D times this eventually to trip E to vent 7 and fill 3, through G past H to extend the cylinder again. When F allows G to trip, 10 vents and 4 fills, past shuttle I to re-turn the cylinder. In the illustrated circuit, air also enters 9, delayed by choke J, finally to trip and reset A to vent the systems ready for the next cycle.

Fig. 4d is obvious, as a double-piloted valve is shifted one way or the other by a momentary signal. Fig. 4e illustrates one use where, when A is actuated to put a momentary signal in line 2, C shifts to put the supply to X. B will put the supply to Y.

Fig. 4f shows the lockup relay arrangement where a momentary signal from A into line 1 past shuttle B into line 2 shifts valve C to put supply into 3, 3a, and 3b so that when 1 is vented, air in 2 is replaced and C is locked up by its own supply. E will trip D to vent 3b and unlock the relay.

The AND and OR functions, as obtained by ex-hausting air, have already been explained. Fig. 5a

Figure 4

LIMITED MEMORY (TIME)

A CONTINUOUS INPUT AT 1 PRODUCES A TIMED OUTPUT AT 2 (a)

A MOMENTARY INPUT AT 3 PRODUCES A TIMED OUTPUT AT 2 (b)

(c)

FULL MEMORY

AIR IN 1 SHIFTS 2 TO 3

AIR IN 4 SHIFTS 3 TO 0 AND BLOCKS 2 (d)

PRESS "A" TO SHIFT "C" AND PUT SUPPLY AT X (e)

LOCK-UP AND UNLOCK SYSTEM

(f)

illustrates a limited memory or time function with venting. With air in line 1, A would be tripped and so would D, so that air in 3 is in line 5. If line 1 is vented, D would trip to block line 3 and vent 5. A is also released so that air in 3a flows, timed, through flow control B and eventually, past shuttle C, to trip D and restore the connection between 3 and 5.

Fig. 5b represents a full memory function actuated by exhausting air. With A as shown and air in 1 and 2, B is tripped and 4 is connected to 5. If line 1 is momentarily dropped, A will shift to vent 3, releasing B to block 4 and vent 5.

Fig. 5c represents the use of full memory in producing alternate cycling from one signal valve. When A is actuated, air in 1 flows through C to 2 to shift E, putting air in 5 to extend the cylinder—line 4 is vented. Air in 2 also trips B, so that air in line 5 is blocked. When A is released, venting line 1 and 2, B is released, so that air in 5 shifts C to connect 1 to 3 and 2 to exhaust. On the next trip of A, air is directed into line 3 to shift E, vent 5, and fill 4 to return the cylinder. D blocks 4 until A is released, then shifting C back. A, B, C, and D can all be small valves, whereas E is the power valve of a size sufficient to handle the cylinder involved.

Figure 7 illustrates the use of "full memory" and "OR" to produce a three-station control for handling waste cotton material. When chutes call for material, the air motors are started in the order in which the demands are made, even when all three call for material near the same time. Assume that station I

Figure 5

VENTING LINE 1 TRIPS *C* TO VENT 5 FOR A LIMITED TIME AS SET BY FLOW CONTROL *B*; THEN 3 IS RE-CONNECTED TO 5 **(a)**

MOMENTARY PRESSURE DROP IN 1 SHIFTS "*A*" TO VENT 3 AND TRIP *B* TO VENT 5 **(b)**

ALTERNATE REVERSE CYCLING (CONTINUAL ACTIONS ON *A* ALTERNATES DIRECTION OF PISTON TRAVEL) **(c)**

Figure 6. Logic circuit controls soap machine.

I   TO POWER VALVE

II   TO POWER VALVE

III   TO POWER VALVE

RV1   A   B   SV2

RV3   A   B   SV5

RV5   A   B   SV8

SV1   RV2   A   B   SV4   SV3

RV4   A   B   SV7   SV6

RV6   A   B   SV9

LV1   LV3   LV5

LV2   LV4   LV6

SUPPL

STATION I   STATION II   STATION III

(a)

Figure 7

(b)

(c)

is being filled; station III calls for material, and then station II before station I is filled. With station I operating, both LV–1 and LV–2 have been tripped, so that RV2 is shifted to put air to air motor I. This also puts air past SV5 and SV8 to hold RV3 and RV5 to the left. Air is also directed past SV6 to RV4 and SV9 to RV6. As the chute fills, LV–2 is released, but air on the upper side of SV–1 holds RV2 in place.

Material drops in station III to trip LV–5 to put air to RV6 and, prior to acting on LV–6, material drops in chute II to trip LV–3 and put air to RV4. If the material drops faster in station III, LV–6 will trip to put air to SV–7, but air on the other side of RV6 holds it in place. Air is also directed to SV–3 and RV2 is held (locked up). Air also goes to SV–6 to hold RV4 left.

When station I is filled, LV–1 is released, unlocking RV2 and releasing RV3 and RV5. The latter will shift to start air motor III, so when LV–4 is finally tripped it will have to wait its turn to start air motor II.

Figure 8 is an automatic bag control system. The initial signal clamps the bag in place, opens the hopper door for pouring, closes off at the designated weight, unclamps, and drops the bag. The operator

holds up the empty bag and tramps FV–1 to put air to P1–A to shift P–1 and clamp the bag in place. Air, timed by TD1, finally trips RV1 to put reduced pressure from the regulator to P2–A and P3–A— tripping both valves to fill lines 1 and 2 to open the hopper doors wide. As the weighing arm approaches near filling, LV–1 is tripped to put air past SV–2 to P3–B, which overcomes the lower pressure in P3A so P3 shifts to vent line 2. Air also passes SV3 into line 3 and the right-hand cylinder section, shifting it back to a point where the left cylinder-section rod stops it at two-thirds closed position, reducing the flow. When filled, LV–2 is tripped to put air past SV–1 to P2–B, overcoming lower pressure in P2–A to shift P2 and vent line 1, allowing the three-position cylinder to close off all flow fully. Air also goes to P1–B to vent the clamp cylinder and release the filled bag. RV1 is also tripped to vent P2–A and P3–A, so that when the weighing arm rises and releases LV–1 and LV–2 the hopper remains closed. The clamp cylinder is returned by the reduced pressure from regulator's being on the rod side. This system uses all five major control functions.

Figure 9 is a molding machine control system. To start the cycle, PB–1 puts a momentary signal

Figure 8

Figure 9

to P1–A to shift P–1 and extend cycle 1. This releases LV–1. Cycle 1 actuates LV–2 to put air through RV1 to P2–A and P5–A. P5 shifts to connect P–2 exhaust to P6–A. P2 shifts to extend cycle 2. Air also flows through TD–1 to RV2–A. LV–3 is released. Eventually, pressure in RV2–A overcomes the set pressure in RV2–B, and air goes to P3–A and P4A to shift P3 and P4. This turns on the vibrators and extends cycle 3. Air also also travels to P2B to shift P2, exhausting it through P5, P6, and the controlled exhaust of P6.

The pressure in P6A is much higher than P6B, allowing it to shift to exhaust. When the cylinder and P6A exhaust to a pressure slightly less than that in P6B, P6 shifts to block pressure in cycle 2 (about 5 psi). Incidentally, TD2 delayed the action on P3 and P4.

Although LV–4 was tripped as cycle–2 piston extended, its supply, coming from P3, was not available, so no action occurred. As the piston retracts, LV–4 is tripped, shifting P4 to cut off vibrators, and shifts P5 rapidly to vent the blocked 5 psi in cycle 2 for a rapid release. On the return, LV–3 is tripped and

shifts P–1 to return cycle 1. When cycle 1 returns, it trips LV–1 finally to, timed by TD3, shift P–3 to release cycle 3 and remove the supply from LV–3 and LV–4.

Figures 10a is a plastic injection molding machine. Fig. 10b and 10c illustrate the JIC symbols and the electric-ladder-type diagram. It is a continuous cycling system. Starting when TD4 completes its time and shifts RV–1 to vent line 15, P–3 shifts to cut off the air blast; RV3–B is vented; RV4–B is vented, and RV1–B is vented. Supply flows into line 2 through RV–4 into 3 and, after timing by TD3, P–1 is shifted to extend the clamp cylinder; when the pressure in line 4 exceeds the set pressure (regulator 2) in line 6, RV2 is shifted to put the pressure in 3 into line 9 and 9a through RV3 into line 10b, past SV–1 into 10 to shift P–2 and extend the injection cylinder. Air also enters TD1 finally to trip RV3 to vent 10b and 10 to retract the injection cylinder and also put air in line 12. TD2 times out and finally shifts RV4 to vent line 3 and re-trip P–1 to return the clamp cylinder. Air also enters line 13, delayed slightly by C–1, and trips RV1 to vent 2 and fill 15. This trips RV4 to

Figure 10

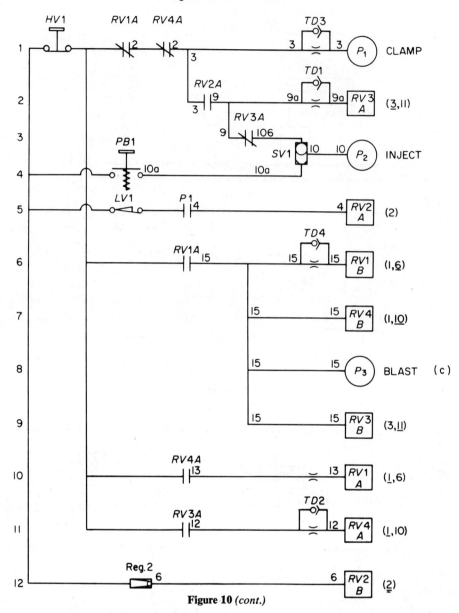

Figure 10 *(cont.)*

vent 13, trips RV3 to vent 12, and trips P–3 to turn on the air blast. TD4 times out to stop the blast and start the cycle again.

Questions

1. What are some of the limitations in using pneumatics for applications requiring large forces?

2. What kind of applications operate most successfully with pneumatic power?

3. Draw the graphical diagram for a pneumatic circuit using linear motion, and list all of the necessary components. Draw the graphical diagram for a pneumatic circuit using rotary motion and list the components necessary.

4. What are some of the advantages of using logic systems for control?

5. What is meant by fluidics, and what are your predictions concerning the future of this field?

THE "HOW OF PNEUMATIC
CONTROL DESIGN"

E. L. HOLBROOK

THE PROBLEM

3 CYLINDERS TO OPERATE CONTINUOUSLY WHEN
TURNED ON WITH A SEQUENCE AS FOLLOWS

CONDITIONS
CYLINDERS ARE HYDRAULIC EACH
WITH A DOUBLE AIR PILOTED 4-WAY VALVE

THERE IS SPACE FOR CAM VALVES

A - CYL. 1 → OUT
B - CYL. 2 → OUT
C - CYL. 3 → OUT
D - CYL. 2 ← IN
E - CYL. 1 ← IN
F - CYL. 2 → OUT
G - CYL. 3 ← IN
H - CYL. 2 ← IN
I - CYL. 1 → OUT ETC.

POSITION SPREAD

CYL 1   CYL 2   TURN CYL. 3

I SET UP WHAT IS KNOWN - (SHADED)

II INSERT CAM VALVES

III FILL IN ACTION

START VALVE

HP 1   CYL. 1   LV 5   LV 1

REDIRECT   OR

HP 2   CYL. 2   LV 4   LV 2

OR

REDIRECT

HP 3   CYL. 3   LV 6   LV 3

SEE SHEET 2 FOR
EXPLANATION

317

HOW OF PNEUMATIC CONTROL DESIGN

E. L. HOLBROOK

(I) SETTING UP WHAT IS KNOWN MEANS THE
CYLINDERS & VALVES -- HP1, HP2, HP3 & CYLS 1, 2 & 3.

(II) 6 CAM VALVES CAN BE INSERTED. WE MIGHT HAVE USED
8 (2 AT LV4 POS. & 2 AT LV2 POS.) - THEN MADE
THEM ALTERNATELY "LIVE" OR "DEAD" - HOWEVER, THERE
WAS A SPACE LIMITATION AND, IN ANY EVENT, IT
WOULD NOT EASE THE PROBLEM

(III) FILL IN THE ACTION

A	THE "START VALVE" CAN FILL LINE 1 TO TRIP HP1 - CYL 1 OUT →
B	LV 1 FILLS LINE 2 TO TRIP HP2 ---- CYL. 2 OUT →
C	LV 2 FILLS LINE 3 TO TRIP HP 3 ---- CYL 3 OUT →
D	LV3 FILLS LINE 4 TO TRIP HP2 ---- CYL. 2 IN ←
E	LV5 FILLS LINE 5 TO TRIP HP 1 ---- CYL. 1 IN ←
F	LV6 FILLS LINE 6 TO TRIP HP2 ---- CYL. 2 OUT →
G	LV2 FILLS LINE 7 TO TRIP HP3 ---- CYL. 3 IN ←
H	LV6 FILLS LINE 8 TO TRIP HP2 ---- CYL. 2 IN ←
I	LV5 FILLS LINE 9 TO TRIP HP 1 ---- CYL. 1 OUT → AGAIN.

THE ABOVE, OBVIOUSLY, CANNOT WORK DUE TO INTERFERING
SIGNALS.
LV2 & LV5 MUST, ALTERNATELY, SHIFT HP2 — (OUT) NEED OR - A SHUTTLE
LV6 & LV3   "   )   "   ,   " HP2 (IN)   "   "   — "   "
LV2 MUST OP. HP3 OUT & IN — NEED RE-DIRECTED SIGNAL,
LV5   "   "   HP1   "   "   "   —   "   "   "   "   )
CERTAIN LINES MUST BE CHANGED FROM CONTINUOUS TO
MOMENTARY SIGNALS TO FREE POWER VALVES TO SHIFT.
LINE 9 AND LINE 5 TO FREE HP1 (OUT & IN)
LINE 2 AND LINE 6 TO FREE HP2 (OUT)
LINE 4 AND LINE 8 TO FREE HP2 (IN)

NEXT STEP - INSERT ALL VALVES THEN LOCATE.
POINTS TO SIGNAL RE-DIRECT VALVES TO POSITIONS
REQUIRED AT THE PROPER TIME FOR ACTION

318

TV1 (START)

ONE SHOT CYCLE

A AIR COMES IN (a) OUT
  AT (c) BUILDS UP
  IN (b)

B BUILD UP IN (b) TRIPS VALVE
  TO BLOCK (a) AND VENT (c)

SINCE ALTHOUGH LV4 IS TRIPPED (AIR IN LINE 9) RV2 (ONE-SHOT) WOULD
VENT LINE 9(a) - NO AIR FOR INITIAL START - INSERT PBI & SHUTTLE

A MOMENTARY TRIP OF PBI, PASSED SV3, THRU TVI TO HPI - CYL 1 OUT →

B CYL 1 TRIPS LVI, AIR IN 2, THRU RV4, 2 (a), PASSED SVI TO
         HP2 (RV4 TRIPS TO VENT 2(a) - - - - - CYL 2 OUT →

C LV2 TRIPPED BY CYL. 2, AIR IN LINE 3 (&7) THRU
         RV7 TO LINE 3(a), HP3 -   - - - - - CYL 3 OUT →

D LV3 TRIPPED, AIR IN LINE 4, THRU RV5, LINE 4(a)
         PASSED SV2 TO HP2 (RV5 VENTS 4(a) CYL. 2 IN ←

E LV4 TRIPPED, AIR IN 5(&9) THRU RV2, THRU RVI (SHOULD HAVE
         BEEN SHIFTED) TO LINE 5, HPI - - - - CYL 1 IN ←

F LV5 TRIPPED, AIR IN 6, THRU RV3 TO 6(a), PASSED SVI
         TO HP2 (RV3 TRIPS TO VENT 6(a)) - - CYL. 2 OUT →

G LV2 TRIPPED, AIR IN LINE 7 (&3), THRU RV7 (SHOULD HAVE BEEN
         SHIFTED) TO LINE 7(a) TO HP3   - - - - CYL. 3 IN ←

H LV6 TRIPPED, AIR IN 8, THRU RV6 TO 8(a) PASSED SV2
         TO HP2 (RV6 SHIFTS TO VENT 8(a)) - - - - CYL. 2 IN ←

I LV4 TRIPPED, AIR IN 9 (&5) THRU RV2 TO 9(a) (&5(a)), THRU
         RVI (SHOULD HAVE BEEN SHIFTED) TO 9(a)
         PASSED SV3, THRU OPEN TVI (CLOSED IF STOP REQUIRED) TO HPI - CYL. 1 OUT →

NOTE - SIGNALS NEEDED TO SHIFT, CORRECTLY, RVI & RV2
AND A CHECK TO REMOVE UNNECESSARY VALVING.
① COULD USE LINE I TO SHIFT RVI TO CONN. 5&9(a) TO 5(a)
② COULD USE LINE 4(a) TO SHIFT RV7 TO CONN. 3&7 TO 7
③ CAN USE ONE VALVE IN LINE 6&2 (BEYOND SVI) IN PLACE
         OF RV3 & RV4
④ CAN USE LINE 6 (OR 6(a)) TO RESHIFT RVI
⑤ CAN USE LINE 8 TO RESHIFT RV7
⑥ SINCE BOTH SIGNALS (FROM LV6 & LV3) MUST FREE RV7,
   MUST MAINTAIN SEPARATE ONE-SHOT FUNCTIONS (RV5 & RV6)

MAKE ABOVE CHANGES & CHECK

RETURN TO PAGE 4 & THE
A TO I EXPLANATION

A - ADD "AIR FROM LINE 1
TO 21 TRIPS RV1 TO CONN.
5 & 9(a) TO 5(a)"
[RV3 GIVES ONE-SHOT CYCLE

B NOW TO BOTH LINES
6 & 2 AFTER PASSING
SHUTTLE SV1 (RV4 REMOVED)

C OK AS IS

D ADD "AIR FROM 4-a TO
24(a) TO SHIFT RV7
AND CONN. 3 & 7 TO 7(a)"
[CHOKE C-3 GIVES SLIGHT DELAY]

E OK AS IS (RV1 WAS SHIFTED AT A)

F ADD" AIR FROM LINE 6 TO 26 (DELAYED BY CHOKE C-1) RESHIFTS
RV1 TO CONN. 5 & 9 (a) TO 9(a)"

G OK AS IS (RV7 WAS SHIFTED IN D)

H ADD "AIR FROM 8(a) TO 28(a) DELAYED BY CHOKE C-2 RESHIFTS
RV7 TO CONN. 3 & 7 TO 3(a)"

Q.E.D.

E.L. HOLBROOK

Fluidics

Fluidics may be thought of as the technology primarily concerned with the use of *fluids in motion, without moving parts,* which perform sensing control and/or actuation functions. Devices of this type operate on the principles of various fluid phenomena, some of which are the Coanda effect, which is the tendency of a stream to attach itself to a wall; the laminar-to-turbulent transition of flow, and the interaction of flow streams. Devices of this type promise to expand the field of fluid power.

Some of the advantages of fluidic devices are

1. Less expensive than comparable conventional mechanical, fluid power, or electrical means, since fewer components are required.
2. Long life, because there are no moving parts to wear.
3. Essentially insensitive to shock and vibration.
4. Impervious to high and low temperature environments.
5. Not affected by nuclear or electromotive radiation fields.
6. Will tolerate overloading—proper operation restored when the overload is removed.
7. No internal heat generated.
8. High degree of reliability.
9. Compact size.
10. May be fabricated from numerous materials.
11. Explosion-proof.

Limitations of fluidic devices include

1. Response too slow for sophisticated computer systems (millisecond response as compared with micro- or nanosecond response of electronics). Response is limited in gases by sonic choking (approximately Mach 0.7), and in liquids by cavitation velocity. The response time of fluidic devices is, however, more than adequate for present fluid power applications, and even for slow speed computers.
2. Frequency interference caused by high noise or vibration levels may have to be "masked" from the system.
3. Contamination may prove to be troublesome in the more sophisticated systems, where output levels must be maintained within narrow limits.
4. Lack of theoretical and analytical information regarding fluid phenomena. Many present developments must be cut-and-try.
5. Numerous present fluidic devices require quiescent (standby) losses—that is, continuous supply whether system is "on" or "off."

Vortex Amplifier

The vortex amplifier operates on the principle of conservation of momentum. The power flow of this device, operating with no control signal, is from the input duct directly to the output duct. As a signal flow is introduced, the power flow takes the vortex path to the output duct. An action of this type of amplifier is also proportional, giving us another analog device.

Turbulence Amplifier

The turbulence amplifier uses a low velocity laminar stream between two pipes to give an output signal. A very low pressure control signal direct against this laminar stream causes it to become turbulent. When this flow is turbulent, no output signal is recaptured and the output of the turbulence amplifier is negligible. This device is used as a bi-stable digital device.

Focused Jet Amplifier

The focused jet amplifier operating without a control signal will have its power jet flow collect in the output duct, producing an output flow. With the injection of a control signal, the focusing of the power jet is disrupted, thereby causing the power jet flow to the outside of the output duct. This device is also a digital device.

Direct Impact Modulator

The operation of the direct impact modulator uses two opposed jets impacting on one another. A control signal will focus power input P_1 causes of impact to move toward power input P_2. As the line of impact approaches the recapture edge, a proportional rise in output pressure will result. High pressure gains can be achieved by using the direct impact modulator. Pressure gains of 8000 have been obtained by cascading two of these devices.

Induction Amplifier

The induction amplifier uses boundary layer control to give bi-stable logic operation. The power jet enters the device and is acted upon by a control flow which flows tangentially to the power jet. This control flow lowers the pressure in its output duct and draws the power jet flow to this duct. The wall attachment principle maintains this flow until it is acted upon by the other control jet.

Double Leg Elbow Amplifier

Another digital device using a modified form of momentum exchange is the double leg elbow amplifier (DLEA). This device was developed by Giannini Controls Corporation and is considered to be the best flow amplifier at the present time.

As the flow is passing through the curved portion of the active leg, it tends to pull away from the inner wall of the device toward the right output duct. Flow from the passive leg provides another stream, which interacts with the active stream. With no control flow, the passive stream directs the active stream to the left output duct. As a control signal is introduced, the passive flow is unable to force all of the active flow to the left output, allowing a portion of this flow to exit at the right output duct. An increase in control flow gives a proportionally increased output at the right output duct. This device gives high flow gains and high power gains of 360:1.

Stream Interaction Amplifier

One type of proportional fluid amplifier is illustrated in Fig. 1–A. A graph of this stream interaction amplifier is shown in Fig. 1–B. This graph gives the amplifier's performance characteristics.

This amplifier operates on the principle of momentum exchange. As these streams act on one another, the result will be proportional to the momenta of the control streams acting on the power input stream. The power input supplies the power fluid to the output ducts. With no control inputs, the flow is evenly divided to the outputs. When a differential flow is applied in the control ducts, a deflection of the power jet will be away from the control duct with the highest flow. The action of this amplifier makes it an analog device.

Fluidics Terminology

Active A fluidic device which is directly attached to the power supply.

Ambient The broad term covering environmental conditions around the area of the device; i.e., atmospheric pressure or temperature or both.

Analog The general class of fluidic elements or circuits having proportional characteristics.

And Basic logic device which has an output when all control signals are on.

Beam deflector A proportional fluidic device which uses control pressure and/or flow to deflect a fluid jet or beam for the functional operating principle.

Bi-stable A fluidic device having two separate and distinct outputs, such as "off-and-on" or "high-and-low," etc.

Circuits Arrays of interconnected fluidic elements which perform specialized functions or standard functions.

Digital The general class of fluidics or circuits which have "on-off" characteristics.

Digital amplifier A digital element which provides amplification and has negligible hysteresis (i.e., no memory); therefore, a continual input signal is required to maintain a given state of bi-stable.

Diode A fluidic element which has higher resistance to flow in one direction compared to that in the opposite direction (a check valve).

Diverting A type of control where the flow is continuous and the normal output of the device is turned off by diverting the flow to another unused port.

Elements Fluidic devices and/or logic devices which are interconnected to form working circuits. Also the general class of devices used in conjunction with fluidic devices in circuits, such as fluid restrictors and capacitors.

Enclosed proportional amplifier A closed fluidic device which has no communication with the local ambient.

Fanout The number of elements which can be driven by a single element. All elements to be operated at the same supply pressure. Also, elements to be of similar size and to have similar switch points.

Flip-flop A digital fluidic element which has a hysteresis loop of sufficient width so that it has "memory." Its state is changed with an input pulse; a continual input signal is not necessary for it to exhibit by stability.

Flow amplifier An element designed specifically for amplifying flow signals.

Fluidics The general field of fluid no-moving-parts control and related logic devices.

Fluidic capacitor A fluidic impedance element in which pressure drop lags flow into it by essentially 90 deg.

Fluidic resistor The fluidic element which produces a pressure drop as a function of the flow through it.

Fluidic valve A fluidic device which has the end use of diverting or modulating flow.

Hysteresis The total width of hysteresis loop expressed as a percentage of peak-to-peak saturation input signal. Measurement to be at frequencies below those where dynamic effects become significant.

Inches of water A term used to define very low pressures, relating to reading the low pressures on a water column manometer. (27.7 in. H_2O = 1 PSIG.)

Inverter A basic device relating to NOT logic which has an output contrary to its single input signal.

Isolation The characteristic of a fluidic device having more than one control port, where the control signal at one of the control ports in no way affects any of the other control ports. Also, having to do with the prevention of the control pressure at one port's working back to some other portion of the logic circuit.

Laminar A type of flow of a liquid or gas wherein the fluid moves in parallel layers, the velocities of these layers not necessarily being the same.

Logic The science of correct reasoning; science which deals with the criteria of valid thought.

Logic elements The general category of fluidic devices which provide logic functions such as AND, OR, and NOR. They can gate (stop or inhibit) signal transmission with the application, removal, or other combination of input signals.

Modulating A modulating flow device is one which is capable of varying the flow output, but never completely to shut the flow off.

Nor A basic logic device which has an output when all control inputs are off.

Not A basic logic device which has an output signal when the single input signal is "off" and has no output when the signal is "on."

Or A basic logic device which has an output if any of its multiple controls have an input.

Passive A fluidic device which operates on input power derived from the output of another fluidic device.

Power amplifier An element optimized to provide maximum power gain. That is the product of pressure and flow.

Pressure amplifier An element designed specifically for amplifying pressure signals.

Proportional The general class of fluidic elements or circuits having analoging characteristics; i.e., output is proportional to input.

Response time The time delay between the application of an input step signal and the resulting output step signal. (The time measurements for the step signals are to be made at the 50 per cent of final value point.)

Sensor A fluidic element which, in general, senses quantities or media other than those related to fluids and converts them into fluid quantities.

Transducer A device which converts signals in other than fluid media into fluid signals (i.e., pressure switch which converts a pneumatic signal into an electrical signal, or a solenoid valve which converts an electrical signal into a pneumatic signal).

Turbulence amplifier A fluidic digital element which uses for the basic operating principle the laminar-to-turbulent transition of a jet controlled by input signal flow.

Turbulent A type of flow wherein the velocity at any point may vary both in magnitude and in direction—a "mixed-up" type of flow.

Vented proportional amplifier An open fluidic amplifier utilizing vents to establish the ambient pressure in the interaction region.

Vortex A fluidic device which utilizes the pressure drop across a vortex for the modulating principle (whirlpool type).

Wall attachment devices Fluidic devices which use jet attachment to a wall (Coanda effect) for the basic operating principle.

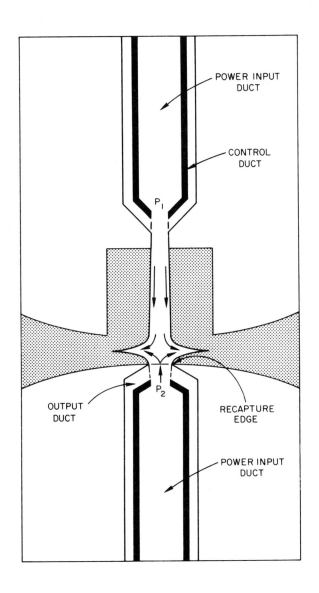

POWER INPUT
DUCT

CONTROL
DUCT

P_1

OUTPUT
DUCT

P_2

RECAPTURE
EDGE

POWER INPUT
DUCT

OUTPUT PRESSURE PSIG

8
7
6
5
4
3
2
1
0

14
12
10
8
$P_1 = 6$

0.25   0.50
INPUT PRESSURE PSIG

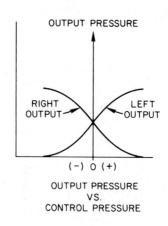

OUTPUT PRESSURE
VS.
CONTROL PRESSURE

SUPPLY MASS FLOW = 4.56 × 10^{-3} LB/SEC

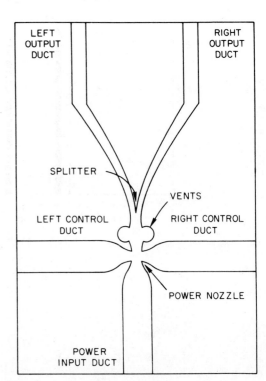

328

FLUIDIC DEVICES CROSS REFERENCE

LOGIC ELEMENT	NOR	AND	NOT	OR	FLIP FLOP	NAND
OUTPUT IF	ALL CONTROL SIGNALS ARE 0	ALL CONTROL SIGNALS ARE 1	SIGNAL CONTROL SIGNAL IS 0	ANY ONE OF THE CONTROLS IS 1	ONE CONTROL IS 1 (HOLDS AFTER THAT IS 0 & UNTIL OPPOSITE CONTROL IS 1	OUTPUT = 0 IF ALL CONTROLS ARE 1
LOGIC SYMBOL (NEMA)						
LOGIC SYMBOL (STD.)	Z Z		Z		FF	Z Z
BOOLEAN ALGEBRA OPERATOR	( ) . ( )	( ) . ( )	(	( ) + ( )	—	) / ( )
ELECTRICAL (RELAY)					CR1 CR2	
ELECTRICAL (SWITCH)						
ASA (J.I.C.) VALVING	P.	P.	P.	P.	P.	P.
FLUIDIC DEVICE TURBULENCE AMPLIFIER						

329

Graphic Symbols for Fluid Power Diagrams

1. Introduction

1.1 General

Fluid power systems are those that transmit and control power through use of a pressurized fluid (liquid or gas) within an enclosed circuit.

Types of symbols commonly used in drawing circuit diagrams for fluid power systems are Pictorial, Cutaway, and Graphic. These symbols are fully explained in the USA Standard Drafting Manual (Ref. 2).

1.1.1 *Pictorial symbols* are very useful for showing the interconnection of components. They are difficult to standardize from a functional basis.

1.1.2 *Cutaway symbols* emphasize construction. These symbols are complex to draw and the functions are not readily apparent.

1.1.3 *Graphic symbols* emphasize the function and methods of operation of components. These symbols are simple to draw. Component functions and methods of operation are obvious. Graphic symbols are capable of crossing language barriers, and can promote a universal understanding of fluid power systems.

Graphic symbols for fluid power systems should be used in conjunction with the graphic symbols for other systems published by the USA Standards Institute (Ref. 3–7 inclusive).

1.1.3.1 Complete graphic symbols are those which give symbolic representation of the component and all of its features pertinent to the circuit diagram.

1.1.3.2 Simplified graphic symbols are stylized versions of the complete symbols.

1.1.3.3 Composite graphic symbols are an organization of simplified or complete symbols. Composite symbols usually represent a complex component.

1.2 Scope and Purpose

1.2.1 *Scope*

This standard presents a system of graphic symbols for fluid power diagrams.

1.2.1.1 Elementary forms of symbols are:

Circles	Triangles	Lines
Squares	Arcs	Dots
Rectangles	Arrows	Crosses

1.2.1.2 Symbols using words or their abbreviations are avoided. Symbols capable of crossing language barriers are presented herein.

1.2.1.3 Component function rather than construction is emphasized by the symbol.

1.2.1.4 The means of operating fluid power components are shown as part of the symbol (where applicable).

1.2.1.5 This standard shows the basic symbols, describes the principles on which the symbols are based, and illustrates some representative composite symbols. Composite symbols can be devised for any fluid power component by combining basic symbols.

Simplified symbols are shown for commonly used components.

1.2.1.6 This standard provides basic symbols which differentiate between hydraulic and pneumatic fluid power media.

1.2.2 *Purpose*

1.2.2.1 The purpose of this standard is to provide a system of fluid power graphic symbols for industrial and educational purposes.

1.2.2.2 The purpose of this standard is to simplify design, fabrication, analysis, and service of fluid power circuits.

1.2.2.3 The purpose of this standard is to provide fluid power graphic symbols which are internationally recognized.

1.2.2.4 The purpose of this standard is to promote universal understanding of fluid power systems.

1.3 Terms and Definitions

Terms and corresponding definitions found in this standard are listed in Ref. 8.

2. Symbol Rules (See Section 10)

2.1 Symbols show connections, flow paths, and functions of components represented. They can indicate conditions occurring during transition from one flow path arrangement to another. Symbols do not indicate construction, nor do they indicate values, such as pressure, flow rate, and other component settings.

2.2 Symbols do not indicate locations of ports, direction of shifting of spools, or positions of actuators on actual component.

2.3 Symbols may be rotated or reversed without altering their meaning except in the cases of: a.) Lines to Reservoir, 4.1.1; b.) Vented Manifold, 4.1.2.3; c.) Accumulator, 4.2.

2.4 Line Technique (See Ref. 1)

Keep line widths approximately equal. Line width does not alter meaning of symbols.

2.4.1 Solid Line

_____

(Main line conductor, outline, and shaft)

2.4.2 Dash Line

___ ___ ___ ___ ___ ___

(Pilot line for control)

2.4.3 Dotted Line

_ _ _ _ _ _ _ _ _ _ _ _

(Exhaust or Drain Line)

2.4.4 Center Line

_____ _ _____ _ _____

(Enclosure outline)

2.4.5 Lines Crossing

(The intersection is not necessarily at a 90 deg angle.)

or

IEC

IEC

2.4.6 Lines Joining

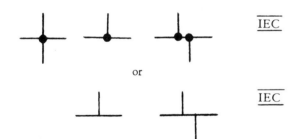

IEC

or

IEC

2.5 Basic symbols may be shown any suitable size. Size may be varied for emphasis or clarity. Relative sizes should be maintained. (As in the following example.)

2.5.1 Circle and Semi-Circle

2.5.1.1 Large and small circles may be used to signify that one component is the "main" and the other the auxiliary.

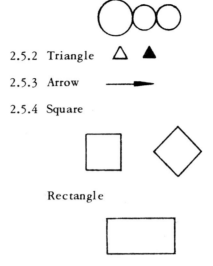

2.5.2 Triangle △ ▲

2.5.3 Arrow ——▶

2.5.4 Square

Rectangle

2.6 Letter combinations used as parts of graphic symbols are not necessarily abbreviations.

2.7 In multiple envelope symbols, the flow condition shown nearest an actuator symbol takes place when that control is caused or permitted to actuate.

2.8 Each symbol is drawn to show normal, at-rest, or neutral condition of component unless multiple diagrams are furnished showing various phases of circuit operation. Show an actuator symbol for each flow path condition possessed by the component.

2.9 An arrow through a symbol at approximately 45 degrees indicates that the component can be adjusted or varied.

2.10 An arrow parallel to the short side of a symbol, within the symbol, indicates that the component is pressure compensated.

2.11 A line terminating in a dot to represent a thermometer is the symbol for temperature cause or effect.

See Temperature Controls 7.9, Temperature Indicators and Recorders 9.1.2, and Temperature Compensation 10.16.3 and 4.

2.12 External ports are located where flow lines connect to basic symbol, except where component enclosure symbol is used.

External ports are located at intersections of flow lines and component enclosure symbol when enclosure is used, see Section 11.

2.13 Rotating shafts are symbolized by an arrow which indicates direction of rotation (assume arrow on near side of shaft).

3.   Conductor, Fluid

3.1 Line, Working (main)

3.2 Line, Pilot (for control)

3.3 Line, Exhaust and Liquid Drain

3.4 Line, sensing, etc. such as gage lines shall be drawn the same as the line to which it connects.

3.5 Flow, Direction of

3.5.1 Pneumatic

3.5.2 Hydraulic

3.6 Line, Pneumatic
Outlet to Atmosphere

3.6.1 Plain orifice, unconnectable

3.6.2 Connectable orifice (e. g. Thread)

3.7 Line with Fixed Restriction

3.8 Line, Flexible

3.9 Station, Testing, measurement, or power take-off

3.9.1 Plugged port

3.10   Quick Disconnect

3.10.1 Without Checks

Connected

Disconnected

3.10.2 With Two Checks

Connected

Disconnected

3.10.3 With One Check

Connected

Disconnected

3.11 Rotating Coupling

4. Energy Storage and Fluid Storage

4.1 Reservoir

Vented

Pressurized

Note: Reservoirs are conventionally drawn in the horizontal plane. All lines enter and leave from above. Examples:

4.1.1 Reservoir with Connecting Lines

Above Fluid Level

Below Fluid Level

4.1.2 Simplified symbol

The symbols are used as part of a complete circuit. They are analogous to the ground symbol

of electrical diagrams. ———|⊢ IEC. Several such

symbols 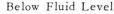 may be used in one diagram to represent the same reservoir.

4.1.2.1 Below Fluid Level

4.1.2.2 Above Fluid Level

(The return line is drawn to terminate at the upright legs of the tank symbol.)

4.1.2.3 Vented Manifold

* Show line entering or leaving below reservoir only when such bottom connection is essential to circuit function.

4.2 Accumulator

4.2.1 Accumulator, Spring Loaded

4.2.2 Accumulator, Gas Charged

4.2.3 Accumulator, Weighted

4.3 Receiver, for Air or Other Gases

4.4 Energy Source
(Pump, Compressor, Accumulator, etc.)

This symbol may be used to represent a fluid power source which may be a pump, compressor, or another associated system.

Hydraulic

Pneumatic

Simplified Symbol

Example:

5. Fluid Conditioners

Devices which control the physical characteristics of the fluid.

5.1 Heat Exchanger

5.1.1 Heater

Inside triangles indicate the introduction of heat.

Outside triangles show the heating medium is liquid.

Outside triangles show the heating medium is gaseous.

5.1.2 Cooler

                or

Inside triangles indicate heat dissipation

(Corners may be filled in to represent triangles.)

5.1.3 Temperature Controller
(The temperature is to be maintained between two predetermined limits.)

              or

5.2 Filter – Strainer

5.3 Separator

5.3.1 With Manual Drain

5.3.2 With Automatic Drain

5.4 Filter – Separator

5.4.1 With Manual Drain

5.4.2 With Automatic Drain

5.5 Dessicator (Chemical Dryer)

5.6 Lubricator

5.6.1 Less Drain

5.6.2 With Manual Drain

6. Linear Devices

6.1 Cylinders, Hydraulic & Pneumatic

6.1.1 Single Acting

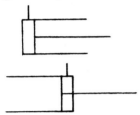

6.1.2 Double Acting

6.1.2.1 Single End Rod

6.1.2.2 Double End Rod

6.1.2.3 Fixed Cushion, Advance & Retract

6.1.2.4 Adjustable Cushion, Advance Only

6.1.2.5 Use these symbols when diameter of rod compared to diameter of bore is significant to circuit function.

(Non-Cushion)

(Cushion, Advance & Retract)

6.2 Pressure Intensifier

6.3 Servo Positioner (Simplified)

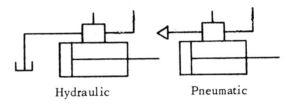

Hydraulic　　　　Pneumatic

6.4 Discrete Positioner

Combine two or more basic cylinder symbols.

7. Actuators and Controls

7.1 Spring

7.2 Manual

(Use as general symbol without indication of specific type; i.e., foot, hand, leg, arm)

7.2.1 Push Button

7.2.2 Lever

7.2.3 Pedal or Treadle

7.3 Mechanical

7.4 Detent

(Show a notch for each detent in the actual component being symbolized. A short line indicates which detent is in use.) Detent may, for convenience, be positioned on either end of symbol.

7.5 Pressure Compensated

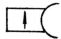

7.6 Electrical

7.6.1 Solenoid (Single Winding)

7.6.2 Reversing Motor

7.7 Pilot Pressure

7.7.1

Remote Supply

7.7.2

Internal Supply

7.7.3 Actuation by Released Pressure

by Remote Exhaust

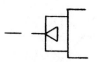

by Internal Return

7.7.4 Pilot Controlled, Spring Centered

Simplified Symbol

Complete Symbol

7.7.5 Pilot Differential

Simplified Symbol

Complete Symbol

7.8 Solenoid Pilot

7.8.1 Solenoid or Pilot

External Pilot Supply

Internal Pilot Supply and Exhaust

7.8.2 Solenoid and Pilot

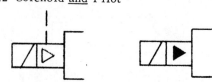

7.9 Thermal

A mechanical device responding to thermal change.

7.9.1 Local Sensing

7.9.2 With Bulb for Remote Sensing

7.10 Servo

(This symbol contains representation for energy input, command input, and resultant output.)

7.11 Composite Actuators (and, or, and/or)

Basic One signal only causes the device to operate.

And One signal <u>and</u> a second signal both cause the device to operate.

Or One signal <u>or</u> the other signal causes the device to operate

And/Or The solenoid <u>and</u> the pilot <u>or</u> the manual override alone causes the device to operate.

The solenoid <u>and</u> the pilot <u>or</u> the manual override <u>and</u> the pilot

The solenoid <u>and</u> the pilot <u>or</u> a manual override <u>and</u> the pilot <u>or</u> a manual override alone.

8. Rotary Devices

8.1 Basic Symbol

8.1.1 With Ports

8.1.2 With Rotating Shaft, with control, and with Drain

8.2 Hydraulic Pump

8.2.1 Fixed Displacement.

8.2.1.1 Unidirectional

8.2.1.2 Bidirectional

8.2.2 Variable Displacement, Non-Compensated

8.2.2.1 Unidirectional

Simplified

Complete

8.2.2.2 Bidirectional

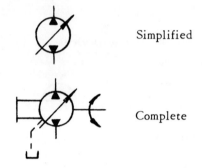

Simplified

Complete

8.2.3 Variable Displacement, Pressure Compensated

8.2.3.1 Unidirectional

Simplified

Complete

8.2.3.2 Bidirectional

Simplified   Complete

8.3 Hydraulic Motor

8.3.1 Fixed Displacement

8.3.1.2 Bidirectional

8.3.2 Variable Displacement

8.3.2.1 Unidirectional

8.3.2.2 Bidirectional

8.4 Pump-Motor, Hydraulic

8.4.1 Operating in one direction as a pump. Operating in the other direction as a motor.

8.4.1.1 Complete Symbol

8.4.1.2 Simplified Symbol

8.4.2 Operating one direction of flow as either a pump or as a motor.

8.4.2.1 Complete Symbol

8.4.2.2 Simplified Symbol

8.4.3 Operating in both directions of flow either as a pump or as a motor.
(Variable displacement, pressure compensated shown)

8.4.3.1 Complete Symbol

8.4.3.2 Simplified Symbol

8.5 Pump, Pneumatic

8.5.1 Compressor, Fixed Displacement

8.5.2 Vacuum Pump, Fixed Displacement

8.6 Motor, Pneumatic

8.6.1 Unidirectional

8.6.2 Bidirectional

8.7 Oscillator

8.7.1 Hydraulic

8.7.2 Pneumatic

8.8 Motors, Engines

8.8.1 Electric Motor

          $\overline{\text{IEC}}$

8.8.2 Heat Engine (E. G. internal combustion engine)

9.   Instruments and Accessories

9.1 Indicating and Recording

9.1.1 Pressure

9.1.2 Temperature

9.1.3 Flow Meter

9.1.3.1   Flow Rate

9.1.3.2   Totalizing

9.2 Sensing

9.2.1 Venturi

9.2.2 Orifice Plate

9.2.3 Pitot Tube

9.2.4 Nozzle

Hydraulic   Pneumatic

9.3 Accessories

9.3.1 Pressure Switch

9.3.2 Muffler

10. Valves

A basic valve symbol is composed of one or more envelopes with lines inside the envelope to represent flow paths and flow conditions between ports. Three symbol systems are used to represent valve types: single envelope, both finite and infinite position; multiple envelope, finite position; and multipe envelope, infinite position.

10.1 In infinite position single envelope valves, the envelope is imagined to move to illustrate how pressure or flow conditions are controlled as the valve is actuated.

10.2 Multiple envelopes symbolize valves providing more than one finite flow path option for the fluid. The multiple envelope moves to represent how flow paths change when the valving element within the component is shifted to its finite positions.

10.3 Multiple envelope valves capable of infinite positioning between certain limits are symbolized as in 10.2 above with the addition of horizontal

bars which are drawn parallel to the envelope. The horizontal bars are the clues to the infinite positioning function possessed by the valve re-represented.

10.4 Envelopes

10.5 Ports

10.6 Ports, Internally Blocked

Symbol System 10.1

Symbol System 10.2

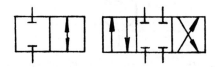

10.7 Flow Paths, Internally Open (Symbol System 10.1 and 10.2)

Symbol   Symbol
System   System
10.1   10.2

10.8 Flow Paths, Internally Open (Symbol System 10.3)

10.9 Two-Way Valves (2 Ported Valves)

10.9.1 On-Off (Manual Shut-Off)

Simplified

Off

On

10.9.2 Check

  Simplified Symbol

Flow to the right is blocked. Flow to the left is permitted)

(Composite Symbol)

10.9.3 Check, Pilot- Operated to Open

10.9.4 Check, Pilot-Operated to Close

10.9.5 Two-Way Valves

10.9.5.1 Two-Position

Normally Closed   Normally Open

10.9.5.2 Infinite Position

Normally Closed   Normally Open

10.10 Three-Way Valves

10.10.1 Two-Position

10.10.1.1 Normally Open

10.10.1.2 Normally Closed

10.10.1.3 Distributor (Pressure is distributed first to one port, then the other)

10.10.1.4 Two-Pressure

10.10.2 Double Check Valve
Double check valves can be built with and without "cross bleed". Such valves with two

poppets do not usually allow pressure to momentarily "cross bleed" to return during transition. Valves with one poppet may allow "cross bleed" as these symbols illustrate.

10.10.2.1 Without Cross Bleed (One Way Flow)

10.10.2.2 With Cross Bleed (Reverse Flow Permitted)

10.11 Four-Way Valves

10.11.1 Two Position

Normal

Actuated

10.11.2 Three Position

(a) Normal

(b) Actuated Left

(c) Actuated Right

10.11.3 Typical Flow Paths for Center Condition of Three Position Valves

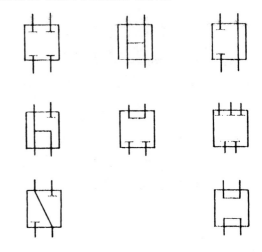

10.11.4 Two-Position, Snap Action with Transition.

As the valve element shifts from one position to the other, it passes through an intermediate position. If it is essential to circuit function to symbolize this "in transit" condition, it can be shown in the center position, enclosed by dashed lines.

Typical Transition Symbol

10.12 Infinite Positioning (Between Open & Closed)

10.12.1 Normally Closed

10.12.2 Normally Open

10.13 Pressure Control Valves

10.13.1 Pressure Relief

Simplified Symbol
Denotes

Normal   Actuated
                 (Relieving)

10.13.2 Sequence

10.13.3 Pressure Reducing

  ISO

10.13.4 Pressure Reducing and Relieving

10.13.5 Airline Pressure Regulator (Adjustable, Relieving)

10.14 Infinite Positioning Three-Way Valves

10.15 Infinite Positioning Four-Way Valves

10.16 Flow Control Valves (See 3.7)

10.16.1 Adjustable, Non-Compensated (Flow control in each direction)

10.16.2 Adjustable with Bypass

Flow is controlled
to the right
Flow to the left by-
passes control

10.16.3 Adjustable and Pressure Compensated With Bypass

10.16.4 Adjustable, Temperature & Pressure Compensated

11. Representative Composite Symbols

11.1 Component Enclosure

Component enclosure may surround a complete symbol or a group of symbols to represent an assembly. It is used to convey more information about component connections and functions. Enclosure indicates extremity of component or assembly. External ports are assumed to be on enclosure line and indicate connections to component.

Flow lines shall cross enclosure line without loops or dots.

11.2 Airline Accessories
(Filter, Regulator, and Lubricator)

Composite

Simplified

11.3 Pumps and Motors

11.3.1 Pumps

11.3.1.1 Double, Fixed Displacement, One

Inlet and Two Outlets

11.3.1.2 Double, with Integral Check Unloading and Two Outlets

11.3.1.3 Integral Variable Flow Rate Control with Overload Relief

11.3.1.4 Variable Displacement with Integral Replenishing Pump and Control Valves

11.3.2 Pump Motor

Variable displacement with manual, electric, pilot, and servo control.

11.4 Valves

11.4.1 Relief, Balanced Type

11.4.2 Remote Operated Sequence with Integral Check

11.4.3 Remote & Direct Operated Sequence with Differential Areas and Integral Check

11.4.4 Pressure Reducing with Integral Check

11.4.5 Pilot Operated Check

11.4.5.1 Differential Pilot Opened

11.4.5.2 Differential Pilot Opened and Closed

11.4.6 Two Positions, Four Connection Solenoid and Pilot Actuated, with Manual Pilot Override.

Simplified Symbol

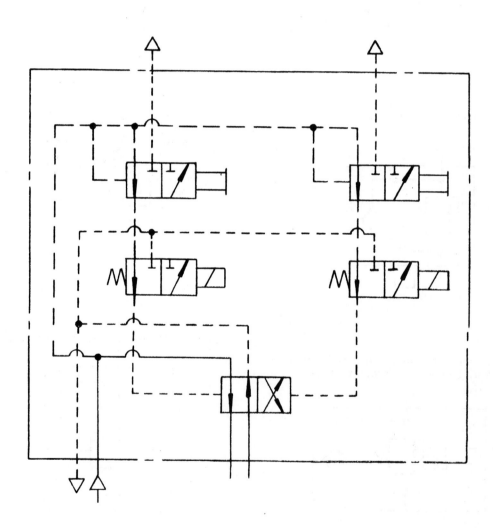

Complete Symbol

11.4.7 Two Position, Five Connection, Solenoid Control Pilot Actuated with Detents and Throttle Exhaust

Symplified Symbol

Complete Symbol

11.4.8 Variable Pressure Compensated Flow Control and Overload Relief

11.4.9 Multiple, Three Position, Manual Directional Control with Integral Check and Relief Valves

11.4.10 Cycle Control Panel, Five Position

11.4.11 Panel Mounted Separate Units Furnished as a Package (Relief, Two Four-Way, Two Check, and Flow Rate Valves)

11.4.12 Single Stage Compressor with Electric Motor Drive, Pressure Switch Control of Receiver Tank Pressure

INDEX

TABLE OF CONVERSION FACTORS

TO CONVERT	INTO	MULTIPLY BY

TO CONVERT	INTO	MULTIPLY BY
A		
atmospheres	ton/sq. inch	.007348
atmospheres	ft of water (at 4°C)	33.90
atmospheres	in. of mercury (at 0°C)	29.92
atmospheres	kg/sq cm	1.0333
atmospheres	kg/sq meter	10,332.
atmospheres	pounds/sq in.	14.70
B		
barrels (U.S. liq.)	gallons	31.5
barrels (oil)	gallons (oil)	42.0
Btu	horsepower-hr	3.931×10^{-4}
Btu	kilowatt-hr	2.928×10^{-4}
Btu/hr	horsepower	3.929×10^{-4}
Btu/min	horsepower	0.02356
C		
centimeters	feet	3.281×10^{-2}
centimeters	inches	0.3937
centimeter-grams	pound-feet	7.233×10^{-5}
centimeters of mercury	atmospheres	0.01316
centimeters of mercury	pounds/sq in.	0.1934
centimeters/sec	feet/min	1.9685
centimeters/sec	feet/sec	0.03281
cubic centimeters	cu feet	3.531×10^{-5}
cubic centimeters	cu inches	0.06102
cubic centimeters	gallons (U.S. liq.)	2.642×10^{-4}
cubic feet	cu cms	28,320.0
cubic feet	cu inches	1,728.0
cubic feet	gallons (U.S. liq.)	7.48052
cubic feet/min	cu cm/sec	472.0
cubic feet/min	gallon/sec	0.1247
cubic inches	cu cm	16.39
cubic inches	cu feet	5.787×10^{-4}
cubic inches	gallons (U.S. liq.)	4.329×10^{-3}
cubic meters	cu cm	10^6
cubic meters	cu feet	35.31
cubic meters	cu inches	61,023.0
cubic meters	gallons (U.S. liq.)	264.2
cubic meters/hr	gpm	4.403
D		
degrees (angle)	radians	0.01745
degrees (angle)	seconds	3,600.0

TO CONVERT	INTO	MULTIPLY BY
degrees/sec	revolutions/min	0.1667
degrees/sec	revolutions/sec	2.778×10^{-3}
dyne/sq cm	atmospheres	9.869×10^{-7}
dyne/sq cm	inch of mercury at 0°C	2.953×10^{-5}
dynes	pounds	2.248×10^{-6}
dynes/sq cm	psi	1.45038×10^{-5}
E		
ergs	dyne-centimeters	1.0
ergs	foot-pounds	7.367×10^{-8}
ergs/sec	Btu/min	$5,688 \times 10^{-9}$
ergs/sec	ft-lb/min	4.427×10^{-6}
ergs/sec	horsepower	1.341×10^{-10}
F		
fathom	meter	1.828804
fathoms	feet	6.0
feet	centimeters	30.48
feet	meters	0.3048
feet of water	atmospheres	0.02950
feet of water	in. of mercury	0.8826
feet of water	kg/sq cm	0.03048
feet of water	pounds/sq in.	0.4335
feet/min	cm/sec	0.5080
feet/min	feet/sec	0.01667
feet/min	km/hr	0.01829
feet	meters/min	0.3048
feet/min	miles/hr	0.01136
feet/sec	cm/sec	30.48
feet/sec	knots	0.5921
feet/sec	meters/min	18.29
feet/sec	miles/hr	0.6818
foot-pounds	Btu	1.286×10^{-3}
foot-pounds	hp-hr	5.050×10^{-7}
foot-pounds	kg-meters	0.1383
foot-pounds/min	Btu/min	1.286×10^{-3}
foot-pounds/min	horsepower	3.030×10^{-5}
G		
gallons	cu cm	3,785.0
gallons	cu feet	0.1337
gallons	cu inches	231.0
gallons (liq.Br.Imp.)	gallons (U.S. liq.)	1.20095
gallons (U.S.)	gallons (Imp.)	0.83267
gallons/min	cu ft/sec	2.228×10^{-3}
gallons/min	cu ft/min	0.13368
grams	pounds	2.205×10^{-3}
grams/cm	pounds/inch	5.600×10^{-3}
grams/cu cm	pounds/cu ft	62.43
grams/sq cm	pounds/sq ft	2.0481
H		
horsepower	Btu/min	42.44
horsepower	foot-lb/min	33,000.
horsepower	foot-lb/sec	550.0
horsepower (metric) (542.5 ft-lb/sec)	horsepower (550 ft-lb/sec)	0.9863
horsepower (550 ft-lb/sec)	horsepower (metric) (542.5 ft-lb/sec)	1.014
horsepower	kilowatts	0.7457
horsepower	watts	745.7
horsepower-hrs	Btu	2,547.
horsepower-hrs	foot-lb	1.98×10^6